John William Draper

Scientific Memoirs

Being experimental contributions to a knowledge of radiant energy

John William Draper

Scientific Memoirs
Being experimental contributions to a knowledge of radiant energy

ISBN/EAN: 9783337094355

Printed in Europe, USA, Canada, Australia, Japan

Cover: Foto ©berggeist007 / pixelio.de

More available books at **www.hansebooks.com**

SCIENTIFIC MEMOIRS

BEING

EXPERIMENTAL CONTRIBUTIONS

TO

A KNOWLEDGE OF RADIANT ENERGY

BY

JOHN WILLIAM DRAPER, M.D., LL.D.

PRESIDENT OF THE FACULTY OF SCIENCE IN THE UNIVERSITY OF NEW YORK
AUTHOR OF "A TREATISE ON HUMAN PHYSIOLOGY" "HISTORY OF THE
INTELLECTUAL DEVELOPMENT OF EUROPE" "HISTORY OF
THE AMERICAN CIVIL WAR" ETC.

NEW YORK
HARPER & BROTHERS, PUBLISHERS
FRANKLIN SQUARE
1878

PREFACE.

During the past forty years I have devoted much time to the experimental investigation of scientific topics, and have published the results in various journals, pamphlets, and the transactions of learned societies. They have been largely disseminated in European languages, and many of the conclusions they have presented have been admitted into the accepted body of scientific knowledge.

It has therefore become desirable for me to collect these scattered memoirs and essays together, and, since they are too voluminous to be published in full, to offer an abridgment or condensation of those that are of less interest. I propose in this book to include only such as are connected with the effects of Radiations or of Radiant energy, these having been distinguished by the American Academy of Science, as manifested by its award to me of the Rumford medal for discoveries in light and heat. A statement of the action of the Academy is annexed.

Besides these, I have several other memoirs on chemical, electrical, and physiological topics, some of them hitherto unpublished. These, for the present, I must reserve.

Among many other subjects treated of in these pages, the reader will find an investigation of the temperature at which bodies become red-hot, the nature of the light they emit at different degrees, the connection between their condition as to vibration and their heat. It is shown that ignited solids yield a spectrum that is continuous, not interrupted. This has become one of the fundamental facts in astronomical spectroscopy. At the time of the publication of this Memoir, no one in America had given attention to the spectroscope, and, except Fraunhofer, few in Europe. I showed that the fixed lines might be photographed, doubled their number, and found other new ones at the red end of the spectrum. The facts thus discovered I applied in an investigation of the nature of flame and the condition of the sun's surface. I showed that under certain circumstances rays antagonize each other in their chemical effect, and that the diffraction spectrum has great advantages over the prismatic, which is necessarily distorted. I attempted to ascertain the distribution of heat in the diffraction spectrum, and pointed out that great ad-

vantages arise if wave-lengths are used in the description of photographic phenomena. I published steel engravings of that spectrum so arranged. I made an investigation of phosphorescence, and obtained phosphorescent pictures of the moon. Up to this time it had been supposed that the great natural phenomenon of the decomposition of carbonic acid by plants was accomplished by the violet rays of light, but by performing that decomposition in the spectrum itself, I showed that it is effected by the yellow. Under very favorable circumstances, I examined the experiments said to prove that light can produce magnetism, and found that they had led to an incorrect conclusion. The first photographic portrait from the life was made by me : the process by which it was obtained is herein described. I also obtained the first photograph of the moon. I made many experiments on and discovered the true explanation of the crystallization of camphor towards the light. When Daguerre's process was published, I gave it a critical examination, and described the analogies existing between the phenomena of the chemical radiations and those of heat. For the purpose of obtaining more accurate results in these various inquiries, I invented the chlor-hydrogen photometer, and examined the modifications that chlorine undergoes in its allotropic states. Since in such researches more delicate thermometers are required than our ordinary ones, I entered on an investigation of the electro-motive power of heat, and described improved forms of electric thermometers. In these memoirs will be found a description of the method made use of for obtaining photographs of microscopic objects, together with specimens of the results. In a physiological digression respecting interstitial movements of substances, I examined the passage of gases through thin films such as soap-bubbles, and the force with which these movements are accomplished, applying the facts so gathered to an explanation of the circulation of the sap in plants, and of the blood in animals. Returning to an inquiry as to the distribution of heat and of chemical force in the spectrum, I was led to conclude, in opposition to the current opinion, that all the colored spaces are equally warm ; and that, so far from one portion — the violet — being distinguished by producing chemical effects, every ray can accomplish special changes. This series of experiments on radiations is concluded in this volume by an examination of the chemical action of burning-lenses and mirrors.

I have endeavored to reproduce these memoirs as they were originally published. When considerations of conciseness have obliged me to be contented with an abstract, it has always been so stated, and the place where the original may be found has been given. Sometimes, the circumstances seeming to call for it, additional matter has been introduced; but this has always been formally indicated under the title of NOTES, or included in parentheses. An instance of the former occurs on page 45, of the latter on page 30. Wood-cuts and their explanation, seldom occurring in the original publications, have been introduced. They have for the most part been obtained from some popular articles published by me in *Harper's Magazine*.

Except in a few instances, I have adhered, in these memoirs, to the chemical nomenclature in use at the time they were written. Though often very weighty reasons may be given that the designations under which substances pass might be made more in accordance with their constitution or properties, and therefore more correct, yet such are the confusions, the inconveniences, the difficulties attending an introduction of new names, that sweeping changes of nomenclature should never be introduced until they have become absolutely indispensable. In Memoir X., which treats of the action of the leaves of plants under the influence of yellow light, I have preferred to retain the term carbonic acid instead of any of its more recent synonyms.

Here and there the reader will detect statements that seem to be contradictory. On examination, however, he will find that this arises from changes which the general progress of science had made necessary. As an illustration of what I mean, I may refer to what is given as regards wave-lengths (from Mosotti) on page 112. These numbers do not agree with the more exact ones of Angström, page 120. It is better in such cases to let the original statements stand.

The pages offered in this volume, though not very numerous, represent a very large amount of work, the occupation of many years. Experimental investigation, to borrow a phrase employed by Kepler respecting the testing of hypotheses, is "a very great thief of time." Sometimes it costs many days to determine a fact that can be stated in a line. The things related in these memoirs have consumed much more than forty years.

Such a publication, therefore, assumes the character of an autobiography, since it is essentially a daily narrative of the occupations of its author. To a reader imbued with the true spirit of philosophy, even the shortcomings easily detectable in it are not without a charm. From the better horizon he has gained he watches his author, who, like a pioneer, is doubtfully finding his way, here travelling on a track that leads to nothing, then retracing his footsteps, and again, undeterred, making attempts until success crowns his exertions. To explore the path to truth implies many wanderings, many inquiries, many mistakes.

Perhaps, then, since this book is a sort of autobiography, its reader will bear with me if I try to make it more complete by here referring to other scientific or historical works in which I have been engaged.

In early life I had felt a strong desire to devote myself to the experimental study of nature; and, happening to see a glass containing some camphor, portions of which had been caused to condense in very beautiful crystals on the illuminated side, I was induced to read everything I could obtain respecting the chemical and mechanical influences of light, adhesion, and capillary attraction. Experiments I soon made in connection with these topics are described in these memoirs. Some of them I used in a Thesis for the degree of Doctor of Medicine in the University of Pennsylvania. My thoughts were thus directed to physiological studies, and I published papers on these topics in the *American Journal of Medical Sciences.* The favorable impression they made caused me to be appointed, in 1836, Professor of Chemistry and Physiology in Hampden Sidney College, Virginia, an appointment which enabled me to convert experimental investigation, thus far only an amusement, into the appropriate occupation of my life.

Several of the memoirs contained in this volume were composed at that time. To them I was indebted, without any application on my part, for an appointment to the Professorship of Chemistry and Physiology in the University of New York. Soon afterwards I published a work on the Forces that Produce the Organization of Plants. The lectures on Physiology I gave at that time I improved from year to year, and at length published them as a treatise on Human Physiology. It was very favorably received by the medical profession.

Among new experiments and explorations on physiological subjects contained in that book may be mentioned the selecting action of membranes; cause of the coagulation of blood; theory of the circulation of the blood; explanation of the flow of sap; endosmosis through thin films; measure of the force of endosmosis; respiration of fishes; action of the organic muscle fibres of the lungs; allotropism of living systems; new observations on the action of the skin; functions of nerve vesicles and their electrical analogies; function of the sympathetic nerve; explanation of certain parts of the auditory apparatus, particularly of the cochlea and semicircular canals; the theory of vision; the theory of muscular contraction.

From the study of individual man it is but a step to the consideration of him in his social relation, and this, accordingly, had been done in the second part of my work on Physiology. But the subject being too extensive to be dealt with satisfactorily in that manner, I published the materials that I had collected in a separate book, under the title of "A History of the Intellectual Development of Europe." The object of this was mainly to point out that the intellectual progress of nations proceeds in the same course as the intellectual development of the individual; that the movement of both is not fortuitous, but under the dominion of law; that the stages of personal development are paralleled by the stages of social development, and, indeed, as palæontology has proved, by the evolution of all animated nature; that there is an ascent of man through well-marked epochs from the most barbarous to the most highly civilized condition. This book was translated into many languages: in some of them several editions of it were issued. Portions of it relating to Mohammedan science were translated into Arabic.

About this time, circumstances led me to deliver before the New York Historical Society a course of lectures on American topics, considered from a similar point of view. These were enlarged, and published under the title of "Thoughts on the Civil Policy of America." This, like the preceding, was extensively translated and circulated. The train of investigation on which I had thus entered led me to a far more serious undertaking—a "History of the American Civil War," which had just then closed. To this, moreover, I was incited by the earnest request of some who had been chief actors in the events, and who very effectively aided me. I had the inestimable advantage of enjoying the friendship of many whose names have now become illustrious in connection with those times. The Secretary of War gave me access to the public documents on both sides, and to him I was indebted for guidance in the description of many of the most important incidents and the course of national policy. To generals who had commanded in great battles and conducted great campaigns I owed information which they alone could impart, and, in like manner, most valuable assistance was given me in special cases by persons eminent in military and civil life. It is often said that the history of any very great social event cannot be correctly written by a contemporaneous author, and that we must wait until passions have subsided and interests ceased for a narrative of the truth. But this is not so. More depends on the impartiality of the writer than on the deadening lapse of time. The best history will always be written by one who has had the best opportunity of getting at the facts, who has had the privilege of the friendship or personal acquaintance of those who have been conspicuous in the events.

No one can consider the intellectual development of Europe without contemplating the forces that have brought that continent to its present social condition—forces that have never ceased to be in active opposition. Under the title of a "History of the Conflict of Religion and Science,"

I endeavored to describe their warfare. This work has passed through a great many editions in America and England. It has been translated into French, Spanish, German, Russian, Italian, Polish, Servian, etc. It finds very many readers in Eastern Europe. And of some of these translations several editions have been issued.

When I thus look back on the objects that have occupied my attention, I recognize how they have been interconnected, each preparing the way for its successor. Is it not true that for every person the course of life is along the line of least resistance, and that in this the movement of humanity is like the movement of material bodies?

To my American reader I need say nothing for the purpose of securing his kind appreciation of this work. I know that he, recognizing the difficulties encountered in such a long series of experiments, will extenuate its imperfections, and regard it as a contribution from this side of the Atlantic to the common fund of human knowledge, and especially to one of its most important departments, at a period when the main subject on which it treats had scarcely attracted scientific attention.

NEW YORK, *January*, 1878.

CONTENTS.

xvi CONTENTS.

MEMOIR IV.

ON THE NATURE OF FLAME AND ON THE CONDITION OF THE SUN'S SURFACE.

MEMOIR V.

ON THE NEGATIVE OR PROTECTING RAYS OF THE SUN.

MEMOIR VI.

ON THE DIFFRACTION SPECTRUM.

MEMOIR VII.

STUDIES IN THE DIFFRACTION SPECTRUM.

MEMOIR VIII.

ON THE PHOSPHORESCENCE OF BODIES.

MEMOIR IX.

ON THE EFFECTS OF HEAT ON PHOSPHORESCENCE.

MEMOIR X.

ON THE DECOMPOSITION OF CARBONIC-ACID GAS BY PLANTS IN THE PRISMATIC SPECTRUM.

MEMOIR XI.

OF THE FORCE INCLUDED IN PLANTS.

B

MEMOIR XVI.

ON THE CHEMICAL CONDITION OF A DAGUERREOTYPE SURFACE.

MEMOIR XVII.

ON SOME ANALOGIES BETWEEN THE PHENOMENA OF THE CHEMICAL RAYS AND THOSE OF RADIANT HEAT.

MEMOIR XVIII.

DESCRIPTION OF THE CHLOR-HYDROGEN PHOTOMETER.

MEMOIR XIX.

ON MODIFIED CHLORINE.

MEMOIR XX.

ON THE ALLOTROPISM OF CHLORINE AS CONNECTED WITH THE THEORY OF SUBSTITUTIONS.

MEMOIR XXI.

ON THE INFLUENCE OF LIGHT UPON CHLORINE, AND SOME REMARKS ON ALCHEMY.

MEMOIR XXII.

ON THE ACTION OF GLASS AND QUARTZ ON THE RADIATIONS THAT PRODUCE PHOSPHORESCENCE.

MEMOIR XXIII.

ON A REMARKABLE DIFFERENCE BETWEEN THE RAYS OF INCANDES-CENT LIME AND THOSE EMITTED BY AN ELECTRIC SPARK.

MEMOIR XXIV.

ON THE ELECTRO-MOTIVE POWER OF HEAT.

MEMOIR XXIX.

ON THE DISTRIBUTION OF CHEMICAL FORCE IN THE SPECTRUM.

MEMOIR XXX.

ON BURNING GLASSES AND MIRRORS—THEIR HEATING AND CHEMICAL EFFECTS.

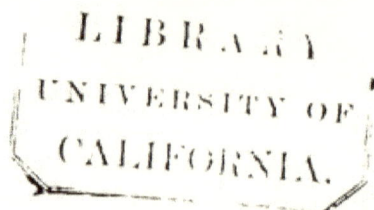

SCIENTIFIC MEMOIRS.

MEMOIR I.

EXAMINATION OF THE RADIATIONS OF RED-HOT BODIES.
THE PRODUCTION OF LIGHT BY HEAT.

From the American Journal of Science and Arts, Second Series, Vol. IV., 1847;
London, Edinburgh, and Dublin Philosophical Magazine, May, 1847; Harper's
New Monthly Magazine, No. 322.

CONTENTS :—*Ascertainment of the temperature at which bodies become self-luminous; it is 977° Fahr.—Proof that all solids begin to shine at the same degree.—The spectrum of incandescent solids has no fixed lines.—The reference spectrum.—Colors of light emitted as the heat increases are in the order of the spectrum.—Frequency of vibration increases with the temperature.—Intensity of the light emitted.—Intensity of the heat radiated.*

ALTHOUGH the phenomenon of the production of light by all solid bodies, when their temperature is raised to a certain degree, is one of the most familiar, no person so far as I know has hitherto attempted a critical investigation of it. The difficulties environing the inquiry are so great that even among the most eminent philosophers a diversity of opinion has prevailed respecting some of the leading facts. Thus Sir Isaac Newton fixed the temperature at which bodies become self-luminous as 635°; Sir Humphrey Davy at 812°; Mr. Wedgwood at 947°; and Mr. Daniel at 980°. As respects the nature of the light emitted, there are similar contradictions. In some philosophical works of considerable repute it is stated that when a solid begins to shine it first emits red and then white rays; in others it is asserted that a mixture of blue and red light is the first that appears.

I have succeeded in escaping or overcoming many of the difficulties of this problem, and have arrived at satisfactory solutions of the main points; and as the experiments now to be described lead to some striking and perhaps unexpected analogies between light and heat, they commend themselves to our attention, as having a bearing on the question of the identity of those principles.   It is known that heretofore I have been led to believe in the existence of cardinal distinctions not only between these, but also other imponderable agents, and I may therefore state that when this investigation was first undertaken it was in the expectation that it would lead to results very different from those which have actually arisen.

The following are the points on which I propose to treat:

1. To determine the point of incandescence of platinum, and to prove that different bodies become incandescent at the same temperature.

2. To determine the color of the rays emitted by self-luminous bodies at different temperatures.   This is done by the only reliable method—analysis by the prism.

From these experiments it will appear that as the temperature rises the light increases in refrangibility; and making due allowance for the physiological imperfection of the eye, the true order of the colors is red, orange, yellow, green, blue, indigo, violet.

3. To determine the relation between the brilliancy of the light emitted by a shining body and its temperature.

Here we shall find that the intensity of the light increases far more rapidly than the temperature.   For example, platinum at 2600° emits almost forty times as much light as it does at 1900°.

The source of light I have employed is in all instances a very thin strip of platinum, 1.35 inch long, and .05 of an inch wide, brought to the temperature under investigation by a voltaic current. Platinum was selected from its indisposition to oxidize, and its power of resisting a high temperature without fusion.

The strip of platinum thus to be brought to different temperatures by an electric current of the proper force was fastened at one end to an inflexible support, and at the other was connected with a delicate lever-index, which enabled me to determine its expansion, and thereby its temperature. For this purpose I have used the coefficient of dilatation of Dulong and Petit. The temperatures here given are upon the hypothesis of the invariability of that coefficient at all thermometric degrees; they are therefore to some extent in error.

In Fig. 1, *a b* represents the strip of platinum, the upper end of which is soldered to a stout and short copper pin, *a*, firmly sunk in a block of wood, *c*, which is immovably fastened to the basis, *d d*, of the instrument. A cavity, *e*, half an inch in diameter is sunk in the block *c*, and into this cavity the pin *a* projects, so that when the cavity is filled with mercury a voltaic current may be passed through the pin and down the platinum. The other extremity of the platinum, *b*, is fastened to a delicate lever, *b f*, which plays on an axis at *g*, the axis working in brass holes supported on a block, *h*. Immediately beneath

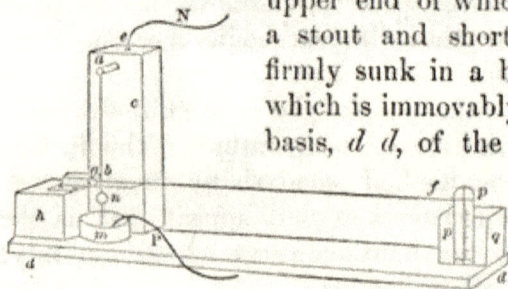
Fig. 1.

the platinum strip, and in metallic communication with it, a straight copper wire dips down into the mercury cup m; on this wire there is a metal ball, n, weighing about 100 grains. The further end of the index plays over a graduated ivory scale, $p\,p$, supported on a block, q; the scale can be moved a little up and down, so as to bring its zero to coincide with the index at common temperatures.

The action of the instrument is readily understood. In the mercury cup e let there dip one of the wires, N, of a Grove's battery of three or four pairs, the other wire, P, being dipped into the cup m. The current passes through the platinum, which immediately expands, the weight n lightly stretching it. The index f moves promptly over the scale, indicating the amount of expansion, and therefore the degree of heat. If the wire N be removed out of its mercury cup e, the platinum instantly becomes cold, and pulls the lever to the zero point.

When the platinum is thin, so as to be quite flexible at the point b, where it is fastened to the index, the movements take place with such promptitude and precision as to leave nothing to be desired. When the heat has been very high and long continued, the limit of elasticity of the platinum is somewhat overpassed, and it suffers a slight permanent extension. But as the ivory scale $p\,p$ can slide up and down a little, the index is readily readjusted to the zero point.

The temperature of the platinum depends entirely on the force of the current passed through it. By intervening coils of brass wire of lengths adjusted beforehand, so as to resist the current to a given extent, any desired temperature may be reached. I found it convenient to intervene in the course of the current a rheostat, so as to be able to bring the index with precision to any degree, notwithstanding slight changes in the force of the voltaic battery.

The following are the dimensions and measures of the instrument I have used: Length of the platinum strip, 1.35 inch; length of the part actually ignited, 1.14 inch; width of ditto, $\frac{1}{20}$ of an inch; length of the index from its centre of motion to the scale, 7.19 inches; distance of the centre of motion of index from the insertion of the platinum at the point b, .22 inch; multiplying effect of the index, 32.68 times; length of each division on the ivory scale, .021 inch. From this it would appear by a simple calculation, using the coefficient of dilatation of platinum given by Dulong and Petit, that each of the divisions here used is equal to 114.5 Fahrenheit degrees. For the sake of perspicuity I have generally taken them at 115°.

The Grove's battery I have employed has platinum plates three inches long and three quarters wide; the zinc cylinders are two inches and a half in diameter, three high, and one third thick. As used in these experiments it could maintain a current nearly uniform for an hour. I commonly employed four pairs.

By the aid of resisting wires of different lengths, or the rheostat, the force of the current in the platinum could be varied, and therefore its temperature. The first attempt was, of course, to discover the degree at which the metal began to emit light.

The platinum and the voltaic battery were placed in a dark room, the temperature of which was 60° Fahr.; and after I had remained therein a sufficient length of time to enable my eyes to become sensible to feeble impressions of light, I caused the current to pass, gradually increasing its force until the platinum was visible. In several repetitions of this experiment it was uniformly found that the index to which the platinum was attached stood at the eighth division when this took place. The metal had therefore dilated $\frac{1}{222}$ of its length; the elevation of its temperature was about 917°, which, added to the existing height of the thermometer, 60°, gave for the temperature of incandescence 977° Fahr.

To the correctness of this number it might be objected that, owing to the narrowness of the metallic strip, it was not well calculated to make an impression on the eye when the light emitted was feeble, and that we ought not to take the dilatations given by the index as representing the uniform temperature of the whole platinum, which must necessarily be colder near its points of support, on account of the conducting power of the metals to which it was attached.

Physiological considerations might also lead to a suspicion that the self-luminous temperature must vary as estimated by different eyes. The experiments of Bouguer, hereafter to be referred to, indisputably show that some persons are much more sensitive to the impressions of light than others. So far as my limited investigation of this matter has gone, I have not, however, found appreciable differences in the estimation of the temperature of incandescence. Different individuals observing the platinum have uniformly perceived it at the same time.

Against the number 977°, it may also be objected that antimony melts at a much lower temperature, and yet emits light before it fuses. If this statement were true, it would lead us to believe that all bodies have not the same point of incandescence. But I think that the experiments of Mr. Wedgwood on gold and earthenware are decisive in this particular; and, moreover, I have reason to believe that the melting-point of antimony is much higher than commonly supposed.

With a view of determining directly whether different

bodies vary in their point of incandescence, I took a clean gun-barrel, and having closed the touch-hole, exposed the following substances in it to the action of a fire: platinum, chalk, marble, fluor-spar, brass, antimony, gas-carbon, lead; each specimen was small: the platinum was in the form of a coil of stout wire.

When one of these bodies was placed in the gun-barrel and the temperature raised, it is clear that any difference in their point of incandescence could be detected by the eye. Thus if the ignition of platinum required a higher degree than that of iron, on looking down the barrel the coil of wire should be dark when the barrel itself had begun to shine; or, if the platinum was incandescent first, the wire should be seen before the barrel had become visibly hot, and these results might be corroborated by observing the inverse phenomena, when the barrel was taken from the fire and suffered to cool.

In Fig. 2, a b is the gun-barrel passing through a hole, c, of suitable size in the side of a stove. At the bottom of the barrel, b, the substances to be examined are placed. Their ignition is observed by looking in at the projecting end, a.

With respect to platinum, brass, antimony, gas-carbon, and lead, they all became incandescent at the same time as the iron barrel itself. I could not discover the slightest dif-

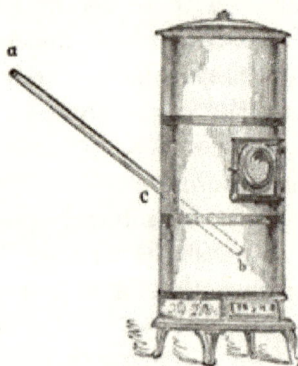

Fig. 2.

ference among them, either in heating or cooling; and it is worthy of remark that the lead was, of course, in the liquid condition. But the chalk and marble were visible before the barrel was red-hot, emitting a faint white light; and the fluor-spar still more strikingly so, its light being of a beautiful blue; and even when the barrel had become bright red I could still see the spar, which had decrepitated to a coarse powder, by its faint blue rays. In these cases, however, it was not incandescence, but phosphorescence that was taking place. I infer, then, that all solids, and probably melted metals, begin to shine at the same thermometric point.

(When phosphorescent substances are to be examined, they must be first exposed to a high temperature and carefully guarded from access of light until they are placed in the gun-barrel. A diamond which, among other bodies, had been thus tried, would recover its quality of phosphorescing by a very short access of light after it had been cooled, but if that had been carefully avoided, it began to shine at the same time as other specimens with which it was placed in the barrel.)

The temperature of incandescence seems to be a natural fixed point for the thermometer; and it is very interesting to remark how nearly this point coincides with 1000° of the Fahrenheit scale when Laplace's coefficient for the dilatation of platinum is used. Upon that coefficient the point of incandescence is 1006° Fahr.

In view of these considerations, and recollecting that the number given by Daniel is 980°, and that of Wedgwood 947°, I believe that 977° is not very far from the true temperature at which solids begin to shine. It is to be understood, of course, that this is in a very dark room.

I pass now to the second proposition. The rays emitted by the incandescent platinum strip were received on a flint-glass prism, placed so as to give the minimum deviation, and, after dispersion, viewed in a small telescope. A movement could be given to the telescope, which was read off on a graduated circle. However, instead of bringing the parts of the spectrum under measurement to coincide with the cross wires in the field of the instrument, it was found more satisfactory to determine them by bringing them to one or other of the edges of the field—a process by which the extreme rays could be better ascertained, their faint light being thus more easily perceived in the darkness by which it was surrounded. It would scarcely be possible to see them accurately while the rest of a bright spectrum was in view.

In Fig. 3, $a\,b$ is the ignited platinum strip, c the prism, $d\,e$ the telescope, moving upon the centre of a graduated circular table, $f\,f$.

As it was absolutely necessary to have fixed points of reference, that all the observations

Fig. 3.

might be brought to a common standard of comparison, and as there are no fixed lines in the light of incandescence such as are in the sunshine and daylight, I therefore previously determined the position of the fixed lines in a spectrum formed by a ray of reflected daylight which passed through a fissure $\frac{1}{20}$ of an inch wide, and one inch long, occupying exactly the position subsequently to be occupied by the incandescent platinum. In Fig. 4, 1 represents the result.*

(I expected to use the Fraunhofer lines for this purpose, and was not a little surprised to find that they are not to be seen in the spectrum of ignited solid bodies.

---

* The letters used to indicate the fixed lines are those employed at that time, 1847.

Thus was discovered one of the fundamental facts in spectrum analysis, a fact that has become of the highest importance in astronomy, as furnishing a means for determining the physical condition of the heavenly bodies, and a test for the nebular hypothesis. An ignited solid will give a continuous spectrum, or one devoid of fixed lines; an ignited gas will give a discontinuous spectrum, one broken up by lines or bands or spaces.

About twenty years subsequently to this discovery, Mr. Huggins (1864) made an examination of a nebula in the constellation of Draco. It proved to be gaseous. Subsequently, of sixty nebulæ examined, nineteen gave discontinuous or gaseous spectra, the remainder continuous ones.

It may therefore be admitted that physical evidence has through this means been obtained demonstrating the existence of vast masses of matter in a gaseous condition, and at a temperature of incandescence. The nebular hypothesis of Laplace and Herschel has thus a firm basis.)

The strip of platinum was now placed in the position of the slit which had given the spectrum represented at 1, and its temperature was raised by the passage of a voltaic current. Though the metal could be distinctly seen by the naked eye when the temperature had reached about 1000° Fahr., yet the loss of light

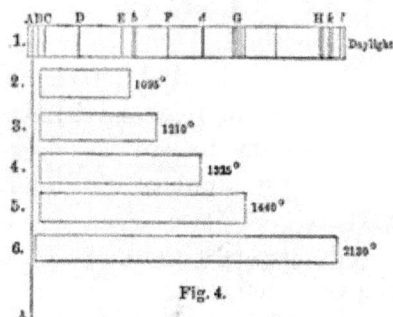

Fig. 4.

Spectra of daylight and of incandescent platinum at different temperatures.

in passing the prism and telescope was so great that it was necessary to carry the temperature to 1210° before a satisfactory observation could be made. At this degree the spectrum extended from the position of the fixed line B in the red almost as far as the line F in the green, the colors present being red, orange, and a tint which may be designated as gray. There was nothing answering to yellow. The rays first visible through this apparatus may therefore be designated as red and greenish-gray; the former commencing at the line B, and the latter continuing to F, as at 3.

The voltaic current was now increased, and the temperature rose to 1325°. The red end of the spectrum remained nearly as before, but the more refrangible extremity reached the position of the little fixed line *d*. Traces of yellow were now visible, and, with a certain degree of distinctness, the red, orange, yellow, green, and a fringe of blue could be seen; 4 shows the result.

The temperature was now carried to 1440°. The red extremity appeared to be advancing towards the line A ; the blue had undergone a well-marked increase. It reached considerably beyond the line G, as shown at 5.

On bringing the platinum to 2130°, all the colors were present, and exhibited considerable brilliancy. Their extent was somewhat shorter than that of the daylight spectrum, as seen at 6.

Having thus by repeated experiments ascertained the continued extension of the more refrangible end as the temperature rose, it became necessary to obtain observations for degrees below 1210, the limit of visibility through the telescope. I therefore carried the prism nearer to the platinum, and looking with the unassisted eye directly through it at the refracted image, found that it could be distinctly seen at a temperature as low as 1095°. Under these circumstances, the total length could not be compared by direct measurement with the other observations, and the result given at 2 is as correct as could be obtained. The colors were red and greenish-gray.

C

The gray rays emitted by platinum just beginning to shine appear to be more intense than the red; at all events, the wires in the field of the telescope are more distinctly seen upon them than upon the other color. The designation of gray may be given them, for they appear to approach that tint more closely than any other, and yet it is to be remarked that they are occupying the position of the yellow and green regions.

Already we have encountered a fact of considerable importance. The conclusion that as the temperature of a body rises it emits rays of increasing refrangibility has obviously to be taken with a certain restriction. Instead of first the red, then the orange, then the yellow rays, etc., in succession making their appearance, in which case the spectrum should regularly increase in length as the temperature rises, we here find that at the very first moment it is visible to the eye it reaches from the fixed line B nearly to F, that is to say, it is equal to about two thirds of the whole length of the diffraction spectrum, and almost one half of the prismatic.

It is to be remarked that while the more refrangible end undergoes a great expansion, the other extremity exhibits a corresponding though a less change. As very important theoretical conclusions depend on the proper interpretation of this fact, it must not be forgotten that to a certain extent it may be an optical deception, arising from the increased brilliancy of the light. While the rays are yet feeble, the extreme terminations may be so faint that the eye cannot detect them, but as the intensity rises they become better marked, and an apparent elongation of the spectrum is the consequence.

It is agreed among optical writers that to the human eye the yellow is the brightest of the rays. In the prismatic spectrum the true relationship of the colors is not perceived, because the less refrangible are crowded together, and the more refrangible unduly spread out. But in the diffraction spectrum, where the colors are arranged side by side in the order of their wave-lengths, the centre is occupied by the most luminous portion of the yellow, and from this point the light declines away on one side in the red, and on the other in the violet, the terminations being equidistant from the centre of the yellow space.

Now if the rays coming from shining platinum were passed through a piece of glass on which parallel lines had been ruled with a diamond point, so as to give a diffraction spectrum, even admitting the general results of the foregoing experiments to be true, viz., that as the temperature rises rays of a higher refrangibility are emitted, it is obvious that it by no means follows that the ray first visible should be the extreme red. Our power of seeing that depends on its having a certain intensity. Even when it has assumed the utmost brilliancy which it has in a solar beam, it is barely visible. We ought, therefore, to expect that rays of a higher refrangibility should be first seen, because they act more energetically on our organ of vision; and as the temperature rises, the spectrum should undergo a partial elongation in the direction of its red extremity.

I may here remark that the general result of these experiments coincides exactly with that of M. Melloni respecting heat at lower thermometric points. In his second Memoir (Taylor's "Scientific Memoirs," vol. i., p. 56) he shows that when rays from copper at 390° and from incandescent platinum are compared by transmission through a rock-salt prism, as the temperature rises the refrangibility of the calorific emanations correspondingly increases. Those who regard light and heat as the same agent will therefore see in this coincidence another argument in favor of their opinion.

In view of the foregoing facts, I conclude that *as the temperature of an incandescent body rises, it emits rays of light of an increasing refrangibility,* and that the apparent departure from this law, discovered by an accurate prismatic analysis, is due to the special action of the eye in performing the function of vision. And as the luminous effects are undoubtedly owing to a vibratory movement executed by the molecules of the platinum, it seems from the foregoing facts to follow that *the frequency of those vibrations increases with the temperature.*

In this observation I am led by the principle that " to a particular color there ever belongs a particular wavelength, and to a particular wave-length there ever belongs a particular color;" but in the analysis of the spectrum made by Sir D. Brewster by the aid of absorption media, this principle is indirectly controverted, that eminent philosopher showing that red, yellow, blue, and consequently white light, exist in every part of the spectrum. This must necessarily take place when a prism which has a refracting face of considerable magnitude is used; for it is obvious that a ray falling near the edge and one falling near the back, after dispersion, will depict their several spectra on the screen; the colors of the one not coinciding with, but overlapping the colors of the other. In such a spectrum there must undoubtedly be a general commixture of the rays; but may we not fairly inquire, whether, if an elementary prism were used, the same facts would hold good; or if the anterior face of the prism were covered by a screen, so as to expose a narrow fissure parallel to the axis of the instrument, would there be found in the spectrum it gave every color in every part, as in Sir David Brewster's original experiment? M. Melloni has shown how this very consideration complicates the phenomena of radiant heat; and it would seem a very plausible suggestion that the effect here pointed out must occur in an analogous manner for the phenomena of light.

I next pass to the third branch of this investigation—to examine the relation between the temperatures of self-luminous bodies and the intensity of the light they emit, premising it with the following considerations:

The close analogy which has been traced between the phenomena of light and radiant heat lends countenance to the supposition that the law which regulates the escape of heat from a body will also determine its rate of emission of light. Sir Isaac Newton supposed that while the temperature of a body rose in an arithmetical progression, the amount of heat escaping from it increased in a geometrical progression. The error of this was subsequently shown by Martin, Erxleben, and Delaroche, and finally Dulong and Petit gave the true law: " When a hot body cools in vacuo, surrounded by a medium the temperature of which is constant, the velocity of cooling for excess of temperature in arithmetical progression increases as the terms of a geometrical progression, diminished by a constant quantity." The introduction of this constant depends on the operation of the theory of the exchanges of heat; for a body when cooling under the circumstances here supposed is simultaneously receiving back a constant amount of heat from the medium of constant temperature.

While Newton's law represents the rate of cooling of bodies, and therefore the quantities of heat they emit when the range of temperature is limited, and the law of Dulong and Petit holds to a wider extent, there are in the present inquiry certain circumstances to be taken into account not contemplated by those philosophers. Dulong and Petit, throughout their memoir, regard radiant heat as a homogeneous agent, and look upon the theory of exchanges, which is indeed their starting-point and guide, as a very simple affair. But the progress of this department of knowledge since their time has shown that precisely the same modifications found in the colors of light occur also for heat; a fact conveniently designated by the phrase "ideal coloration of heat," and, further, that the wave-length of the heat emitted depends upon the temperature of the radiating source. It is one thing to investigate the phenomena of the exchanges of heat-rays of the same color, and another when the colors are different. A complete theory of the exchanges of heat must include this principle, and, of course, so too must a law of cooling applicable to any temperature.

There is another fact to some extent considered by Dulong and Petit, but not of such weight in their investigations, where the range of temperature was small, as in these where it rises as high as nearly 3000° Fahr. This is the difference of specific heat of the same body at different temperatures. At the high temperatures herein employed, there cannot be a doubt that the capacity of platinum for heat is far greater than that at a low point. This, therefore, must affect its rate of calorific emission, and probably that for light also.

From these and similar considerations we should be led to expect that as the temperature of an incandescent solid rises, the intensity of the light emitted increases very rapidly.

I pass now to the experimental proofs which substantiate the foregoing reasoning.

The apparatus employed as the source of the light and measure of the temperature was the same as in the preceding experiments—a strip of platinum brought to a known temperature by the passage of a voltaic current of the proper force, and connected with an index which measured its expansion.

The principle upon which the intensities of the light

were determined was that originally described by Bouguer, and subsequently used by Masson. After many
experiments, I found that it is the most accurate method
known.

Any one who will endeavor to determine the intensities of light by Rumford's method of contrasting shadows, or by that of equally illuminated surfaces, will find,
when every precaution has been used, that the results
of repeated experiments do not accord. There is, moreover, the great defect that when the lights differ in color,
it is impossible to obtain reliable results except by resorting to such contrivances as that described in the
Philosophical Magazine, August, 1844.

Bouguer's principle is far more exact; and where the
lights differ in color, that difference actually tends to
make the result more correct. As it is not generally
known, I will indicate the nature of it briefly :

Let there be placed at a certain distance from a sheet
of white paper a candle, so arranged as to throw the
shadow of an opaque body, such as a rod of metal, on
the sheet. If a second candle be placed also in front of
the paper and nearer than the former, there is a certain
distance at which its light completely obliterates all
traces of the shadow. This distance is readily found,
for the disappearance of the shadow can be determined
with considerable exactness. When the lights are equal,
Bouguer ascertained that the relative distances were as
1 : 8, and therefore inferred correctly that in the case of
his eye the effect of a given light was imperceptible
when it was in presence of another sixty-four times as
intense. The precise number differs according to the
sensibility of different eyes, but for the same organ it is
constant.

Upon a paper screen I threw the shadow of a rod of
copper, which intercepted the rays of the incandescent

platinum; then taking an Argand lamp, surrounded by
a cylindrical metal shade, through an aperture in which
the light passed, and the flame of which I had found by
previous trial would continue for an hour of almost the
same intensity, I approached it to the paper sheet, until
the shadow cast by the copper disappeared. The distance at which this took place was then measured, and
the temperature of the platinum determined.

The temperature of the platinum was now raised, the
shadow became more intense, and it was necessary to
bring the Argand lamp nearer before it was effaced.
When this took place, the distance of the lamp was
again measured, and the temperature of the platinum
again determined.

In Fig. 5, $a\,b$ is the strip of ignited platinum. It casts
a shadow, h, of the metal rod e on the white screen $f\,g$;
c is a metallic cylinder containing an Argand lamp, the
light of which issues through an aperture, d, and extinguishes the shadow on the screen.

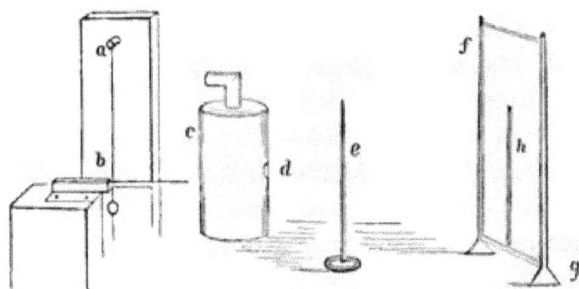

Fig. 5.

In this manner I obtained several series of results, one
of which is given in the following table. They exhibited a more perfect accordance among each other than I
had anticipated.

The intensity of the light of the platinum is of course
inversely proportional to the square of the distance of

Table of the Intensity of Light emitted by Platinum at Different Temperatures.

Temperature of the platinum.	Distance of Argand lamp.		Mean.	Intensity of light.
	Experiment I.	Experiment II.		
980°	. . . .	. . . .	. . . .	0.00
1900	54.00	54.00	54.00	0.34
2015	39.00	41.00	40.00	0.62
2130	24.00	24.00	24.00	1.73
2245	18.00	19.00	18.50	2.92
2360	14.50	15.50	15.00	4.40
2475	11.50	12.00	11.75	7.24
2590	9.00	9.00	9.00	12.34

the Argand lamp at the moment of the extinction of the shadow.

In this table the first column gives the temperatures under examination in Fahrenheit degrees; the second and third the distances of the Argand lamp from the screen in English inches, in two different sets of experiments; the fourth the mean of the two; and the fifth the corresponding intensity of the light.

The results thus obtained proved that the increase in the intensity of the light of the ignited platinum, though slow at first, became very rapid as the temperature rose. At 2590° the brilliancy of the light was more than thirty-six times as great as it was at 1900°.

Thus, therefore, the theoretical anticipation founded on the analogy of light and heat was completely verified, the emission of light by a self-luminous solid as its temperature rises being in greater proportion than would correspond to mere difference of temperature.

To place this in a more striking point of view, I made some corresponding experiments in relation to the heat emitted. No one thus far had published results for high temperatures, or had endeavored to establish through an extensive scale the principle of Delaroche, that "the quantity of heat which a hot body gives off in a given time, by way of radiation to a cold body situated at a distance, increases, other things being equal, in a progression more rapid than the excess of the temperature of the first above that of the second."

As the object thus proposed was mainly to illustrate the remarkable analogy between light and heat, the experiments now to be related were arranged so as to resemble the foregoing; that is to say, as in determining the intensities of light emitted by a shining body at different temperatures, I had received the rays upon a screen placed at an invariable distance, and then determined their value by photometric methods, so in this case I received the rays of heat upon a screen placed at an invariable distance, and measured their intensity by thermometric methods. In this instance the screen employed was, in fact, the blackened surface of a thermo-electric pile. It was arranged at a distance of about one inch from the strip of ignited platinum, a distance sufficient to keep it from any disturbance from the stream of hot air arising from the metal; care was also taken that the multiplier itself was placed so far from the rest of the apparatus that its astatic needles could not be affected by the voltaic current igniting the platinum, or the electro-magnetic action of the wires or rheostat used to modify the degrees of heat.

In Fig. 6, $a\ b$ is the ignited platinum strip, c the thermo-electric pile, $d\ d$ the multiplier.

The experiments were conducted as follows: The needles of the thermo-multiplier standing at the zero of their scale, the voltaic current was passed through the platinum, which immediately rose to the corresponding temperature, and radiated its heat to the face of the pile. The instant this current passed, the needles of the multiplier moved, and kept steadily advancing on the scale. At the close of one minute the deviation of the needle and the temperature of the platinum were si-

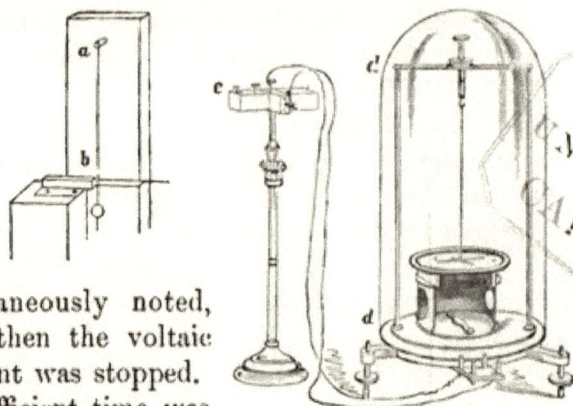

Fig. 6.

multaneously noted, and then the voltaic current was stopped.

Sufficient time was now given for the needles of the multiplier to come back to zero. This time varied in the different cases, according to the intensity of the heat to which the pile had been exposed; in no instance, however, did it exceed six minutes, and in most cases was much less. A little consideration will show that the usual artifice employed to drive the needles back to zero by warming the opposite face of the pile was not admissible in these experiments.

The needles having regained their zero, the platinum was brought again to a given temperature, and the experiment conducted as before. The following table exhibits a series of these results.

In this table the first column gives the temperatures of the platinum in Fahrenheit degrees; the second and third, two series of experiments expressing the arc passed over by the needle at the close of a radiation lasting one minute, each number being the mean of several successive trials, and the fourth the mean of the two. It therefore gives the radiant effect of the incandescent platinum on the thermo-multiplier for the different temperatures.

Of course it is understood that I here take the angu-

Table of the Intensity of Radiant Heat emitted by Platinum at Different Temperatures.

Temperature of the platinum.	Intensity of heat emitted.		Mean.
	Experiment I.	Experiment II.	
980°	.75	1.00	.87
1095	1.00	1.20	1.10
1210	1.40	1.60	1.50
1325	1.60	2.00	1.80
1440	2.20	2.20	2.20
1555	2.75	2.85	2.80
1670	3.65	3.75	3.70
1785	5.00	5.00	5.00
1900	6.70	6.90	6.80
2015	8.60	8.60	8.60
2130	10.00	10.00	10.00
2245	12.50	12.50	12.50
2360	15.50	15.50	15.50

lar deviations of the needle as expressing the force of the thermo-electric current, or, in other words, as being proportional to the temperatures. This hypothesis, it is known, is admissible.

It therefore appears that if the quantity of heat radiated by platinum at 980° be taken as unity, it will have increased at 1440° to 2.5, at 1900° to 7.8, and at 2360° to 17.8, nearly. The rate of increase is, therefore, very rapid. Further, it may be remarked, as illustrative of the same fact, that the quantity of heat radiated by a mass of platinum in passing from 1000° to 1300° is nearly equal to the amount it gives out in passing from common temperatures up to 1000°.

I cannot here express myself with too much emphasis on the remarkable analogy between light and heat which these experiments reveal. The march of the phenomena in all their leading points is the same in both cases. The rapid increase of effect as the temperature rises is common to both.

It is not to be forgotten, however, that in the case of light we necessarily measure its effects by an apparatus which possesses special peculiarities. The eye is insensible to rays not comprehended within certain limits of refrangibility. In these experiments it is requisite to raise the temperature of the platinum almost to 1000° before we can discover the first traces of light. Measures obtained under such circumstances are dependent on the physiological action of the visual organ itself, and hence their analogy with those obtained by the thermometer becomes more striking, because we should scarcely have anticipated that it could be so complete.

Among writers on Optics it has been a desideratum to obtain an artificial light of standard brilliancy. The preceding experiments furnish an easy means of supplying that want, and give us what might be termed a "unit lamp." A surface of platinum of standard dimensions, raised to a standard temperature by a voltaic current, will always emit a constant light. A strip of that metal, one inch long and $\frac{1}{20}$ of an inch wide, connected with a lever by which its expansion might be measured, would yield at 2000° a light suitable for most purposes. Moreover, it would be very easy to form from it a photometer by screening portions of the shining surface. An ingenious artist would have very little difficulty, by taking advantage of the movements of the lever, in making a self-acting apparatus in which the platinum should be maintained at a uniform temperature, notwithstanding any change taking place in the voltaic current.

UNIVERSITY OF NEW YORK, *Feb.* 27, 1847.

NOTE.—The experiments related in the foregoing pages were made by me between 1844 and 1847. They were published in May in the latter year.

In the following July, M. Melloni, who was at that time recognized as the chief authority on the subject of radiant heat, read before the Royal Academy of Sciences at Naples a memoir entitled "Researches on the Radiations of Incandescent Bodies and on the Elementary Colors of the Solar Spectrum." This was translated into French from the Italian, and published in the *Bibliothèque Universelle* of Geneva. It was also translated into English, and published both in England and America.

M. Melloni commences his memoir as follows: " Among the more recent scientific publications will be found a memoir by the American professor J. W. Draper 'On the Production of Light by Heat,' which appears to me to merit the attentive consideration of those who interest themselves in the progress of the natural sciences. The author treats in a very ingenious manner some questions allied to my own researches on light and radiant heat. In reading this interesting work, several ideas have presented themselves to me, which I have submitted to the test of experiment. I believe that an analysis of the memoir of M. Draper, accompanied with a brief account of what I have done, will not be without interest.

" Every one knows that heat, when it accumulates in bodies, at last renders them *incandescent*, that is to say, more or less luminous and visible in the dark. Is the temperature necessary to produce this state of incandescence always the same, or does it vary with the nature of the body ? In either case, what is its degree, and what is the succession of colored lights emitted by a given substance when brought to temperatures more and more elevated ? Finally, what is the relation that subsists at different periods of incandescence between the temperature and the quantity of light and of heat emitted by a body ?"

Melloni then describes the apparatus and the processes I had used to determine the thermometric degree of incandescence and its uniformity for different substances.

He dwells on the fact that melted metals, such as lead, have the same point of ignition. He agrees in excepting the phenomena of phosphorescence, and those in which light is developed in chemical combinations. He remarks that "some philosophers of the highest eminence, among them M. Biot, suppose that the first light disengaged by incandescent bodies is blue, and they have accounted for this on the principle of a theory now universally abandoned. But these cases," he adds, "ought to be carefully distinguished from incandescence properly speaking, which arises directly and solely from an elevation of temperature in the body, and which always commences with a red light.

"As to the exact degree of this temperature, the objections which might be raised against the mode employed by M. Draper are of very little importance. If we compare the results at which he arrives with those that have been obtained by Wedgwood and Daniel, the difference is only 30° in excess in the first case, and 3° too little in the second. The differences are much greater when compared with the deductions of Davy and Sir I. Newton, which gave 812° and 635°, respectively. But those numbers, and especially the latter, were obtained by methods too imperfect to be trustworthy. Consequently the number 977° Fahr., given by M. Draper, must approach very closely the degree of heat which produces the first incandescence of bodies."

Melloni then describes the method I had resorted to for investigating the nature of the colors which are developed by an ignited body as its temperature is increased.

He dwells on the employment of a reference spectrum, which was resorted to in consequence of the spectrum of a solid having no fixed lines—a discovery which has become of the utmost value in astronomical spectrum analysis. He states the results given in the foregoing pages, and adds: "In other words, the spectrum of the strip of platinum which corresponds to the red extremity of the prismatic spectrum is at first very short, and contains only the less refrangible colors; but as the temperature rises, the spectrum of incandescence extends towards the violet extremity, obtaining the more refrangible tints, and at last acquiring all the colors and all the extent of the solar spectrum, except the terminal rays at the two extremities, which escape the observer evidently on account of their extreme feebleness. The same cause (insensibility due to a want of luminous energy) makes the first spectrum appear at the red end a little shorter than the last, because the less refrangible rays of that color are, as is well known, so feeble even in the solar spectrum that we are unable to perceive them, unless they are isolated in a place that is totally dark. Much more, therefore, ought they to remain invisible to the observer when the spectrum arises from luminous agencies so little energetic as are those of the first periods of incandescence.

"To a perfectly sensitive eye the variations of length would evidently have taken place in the direction of the more refrangible rays only, and all the spectra would have commenced at the extreme limit of the red rays.

"It results from all these observations that when the incandescence of a body becomes more and more vivid and brilliant by the elevation of its temperature, there is not only an augmentation in the intensity of the resulting light, but also in the variety of elementary colors which compose it; there is, too, an addition of rays so much the more refrangible as the temperature of the incandescent body is higher. In this there is, therefore, established an intimate analogy between the progressive development of light and that of heat. Indeed," M. Melloni adds, " as soon as I had convinced myself of the immediate transmission of every variety of radiant heat through rock-salt, I availed myself of that valuable property to study the refraction of heat from various sources, and I discovered that radiations coming from those of a high temperature contain elements more refrangible than those which are derived from sources that are not so hot."

M. Melloni then passes to a criticism of the methods I had used in investigating the law of the increase of the luminous and calorific radiations, according as the temperature of the source of heat is elevated.

He adds: "The method invented by Bouguer to determine the relative intensities of different luminous sources, and employed by Draper to measure the quantities of light emitted by a strip of platinum brought to different degrees of incandescence, is the only one by which we could hope for a successful result. The method of the equality of shadows, well known under the name of Rumford's method, would have furnished in the researches of the learned American uncertain data, on account of the difficulty of establishing an exact comparison between the accidental green tint introduced into the shadow enlightened by the yellow rays of the lamp and the red light emitted by the ignited metal. As to the measures of the radiant heat, they were determined by the aid of the thermo-multiplier, that admirable instrument which has revealed to science so many new properties of calorific radiations, and which still is rendering eminent services in the hands of able chemists far beyond the Alps.

"The numbers obtained by M. Draper show evidently that the augmentations of both light and heat, though feeble at first, become very rapid at last, from which it results that the radiations both of light and heat follow in the *progression of quantity* the same analogy that we

D

have just observed in the *progression of quality*. These researches then conduct, as do others heretofore known on light and radiant heat, to a perfect analogy between the general laws which govern these two great agents of nature."

The law of the radiation of heat, as illustrated by the foregoing experiments with an ignited strip of platinum, has been applied in recent discussions respecting the age of the earth. Geological evidence has satisfactorily established that the temperature of the earth was formerly much higher than now, and the decline that has happened could only have taken place by radiation into space. Considering how slow the cooling now is—a scarcely perceptible fraction of a degree in the course of many centuries—it would seem that to accomplish the whole descent, if even we go no further back than the paleozoic era, an amazing lapse of time would be required. And if we accept the nebular hypothesis, since the original temperature must have been at least that of the surface of the sun, the time must be correspondingly extended. Even if numbers could be given, the imagination would altogether fail to appreciate them.

But we have here experimental proof that the higher the temperature of a body, the more rapidly it cools. A descent through a given number of degrees is more quickly made when a body is at a high than when it is at a low temperature. Anciently the cooling of the earth was more rapid than it is now. Not that there was any change or breach in the general law under which the operation was taking place, for the same mathematical expression applies to all temperatures, no matter how high or how low they may be. Mr. Croll, in his recent researches on the distribution of heat over the globe, points out the bearing of these experiments.

Our estimate of the age of the earth, as deduced from the cooling she has undergone, must therefore, in view of these considerations, be diminished—a result insisted upon by many recent authors. Too much weight must, however, not be given to this conclusion, since it ought to be borne in mind that the cooling was not taking place by radiation into space from the earth alone as a solitary body. She was in presence of a high extraneous temperature, which diminished her speed of cooling, and correspondingly increased the time.

Though the problem of the age of the earth, as investigated through the changes of her temperature, may not at present be capable of exact solution, it must be admitted that the time required to bring her heat to its present degree must have been inconceivably long.

•

MEMOIR II.

SPECTRUM ANALYSIS OF FLAMES.   PRODUCTION OF LIGHT
BY CHEMICAL ACTION.

From the American Journal of Science, Second Series, Vol. V., 1848; Philosophical Magazine, London and Edinburgh, Feb., 1848; Harper's New Monthly Magazine, No. 323.

CONTENTS :—*Spectrum analysis of a candle flame.—Examination of various other flames.—Spectrum analysis of the light of a burning solid.—Flames consist of a succession of shells.—Spectrum of cyanogen.—Combustion of flames in oxygen.—Effect of air in the interior of a flame.—The blowpipe cone.—General result, the more energetic the chemical action the higher the refrangibility of the resulting light; the vibrations increase in frequency as the chemical action is more violent.—Fraunhofer's fixed lines.*

THE production of light and heat by the combustion of various bodies is, of all chemical processes, that which ministers most to the comfort and well-being of man. By it the rigor of winter is abated, and night made almost as available for our purposes as day.

One would suppose that, of a phenomenon on which so much of our personal and social happiness depends, and which must have been witnessed by every one, all the particulars ought to have been long ago known. Among scientific men its importance has been universally recognized. The earlier theories of chemistry, such as those of Stahl and Lavoisier, are essentially theories of combustion.

It is nevertheless remarkable how little positive knowledge, until quite recently, was possessed on this subject. Some chemists thought that the light emitted by flames is due to electric discharges; others, regarding light and heat as material bodies which can be incorporated or united with ponderable substances, supposed that they are disengaged as chemical changes go on. In this confusion of opinions a multitude of interesting and hitherto unanswered questions present themselves. It is known that different substances, when burning, emit rays of different colors. Thus sulphur and carbonic oxide burn blue, wax yellow, and cyanogen lilac. What are the chemical conditions that determine these singular differences? How is it that, by changing the conditions of combustion, we can vary the nature of the light? We turn aside the flame of a candle by means of a blow-pipe, and a neat blue cone appears. Why does it shine with a blue light?

Such inquiries might be multiplied without end; but a little consideration shows that their various answers depend on the determination of a much more general problem, viz., Can any connection be traced between the chemical nature of a substance, or the conditions under which it burns, and the nature of the light it emits? It is to the discussion of that problem that this memoir is devoted.

Sir H. Davy has already furnished us with two important facts in relation to the nature of flame: 1st, All common flames are incandescent shells, the interior of which is dark; 2d, the relative quantity of light emitted depends upon the temporary disengagement of solid particles of carbon.

It is only by a very general examination of the light arising from various solids, vapors, and gases, when burning, that we can expect to obtain data for a true theory of combustion. This is what I shall endeavor to furnish on the present occasion.

As was foreseen by all the older chemists, the true theory of combustion, whatever it may prove to be, must necessarily be one of the fundamental theories of chemistry. It must include the nature of all chemical changes whatsoever. The subject is therefore not alone interesting in a popular sense, but of great importance in its scientific connections.

I. *Prismatic analyses of the flames of various vapors and gases; proving that they yield all the colors of the spectrum.*

I commenced this investigation of the nature of flame and of combustion generally by an optical examination of various bodies in the act of burning. Some authors have asserted that certain flames yield monochromatic light. It is necessary to verify this assertion, if true, or set it aside if false.

The instrumental arrangement I resorted to for the determination of the structure of a flame may be thus described: The rays of the flame of which the examination was to be made passed through a horizontal slit, one thirtieth of an inch wide and one inch long, in a metallic screen placed near to the flame, and were received at a distance of six or eight feet on a flint-glass prism, the axis of which was parallel to the slit. After passing the prism, they entered a telescope, which had a divided micrometer and parallel wires in its eye-piece. Through this telescope the resulting spectrum was viewed. In this form of spectroscope no collimating lens was used.

In Fig. 7, *a* is the candle, or flame which is to be examined, *b* the screen with a horizontal slit, *c* the prism, *d d* the telescope.

If it be the flame of a lamp of any kind that is to be examined, by using a movable stand we are able to raise or lower it, and thus analyze different *horizontal elements* in its lower, its middle, or its upper parts at pleasure.

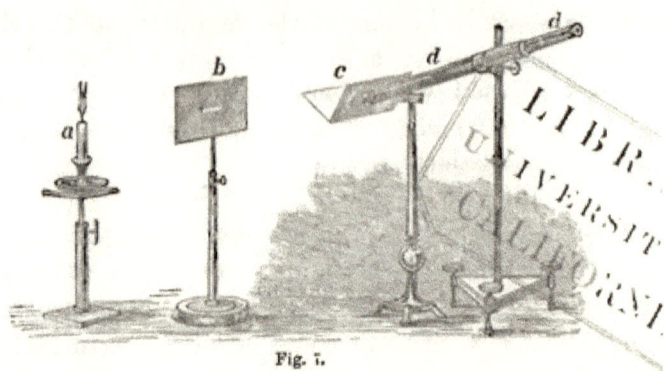

Fig. 7.

If, instead of a horizontal, we wish to examine a *vertical element* of the flame, the slit and the prism must, of course, be set vertically. The former mode possesses great advantages, as will be presently pointed out. It is to be understood in all cases that the eye-piece of the telescope is adjusted to give a sharp image of the slit, and the prism is at its angle of minimum deviation.

By this arrangement I have examined a great number of different flames, as those of oil, alcohol, solutions of boracic acid and nitrate of strontian in alcohol, phosphorus, sulphur, carbonic oxide, hydrogen, cyanogen, arseniuretted hydrogen, etc. Among these, it will be noticed that different colors occur. Oil gives a yellow flame, alcohol a pale blue, boracic acid green, strontian red, phosphorus yellowish · white, sulphur and carbonic oxide blue, hydrogen pale yellow, cyanogen lilac, arseniuretted hydrogen white, etc.

Notwithstanding this diversity of color, all these flames, as well as many others I have tried, yield the same result: every prismatic color is found in them. Even in those cases where the flame is very faint, as in alcohol and hydrogen gas, not only may red, yellow, green, blue, and violet light be traced, but even bright Fraunhoferian lines of different colors.

This observation holds good for those flames reputed to be monochromatic; for example, alcohol burned from a wick imbued with common salt. It is not only a yellow light which is evolved; the other colors plainly, though more faintly, appear.

All flames, no matter what their special colors may be, evolve all the prismatic rays. Their special tints arise from the preponderance of one class of rays over another; thus in cyanogen the red predominates, and in sulphur the blue.

(Later experiments have proved that the spectrum thus containing all the prismatic colors, and acting as a background to the bright fixed lines, is not due, except indirectly, to the flame under examination, but to other causes, not here taken into account, especially the accidental presence of daylight, combustion of dust in the air, etc. The statements here given are, however, in accordance with observations actually made. These are the effects that will be seen in a partially illuminated room such as the laboratory in which these experiments were made. In a dark room the spectrum background disappears.)

The production of light in the case of flames is thus proved to be a very complex phenomenon. The chemical conditions under which the burning takes place are likewise very complex. The combustible vapor is surrounded on all sides by atmospheric air; diffusion occurs, and rapid currents are established by the high temperature. Such circumstances complicate the result; and it is only by observing the burning of an elementary solid, in which most of these disturbances are cut off, that we can hope to effect a proper resolution of the problem.

II. *Prismatic analysis of the light of an elementary*

solid burning at different temperatures; proving that as the temperature rises the more refrangible rays appear.

I took from the fire a piece of burning anthracite coal —the fuel ordinarily used in domestic economy in New York—and which from its compactness, the intense heat it evolves, and other properties, appears to be well fitted for these investigations. This coal was placed on a support so as to present a plane surface to the slit in the metal screen. The rays coming from it and passing the slit were received on the flint-glass prism, and viewed through the telescope.

When the coal was first taken from the fire, and was burning very intensely, on looking through the telescope all the colored rays of the spectrum were seen in their proper order. I had previously passed through the slit a beam of sunlight reflected from a mirror, so as to have a reference spectrum with fixed lines. Now when the coal was burning at its utmost vigor, the spectrum it gave did not seem to differ, either as respects length or the distribution of its colors, from the spectrum of sunlight; but as the combustion declined and the coal burned less brightly, its spectrum became less and less, the shortening taking place first at the more refrangible extremity, one ray after another disappearing in due succession. First the violet became extinct, then the indigo, then the blue, then the green, until at last the red, with an ash-gray light occupying the place of the yellow, was alone visible, and presently this also went out.

From numerous experiments of this kind, I conclude that *there is a connection between the refrangibility of the light which a burning body yields and the intensity of the chemical action going on, and that the refrangibility always increases as the chemical action increases.* It might, perhaps, be objected that, in the form of experiment here introduced, two totally different things are confounded, and that the burning coal not only gives forth its rays as a combustible body, strictly speaking, but also as an incandescent mass.

To avoid this objection as far as possible, and also to reach a much higher temperature than could have been otherwise obtained, I threw a stream of oxygen gas on that part of the anthracite which was opposite the slit; but my expectations were disappointed, for, instead of the combustion being increased, the coal was actually extinguished by the jet playing on it. I therefore replaced the anthracite with a flat piece of well-burned charcoal, kindled at the portion opposite the slit, and throwing a stream of oxygen on this part, the combustion was greatly increased. A spectrum rivalling that of the sunbeam in brilliancy was produced, all the colors from the extreme red to the extreme violet being present.

On shutting off the supply of oxygen, the combustion, of course, declined, and while this was going on the violet, the indigo, the blue, the green, etc., faded away in succession. By merely turning the gas on or off, the original colors could be re-established or made to decline. It was very interesting to see with what regularity, as the chemical action became more intense, the more refrangible colors were developed, and how, as it declined, they disappeared in due succession; the final tint being red and that ash-gray in the position of the yellow which has been described in the preceding memoir.

In the form of experiment here made the combustion is, of course, merely superficial; and the rays come from the charcoal not as an incandescent, but as a burning body.

III. *Of the constitution of flames; proving that they*

consist of a series of concentric and differently colored shells.

I regard the foregoing experiments as affording the means of explanation of the much more complicated phenomena of flames, and proceed to inquire whether the principle I have just brought forward of the co-ordinate increase of refrangibility and of chemical action will hold good; premising the experiments now to be detailed with the following considerations.

All common flames, as is well known, consist of a thin shell of ignited matter, the interior being dark, the combustion taking effect on those points only which are in contact with the air. From the circumstances under which the air is usually supplied, this ignited shell cannot be a mere mathematical superficies, but must have a sensible thickness. If we imagine it to consist of a series of cone-like strata, it is obvious that the phenomena of combustion are different in each. The outer stratum is in contact with the air, and there the combustion is most perfect; but by reason of the rapid diffusion of gases into one another, currents, and other such causes, the atmospheric air must necessarily pervade the burning shell to a certain depth, and in the successive strata, as we advance inward, the activity of the burning must decline. On the exterior stratum oxygen is in excess, at the interior the combustible vapor, and between these limits there must be an admixture of the two, which differs at different depths. Admitting the results of the foregoing experiments with anthracite coal and charcoal to be true, viz., that as the combustion is more active, rays of a higher degree of refrangibility are evolved, it follows that *each point of the superficies of such a flame must yield all the colors of the spectrum,* the violet coming from the outer strata, the yellow from the intermediate, the red from those within. If we could isolate an ele-

mentary horizontal section of such a flame, it would exhibit the appearance of a rainbow ring, and when those
compound rays are received on the face of a prism, the
constituent colors are parted out by reason of their different refrangibility, and the eye is thus made sensible
of their actual existence.

When thus by the aid of a prism we analyze the light
coming from any portion of the superficies of a flame, we
in effect dissect out in a convenient manner and arrange
together side by side rays that have come from different
strata of the burning shell. These, without the prism,
would have pursued the same normal path, and produced a commixed effect as white light on the eye, but
with it are separated transversely, and each becomes
perceptible.

It is immaterial whether we impute the light emitted
by an ordinary flame to the liberation of solid particles
of carbon in an ignited condition, and becoming hotter
and hotter as they pass outwardly towards the surface,
or consider these particles to be in a state of combustion. The experiments of an ignited wire in one case,
and of charcoal in presence of oxygen in the other, lead
to the same explanation. We are not to suppose that
it is simply a gas which is burning; we are examining
the light emitted by an incandescent solid—the carbon
particles that for the moment are set free.

(This explanation, that the luminosity of a flame is
due to the temporary extrication of solid carbon, was
given by Sir H. Davy. It has been called in question
by Frankland. Experiments and criticisms have since
been offered by Deville, Knapp, Stein, Blockmann, and
others, but Davy's theory still remains substantially unaffected. This is the conclusion to which Heumann has
come in his recently published researches on luminous
flames (1876).)

It might be supposed that in the familiar instance of an oil lamp, if we put any check on the supply of the air, and thereby check the intensity of the combustion, we ought to produce a flame emitting rays of light the refrangibility of which becomes less and less, and which, from their being originally white, should pass through various shades of orange, and end in a dull red. But the compound nature of the burning vapor interferes with that result; for when a certain point is gained the hydrogen, for the most part, alone burns, the carbon being set free as smoke, and such a flame cannot support itself in strict accordance with the principle given.

We must, then, search for other conditions under which carbon is found which are free from this difficulty. Two at once present themselves: they are carbonic oxide and cyanogen gas. In the former the carbon is already united with half the oxygen required for maximum oxidation: its complete combustion can therefore be carried on with a limited supply of atmospheric air; in the latter the carbon is united with nitrogen, which during the combustion is set free, and interferes with the process by cutting off the more complete access of the atmosphere.

In place of the burning coal of the former experiments I substituted a jet-pipe through which the various gases might be made to pass, and the rays emitted by their flames enter the telescope after passing through the slit and prism. In this arrangement the slit should be horizontal and not vertical. So far from it being immaterial which of the two positions is selected, very great advantages arise from the former. If the slit be vertical, the prism, it is true, will separate the constituent colors from one another, but it fails to show their relative positions. If it be horizontal, the relative positions of the different colors can be demonstrated, and it can be proved that a horizontal section of a flame is in reality, as has been al-

ready remarked, a colored ring, the red being the innermost color, and the violet outside; for if this be the
order in which the colors occur, the red ring must necessarily have a less diameter than the green, and the green
less than the violet; and when the prism, set in a horizontal position, separates those colors from each other,
the sides of the resulting spectrum ought not to be parallel, but inclined to one another, the breadth being least
in the red and increasing towards the violet end.   This
increasing breadth proves that the constituent colored
shells of the flame envelop each other, the violet being
outermost, and therefore broadest. This valuable indication would be wholly lost if the slit were vertical.

This being understood, I may illustrate the facts now
to be brought forward by an example of the prismatic
analysis of a horizontal element of the flame of a spirit-
lamp; it being understood that the prism is at its angle
of minimum deviation, and the spectrum seen through
the telescope. All the prismatic colors, in their proper
order, are visible; the sides of the spectrum not being parallel, the inclination being quite rapid towards the red
extremity, the rays of which come from the interior of
the flame, where the diameter is less. Mere inspection is
sufficient to show the rapid approach of the red sides to
each other, and I satisfied myself that even in the more
refrangible regions there is the same want of parallelism,
by rotating the telescope on its vertical axis so that the
vertical wires in its eye-piece might coincide with first
one and then the other side of the spectrum.   It will be
understood that I took the proper precaution not to be
deceived by a partial want of achromaticity in the telescope, which might have led to a mistake.

But further, the yellow space of such a spirit-flame
spectrum is crossed by a bright fixed line—Brewster's
monochromatic ray. It is a beautiful example of the

principles just pointed out in this method of horizontal analysis, being of much greater width than the rest of the spectrum, and recalling to the imagination the appearance of Saturn's ring when nearly closed and seen through a telescope of moderate power. This ray, from its superior breadth, must necessarily come from that pale, tawny light which invests the bright part of the flame. This, which is readily seen when the flame is large, envelops the middle and upper parts, but cannot so easily be detected low down. It is to be attributed to the carbonic acid and steam that have risen at a high temperature in the burning shell, and are escaping at a degree above that of incandescence into the air, and are mingled with oxygen diffusing from the air into them. A similar tawny cloak surrounds the upper part of the flame of a candle; it answers to the oxidizing flame of the blow-pipe, and yields Brewster's monochromatic yellow light.

(A few years subsequently this yellow ray was discovered by Swan to be due to sodium. It is now known as the lines D. At the date of this memoir it was not suspected that sodium is so universally diffused.)

IV. *Explanation of the nature of colored flames, showing, for example, why carbonic oxide burns blue, and cyanogen red.*

To return now to carbonic oxide and cyanogen. Fig. 8, No. 1, represents the solar spectrum with its fixed lines; No. 3 represents the spectrum of carbonic oxide burning in the air. It begins in the red region short of the fixed line C, and terminates between the lines G and H. It yields, therefore, rays of every color, and this in accordance with the principles we have laid down; but when the relative quantity and force of the rays are estimated in comparison with the sunlight spectrum, the

Spectra of Various Flames.

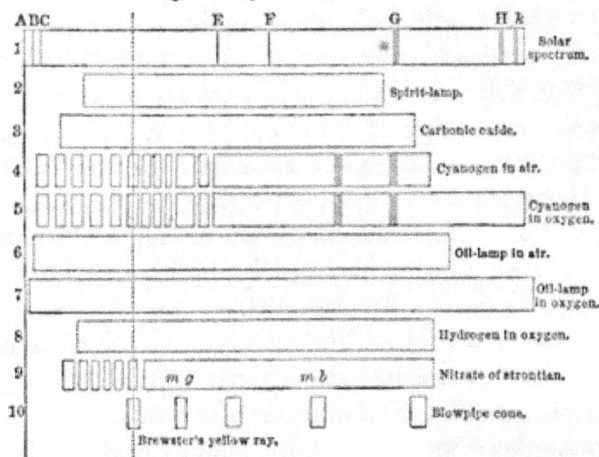

Fig. 8.

red and orange are deficient, and the more refrangible colors predominate, and, indeed, it is the excess of these that gives the flame its characteristic blue tint. This agrees with what has been observed as to anthracite and charcoal; for with carbonic oxide a limited supply of oxygen can bring about the maximum chemical action, and therefore liberate in abundance rays of maximum refrangibility.

This condition of things is inverted in the case of cyanogen. It is the nature of its flame to be enveloped, as it were, in a sheet of nitrogen arising from its own burning, and this necessarily impedes the access of air and checks the intensity of the chemical change: a check which is at once betokened by the emission of a predominant number of rays of low refrangibility or of a red color.

But there is a striking difference in the chemical conditions under which carbonic oxide and cyanogen burn. In the case of the former the whole gas is combustible, in the latter the carbon alone, and we have, in reality, introduced an incombustible element into the flame; for as the carbon burns the incombustible nitrogen is set free. It occurred to me, in selecting the gas for experiment, that this condition should impress a physical characteristic on the flame. I thought it was not impossible that dark lines in its spectrum might be the result, because there must be a peculiar arrangement of the burning strata which together make up the shell of the flame, every two atoms of carbon setting free one of nitrogen. I did not know, until subsequently, that this flame had already been examined by Faraday. Having therefore confined some cyanogen, made from cyanide of mercury, in a glass gas-holder filled with a saturated solution of common salt, I burned it from the jet-pipe, and found that what I had surmised was actually the fact. There was a spectrum so beautiful that it is impossible to describe it by words, or depict it in colors. It was crossed throughout its extent by black lines, separating it into well-marked divisions. I could plainly count four great red rays of definite refrangibility, followed by one orange, one yellow, and seven green rays; while in the more refrangible spaces were two extensive groups of black lines, recalling somewhat from their position, but greatly exceeding in extent, Fraunhofer's lines G and H in the sun-rays. I shall return to the consideration of this spectrum and to the relation of fixed lines presently, here only making the remark that the burning of cyanogen, both as respects the color of the light and the occurrence of fixed lines, is a direct consequence of the principle I am establishing.

The unassisted eye detects two well-marked regions in the cyanogen flame: a greenish-gray stratum on the outside, and a lilac-colored nucleus within. Decomposed by the prism, a horizontal element of this flame shows that the exterior shell contains all the prismatic colors,

except, perhaps, the yellow; but the green, the blue, and
the violet greatly predominate. The interior lilac flame
is the source of the bright spectrum with fixed lines just
described.

V. *Continuation of the same principle in the case in
which combustion is carried on in oxygen gas instead of
atmospheric air.*

If the principle that high refrangibility is connected
with intense chemical action be true, it must hold good
when the nature of the atmosphere in which the burning
is carried forward is changed. If, instead of being the
common air, it is oxygen gas, we ought to be able to
foresee the result. Carbonic oxide, when made to burn
in that gas, should not change its tint; because if the air
can carry on the process to its maximum effect, oxygen
can do no more. But the result should be just the reverse with cyanogen, which, if made to burn in oxygen,
should be capable of emitting rays of higher refrangibility.

Foreseeing this result, I submitted the two gases to
experiment, and first arranged the carbonic oxide so
that its spectrum might be examined in the telescope
as already described; then causing a clean bell-jar full
of oxygen to be inverted over it, the flame diminished
somewhat in size, emitted a slight crackling sound, *but
retained its color unchanged.* Its spectrum appeared precisely the same both as respects extent and the distribution of color, whether the burning took place in oxygen
gas or in atmospheric air.

If cyanogen be made to burn in oxygen, we should
expect that it would lose to a great extent its characteristic lilac tint, and emit a whiter light. It was therefore
very interesting to find that the moment the flame was
immersed in oxygen it lost much of its pinkish color,

and became of a dazzling brilliancy; and on examination through the telescope, though all the colors had increased in brightness, the most remarkable effect took place among the extreme refrangible rays. Far out of the limits of the ordinary spectrum a ray of great purity and force was developed, as represented in the Fig. at No. 5. Its color is violet.

I have made similar experiments on many other flames besides those here mentioned. It is not necessary to relate them in detail, for they give the same results. In every instance of combustion in the air, when the flame is bright enough, all the colors are visible; and when the combustion takes place in oxygen they are increased in intensity. With hydrogen gas and alcohol the light is so feeble that the eye cannot catch the terminal rays; but as soon as the combustion is made in oxygen the red and the violet both appear, the latter, however, predominating. Several of these spectra both in air and oxygen are represented in Fig. 8. In No. 9, the letters $m\ g$ and $m\ b$ indicate a maximum of green and blue light in the form of bright lines.

It does not require the use of a prism to satisfy one's self of the change of tint that flames exhibit when the chemical action increases. In reality it is only necessary to compare by eye-sight the color of the light emitted in air and in oxygen gas. In the latter case rays of a higher refrangibility uniformly arise.

On the evidence furnished by the foregoing experiments I regard a common flame as consisting of a shell of ignited matter in which combustion is going on with different degrees of rapidity at different depths, being most rapid at the exterior, where there is a more perfect contact with the atmosphere, and diminishing inward. In a horizontal section, the interior space consisting of unburned vapor is black; this is surrounded by a ring

where the combustion is incipient, and from which red
light issues; then follow orange, yellow, green, blue,
indigo, and violet circles in succession, the production
of each of these tints being dependent on the rapidity
with which chemical action is going forward—that is,
on the amount of oxygen present—the tints gradually
shading off into one another and forming, as I have said,
a circular rainbow. An eye placed on the exterior of
such a flame sees all the colors conjointly, and from their
general admixture arises the predominant tint.

An examination of the flame of a candle *vertically*
confirms this conclusion, for the red projects on the top
of the flame, and the blue towards the bottom.

From this, which may be regarded as the normal
flame, the flame of cyanogen differs. It must consist of
as many concentric shells as the prism separates it into
regions of definite refrangibility. Its interior part is
therefore divided into four red layers, followed by one
of orange, one of yellow, seven of green, etc. There are
two great inactive spaces towards the outside of the
flame, corresponding to the two great groups of fixed
lines. Perhaps through all these inactive parts the incombustible nitrogen chiefly escapes.

VI. *Effects of the introduction of air into the interior
of a flame, producing the destruction of the red and
orange strata, and converting them into violet.*

It now becomes a curious subject to determine what
takes place when an ordinary flame is disturbed by the
introduction of air into its interior. When a blow-pipe
jet is thrown through the flame of an oil-lamp, the sharp
blue cone which forms indicates, on the principles here
set forth, that the combustion is much more active. But
if the colors of the common flame come from different
depths, the red being the innermost, it is clear that the

introduction of a jet of air by a blow-pipe should make the combustion rapid where before it was slowest, and the less refrangible colors ought to be destroyed. A prismatic analysis should exhibit the spectrum of a blow-pipe flame without any red or orange.

In this examination no slit was required, as in the former experiments, for the cone itself, when at a distance of six or eight feet, was narrow enough for the purpose. It yielded a very extraordinary spectrum. As I anticipated, all the red rays were gone; not a vestige of either them or of the orange could be found. But the spectrum was divided into five well-marked regions, separated from one another by dark spaces. There were five distinct images of the blue cone: one yellow, two green, one blue, and one violet. In Fig. 9 this result is represented.

This experiment may be verified without a telescope. On looking through a prism, set horizontally at its angle of minimum deviation, at a blow-pipe cone some six or eight feet distant, there will be seen a spectrum of that part of the flame which does not join in the production of the blue cone. It contains, of course, all the prismatic colors. But projecting from this are five colored images of the cone—one yellow, two green, one blue, and one violet. They are entirely distinct from one another, and are parted by dark spaces.

Such is the effect of introducing air into the interior of a flame, and destroying those strata that yield the red and orange colors. The effect of a blow-pipe is to produce two strata of blue light, one being external, the other internal; also two strata of green, one again external, the other internal, and the escaping products

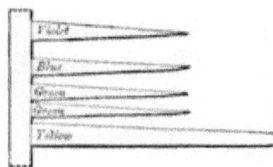

Fig. 9.

of combustion, steam and carbonic acid, mingled with
atmospheric air, constitute the oxidizing flame, which
envelops the blue cone, and emits Brewster's monochromatic yellow light. That the yellow light comes
from this flame is proved by the greater length of its
image.

VII. *Physical cause of the production of light by chemical action.*

Do not the various facts here brought forward prove
that chemical combinations are attended by a rapid
vibratory motion of the particles of the combining
bodies, which vibrations become more frequent as the
chemical action is more intense?

The burning particles constituting the inner shell of
a flame are executing about four hundred billions of
vibrations in one second; those in the middle about
six hundred billions, and those on the exterior, in contact with the air, about eight hundred billions in the
same time. The quality of the emitted light, as respects its color, depending on the frequency with which
these vibrations are accomplished, increases in refrangibility as the energy of the chemical action becomes
greater.

The parts of all material bodies are in a state of
incessant vibration; that which we call *temperature* depends on the frequency and amplitude of these vibrations conjointly. If by any process, as by chemical
agencies, we increase that frequency to between four
and eight hundred billions of vibrations in one second,
ignition or combustion results. In the case of the former of these numbers, the temperature is 977° Fahr.
At this temperature the waves propagated in the ether
impress the organ of vision with a red light. *This also
is the temperature of the innermost shell of a flame.* If

the frequency of vibration still increases, the temperature correspondingly rises, and the light successively becomes orange, yellow, green, blue, etc., and this condition obtains in the successive strata of a flame, as we pass from its interior to its exterior surface.

The general principle at which I thus arrive, as the result of this experimental investigation, viz., that there is a connection between the energy with which chemical affinity is satisfied and the refrangibility of the resulting light, assumes the position of a simple consequence of the undulatory theory. Is it not very natural, if all chemical changes are attended by vibratory motions in the particles of the bodies engaged, that those vibrations should increase in frequency as the action becomes more violent? But an increased frequency of vibration is the same thing as an increased refrangibility.

VIII. *On the physical cause of Fraunhofer's dark lines.*

Although I have extended this memoir to so great a length, I have omitted many facts which have been made the subject of experiment. I cannot conclude, however, without offering some remarks on the artificial production and cause of Fraunhofer's fixed lines.

It has been stated that I was led to expect the production of these lines in the flame of cyanogen, from considering the circumstances under which its combustion takes place. Returning to this phenomenon, I shall here point out a very remarkable numerical relation existing among the fixed lines of the solar spectrum.

The following table contains Fraunhofer's determination of the wave-lengths of the seven great fixed lines of the spectrum, which are designated by the capital letters of the alphabet from B to H. I have added the wave-length of A from my own experiments.

*Table of wave-lengths corresponding to the eight great fixed lines of
the solar spectrum, the Paris inch being supposed to be divided into
one hundred millions of equal parts.*

A = 2660	E = 1945
B = 2541	F = 1794
C = 2422	G = 1587
D = 2175	H = 1464

An examination of this table proves that—

The wave-length of B is	119 parts less than A ;
"	" C " 238 " "
"	" D " 485 " "
"	" E " 715 " "
"	" F " 866 " "
"	" G " 1073 " "
"	" H " 1196 " "

and these differences of length are obviously very nearly
as the whole numbers 1, 2, 4, 6, 7, 9, 10. This coincidence
is far too striking to be merely accidental. Moreover,
it must not be forgotten that the observed numbers as
determined by Fraunhofer are wholly independent of
any hypothesis.

If the relation of whole numbers were rigorously true,
the numbers in the foregoing table would stand as follows: 119, 238, 476, 714, 833, 1071, 1190.

The wave-length of the most luminous portion of the
spectrum, the centre of the yellow space, is 2060 parts.
If we take this as an optical centre, it will be found that
the great lines are situated symmetrically in relation to
it. E and D are equidistant above and below it; the
same observation applies to G and B, and also to H and
A. The only departure from this symmetry is in the
case of F, which is not symmetrical with C. It will be
understood that I am here speaking of one of those
spectra which are formed when a grating or ruled surface is used. In this the colors are arranged side by
side, according to their wave-length, the centre of the
spectrum, which is its most luminous portion, is occupied

by the centre of the yellow space, and the light terminates at equal distances in the violet and red.

Do not these observations lead us to conclude that the cause, whatever it may be, which produces these fixed lines is periodic in its action?

What that cause in reality is we have not now facts sufficient to determine. I would not affirm that the disengagement of incombustible matter by a flame will always give rise to dark lines. But this is very clear, that in all those cases, as cyanogen, alcoholic solutions of nitrate of strontia, of boracic acid, etc., in which these lines are developed, incombustible matter is uniformly disengaged.

UNIVERSITY OF NEW YORK, *Dec.* 25, 1847.

•

MEMOIR III.

ON INVISIBLE FIXED LINES IN THE SUN'S SPECTRUM DE-TECTED BY PHOTOGRAPHY.

From the Philosophical Magazine, May, 1843.

CONTENTS:—*The photography of Fraunhofer's fixed lines.—The lines in the red due to absorption by the earth's atmosphere.—Nomenclature of the Fraunhofer lines.—Discovery of the invisible fixed lines.—Original map of them.—The new ultra spectral red lines α, β, γ.—Rediscovered by Foucault and Fizeau.—M. E. Becquerel's discovery.—Experiments in 1834.*

WHEN a beam of the sun's light, directed horizontally by a heliostat, is admitted into a dark room, and, passing through a slit with parallel edges, is received on the surface of a flint-glass prism, which refracts it at the angle of minimum deviation, and, after its passage through the prism, is converged to a focal image on a white screen by the action of an achromatic lens, the resulting spectrum is given in great purity, and Fraunhofer's lines are very distinct. If a photographic surface be set in the place of the white screen, it will exhibit the representation of multitudes of dark lines, varying greatly in dimensions.

After several attempts last summer, I succeeded in discovering these lines, and have obtained impressions of them sufficiently perfect.

Before proceeding to the description of the mode to be followed, and of the characters of the lines themselves, I cannot avoid calling attention to the remarkable circumstance which has frequently presented itself to me of a

great change in the *relative visibility* of Fraunhofer's
lines when seen on different occasions. There are times
at which the strong lines seen in the red ray are so feeble that the eye can barely catch them, and then again
they come out as dark as though marked with India-ink
on the paper. During these changes the other lines may
or may not undergo corresponding variations. The same
remark applies equally to the blue and yellow rays. It
has seemed to me that the lines in the red are more visible as the sun approaches the horizon, and those at the
more refrangible end of the spectrum are plainer in the
middle of the day.

(I subsequently substantiated this remark, and satisfied myself that many of the lines in the red are due to
absorption by the earth's atmosphere, and therefore more
distinct with a rising or setting sun. Those in the more
refrangible regions, the blue, the indigo, and the violet,
are due to absorption by the atmosphere of the sun.)

A sunbeam, passing horizontally from a heliostat mirror into a dark room, was received on a metal plate with
a slit in its centre, the slit being formed by a pair of
parallel knife edges, one of which was movable by a
micrometer screw, the instrument being, in fact, the common one used for showing diffracted fringes. The screw
was adjusted so as to give an aperture $\frac{1}{52}$ inch wide, and
the light passing through fell upon an equiangular flint-
glass prism placed at a distance of eleven feet. Immediately on the posterior face of the prism the ray was
received on an achromatic lens, the object-glass of a telescope, and brought to a focus at the distance of six feet
six inches, at which an arrangement was adjusted for exposing white paper screens, on which the greater fixed
lines might be seen, or sensitive plates substituted for
the screens, occupying precisely the same position. The
lines on the screens could, therefore, be compared with

those on the sensitive surfaces as to position and magnitude with considerable accuracy.

In order to identify these lines I have made use of the map of the spectrum published by Professor Powell in the Report of the British Association for 1839. With the apparatus, as above described, they are exceedingly distinct; no difficulty arises in the identification of the more prominent ones. The spectrum with which I have worked occupies upon the screen a space of nearly four inches and a quarter in length from the red to the violet, or, more correctly speaking, from the ray marked in that map A to the one marked *k*. In stating, however, that no difficulty arises in identifying these lines, I ought to add that I am referring to that particular map. In the figure annexed to Sir John Herschel's "Treatise on Light," in the "Encyclopædia Metropolitana," the rays marked G seem to differ from that in the report. But Professor Powell's map being drawn from his personal observations, with reference to these very difficulties, and as it agrees with my own observations and measures, I have employed it, and therefore take the letters he gives. (I may add that in all the earlier of these memoirs his nomenclature of the fixed lines is used. It differs a little from that now employed by spectroscopists.)

It will be understood that the *whole* spectrum and *all* its lines cannot be obtained at one impression. The difficulty is that the different regions of the spectrum act with different power in producing the proper effect. Thus, if on common yellow iodide of silver the attempt were made to obtain all the lines at one trial, it would be found that the blue region would have passed to a state of high solarization, and that all its fine lines were extinguished by being overdone long before any well-marked action could be traced in the less refrangible extremity. It is necessary, therefore, to examine the different regions in succession, exposing the sensitive surface to each for a suitable length of time.

In Fig. 10 I have given on the left side a representation of the larger lines of Fraunhofer; the right side gives them as obtained on a daguerreotype plate which has been iodized to a yellow, brought by the vapor of bromine to a red, and then slightly exposed to the vapor of chloride of iodine. The photograph is so adjusted as to have its lines by the side of those of Fraunhofer which have the same name. It will be seen that there are beyond the red ray three extra spectral lines, which I have marked a, β, γ. These, however, I have only occasionally found, for from the general diminution of effect in that region they do not always come out in a plain and striking manner. None of Fraunhofer's lines in the yellow and green are given, but G and its companions are very strongly impressed, as also the group about i. But by far the most striking in the whole photograph are those marked H and k. Then passing beyond the violet and out of the visible limits of the spectrum, four very strik-

Fig. 10.

ing groups make their appearance. To the first line of each of these, in continuation of Fraunhofer's nomenclature, I gave the designations M, N, O, P. In l there are three lines, in M eight, in N three, in O four, and in P five.

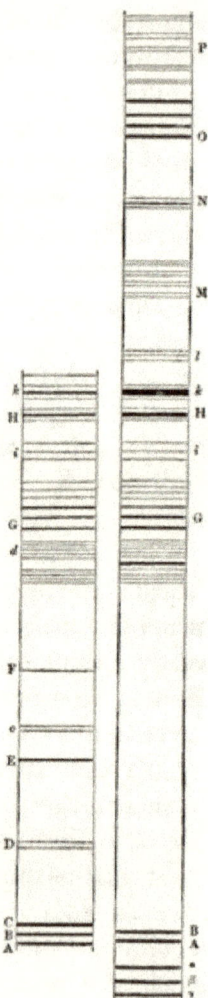

Besides these larger groups, the photographs were

crossed by hundreds of minuter lines, so that it was impossible to count them. If nearly six hundred have been counted between A and H, I should think there must be quite as many between H and P. In speaking of these lines as though they were strong individual ones, the statement is to be taken with some limitation. It is quite likely that each of these bolder lines is made up of a great number that are excessively narrow and close together.

If the absorptive action of the sun's atmosphere be the cause of this phenomenon, that action must take place much more powerfully on the more refrangible and extra-spectral region. The lines exhibited there are bold and strongly developed; they are crowded in groups together.

(Scarcely was the paper from which the foregoing extracts are made published in the *Philosophical Magazine*, when I learned that in France M. E. Becquerel had already photographed the more refrangible lines, and published statements to that effect. But he had not observed those in the less refrangible regions, designated by me a, β, γ.

In fact, the process I was using was one I had recently discovered: it consisted in permitting the daylight to fall along with the sun rays on the photographic surface. The daylight and the sunlight antagonized each other, and these hitherto undiscovered lines made their appearance as positive photographs. The peculiarities of this singular and interesting process I will describe hereafter in one of these memoirs.

In 1846, MM. Foucault and Fizeau, having repeated the experiment thus originally made by me, presented a communication to the French Academy of Sciences. They had observed the antagonizing action above described, and had seen the ultra spectrum heat lines a, β, γ. They

had taken the precaution to deposit with the Academy
a sealed envelope containing an account of their discovery, not knowing that it had been made and published
long previously in America.

Hereupon M. E. Becquerel communicated to the same
Academy a criticism on their paper. In this he remarks:
" M. Draper, in examining the image produced by the
action of the spectrum on plates of iodized silver, announced before those gentlemen the existence of protecting rays antagonizing the action of the solar rays, and
even acting negatively on iodide of silver." He strengthened his views by adding some observations that had
been made by Sir J. Herschel, who did not assent to the
existence of this protecting action, but thought that the
daguerreotype impressions could be explained on Newton's theory of the colors of thin plates.

Herschel had made some investigations on the distribution of heat in the spectrum, using paper blackened on
one side and moistened with alcohol on the other. He
obtained a series of spots or patches, commencing above
the yellow and extending far below the red. Some writers on this subject have considered that these observations imply a discovery of the lines a, β, γ; they forget,
however, that Herschel did not use a slit, but the direct
image of the sun—an image which was more than a
quarter of an inch in diameter, as I know from the specimens he sent me, and which are still in my possession.
Under such circumstances it was physically impossible
that these or any other of the fixed lines should be seen.

As I had thus been unsuccessful in obtaining impressions of the fixed lines D and E—once only I thought I
perceived a line corresponding to Fraunhofer's F, but it
was exceedingly faint and on the whole doubtful—I supposed that this furnished an argument for the physical
independence of the luminous and actinic rays, as they

were subsequently called. The lines D, E, F were, however, afterwards photographed by my son, Henry Draper,
and so the argument fell to the ground.)

In 1834, when my attention was first drawn to these
subjects, and I began to make prismatic analyses by the
aid of sensitive paper, some of my earliest attempts were
directed to the detection of these fixed lines. At that
time I was employing sensitive paper, made with bromide of silver, precisely as has been subsequently done
in Europe—a number of the results were published in
the American journals during the year 1837. In the detection of the fixed lines I failed at that time entirely;
but the bromuretted paper enabled me at that early period, when the attention of no other chemist was as yet
turned to these matters, to trace the blackening action
from far beyond the confines of the violet, down almost
to the other end of the spectrum. I distinctly made out
that the dark rays underwent interference, after the manner of their luminous companions, a result originally due
to Arago, and printed some long papers in proof of the
physical independence of the chemical rays and light and
heat throughout the spectrum.

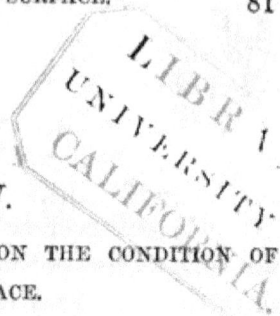

MEMOIR IV.

ON THE NATURE OF FLAME AND ON THE CONDITION OF THE SUN'S SURFACE.

From the American Journal of Science and Arts, Second Series, Vol. XXVI., 1858;
Philosophical Magazine, Feb., 1858.

CONTENTS:—*Dove's experiments on electric light.—Dark lines replaced by bright ones.—Electric spark between metallic surfaces.—The lines depend on the chemical nature of the substance from which the light issues.—They may be used for determining the physical condition of the sun and stars.—Three hypotheses of the condition of the sun's surface examined.*

AMONG the more recent publications on Photo-chemistry, there is one by Professor Dove on the electric light (*Philosophical Magazine*, Nov., 1857) which will doubtless attract the attention of those interested in that branch of science. Examination by the prism, and by absorbing and reflecting colored bodies, leads him to the conclusion that it is necessary to consider the luminous appearance as having two distinct sources: 1st, the ignition or incandescence of the material substances bodily passing in the course of the discharge; 2d, the proper electrical light itself. As respects the former, he illustrates its method of increase from low to high temperatures by supposing a screen to be withdrawn from the red end of the spectrum through the colored spaces successively towards the violet; and that of the latter from the bluish brush to the bright Leyden spark, by a like screen drawn from the violet towards the red.

The true electric light exhibits properties resembling those observed in actual combustions, as though there

F

were an oxidation of a portion of the translated matter
when the spark is taken in air.   The order of evolution
of rays in this instance happens to be the same as in the
second illustration of Professor Dove, that is, from the
violet to the red.   There are certain facts connected
with these appearances of color which are not generally
known, and deserve to be pointed out.

(I then give an abstract of the preceding memoir, and,
after speaking of the production of dark lines in the cyanogen flame, continue as follows:)

In other cases dark lines are replaced by bright ones,
as in the well-known instance of the electric spark between metallic surfaces.   *The occurrence of lines, whether
bright or dark, is hence connected with the chemical nature
of the substance producing the flame.*   For this reason
these lines merit a much more critical examination than
has yet been given to them, for by their aid we may be
able to ascertain points of great interest in other departments of science.   Thus if we are ever able to acquire
certain knowledge respecting the physical state of the
sun and other stars, it will be by an examination of the
light they emit.   Even at present, by the aid of the few
facts before us, we can see our way pretty clearly to certain conclusions respecting the sun.   For since substances
which are incandescent, or in an ignited state, through
the accumulation of heat in them, show no fixed lines,
their prismatic spectrum being uninterrupted from end
to end, it would appear to follow that the luminous condition of our sun, whose light contains fixed lines, cannot
be referred to such incandescence or ignition.   At various times those who have studied this subject have offered different hypotheses: one regarding the sun as a
solid or perhaps liquid mass in a condition of ignition;
another considering the light to be electrical; a third
supposing him to be the seat of a fierce combustion.

Of such hypotheses we have given reasons for declining the first. Prismatic analysis, which demonstrates no resemblance between the light of the sun and that of any form of electric discharges with which we are familiar, enables us in like manner to reject the second; and, upon the whole, facts seem most strongly to prepossess us in favor of the third; in artificial combustions similar fixed lines being observed. If such is to be regarded as the physical condition of the sun, we can no longer contemplate him as an immense mass slowly and tranquilly cooling in the lapse of countless centuries by radiation into space, as so many considerations drawn from other branches of science have hitherto led us to suppose, but he must be regarded as the seat of chemical changes going on upon a prodigious scale, and with inconceivable energy.

If the law designated above, that the more energetic the chemical action in combustion the more refrangible the emitted light, be translated into the conceptions of the undulatory theory, it not only puts us in possession of a distinct idea of the manner in which the combustive union of bodies is accomplished, the quickness of vibration increasing with the chemical energy, but it also enables us to transfer for the use of chemistry some of the most interesting numerical determinations of optics.

UNIVERSITY OF NEW YORK, *Dec.* 10, 1857.

NOTE. — I have thus presented the four preceding memoirs as early contributions to the history of spectrum analysis, and applications of spectroscopic researches to solar physics or astronomical problems. I think that perhaps they will not be less interesting to the scientific reader from the circumstance of their imperfections.

He will not look upon them from the present elevated point of view, but regard them as results gathered with much labor by a pioneer—results that have had something to do with the development of the subject.

The history of science shows that there have sometimes been occasions on which one investigator, writing at an opportune moment, has carried off from his predecessors all the credit they were entitled to, and, perhaps without any definite intention on his part, the world has unjustly awarded it to him. In America, as the foregoing memoirs show, some attention had been paid to the use of the spectroscope long previously to the time when M. Kirchoff occupied himself with the subject.

Thirteen years after the publication of the first of these memoirs, M. Kirchoff (1860), in a memoir regarded at that time as the origin of spectrum analysis, and entitled, "On the Relation between the Radiating and Absorbing Powers of Different Bodies for Light and Heat," published, under the guise of mathematical deductions, many of these facts as discoveries of his own. This memoir appeared in German in Poggendorff's *Annalen*, Vol. CIX., p. 275, and was translated into English in the *Philosophical Magazine*, July, 1860.

Among these deductions are the following. I quote M. Kirchoff's own language:

"If a body (a platinum wire, for example) be gradually heated up to a certain temperature, it only emits rays consisting of waves longer than those of the visible rays. Beyond that point waves of the length of the extreme red begin to appear, and as the temperature rises, shorter and shorter waves are added, so that, for every temperature, rays of a corresponding length of wave are originated, while the intensity of the rays of greater wave-length is increased.

"Whence, applying the same proposition to other bodies, it follows that all bodies, when their temperature is gradually raised, begin to emit waves of the same length at the same temperature," etc.   (Draper, *Phil. Mag.*, Vol. XXX., p. 345.   Berl., 1847.)

"For the same temperature the magnitude (I) is a continuous function of the wave-length, except for such values of the latter as render (I) evanescent.   The truth of this assertion may be concluded from the continuity of the spectrum of a red-hot platinum wire, provided it be admitted that the power of absorption of such a body is a continuous function of the length of the waves of the incident rays."

These, together with other facts, were presented by M. Kirchoff, not as experimental, but as mathematical results. No allusion was made to the fact that the whole subject had been extensively investigated, as shown in the preceding pages, many years before, the only reference to such investigation being that contained, as shown above, in a parenthesis, which in the original is in a foot-note, and even this was omitted in an historical memoir on the subject shortly afterwards published by M. Kirchoff.

As an example of the effect of this, I may quote from the *Cours de Physique de l'École Polytechnique*, of Paris, by Professor Jamin:

"M. Kirchoff has deduced the following important consequences:

"Black bodies begin to emit at 977° Fahr. red radiations, to which are added successively and continuously other rays of increasing refrangibility as the temperature rises.

"All substances begin to be red-hot at the same temperature in the same enclosure.

"The spectrum of solids and liquids contains no fixed lines."

Subsequently, in the *Philosophical Magazine* (April, 1863), a memoir appeared, under the title of "Contributions toward the History of Spectrum Analysis and of the Analysis of the Solar Atmosphere," by M. Kirchoff. In this all allusion to the foregoing memoirs is avoided.

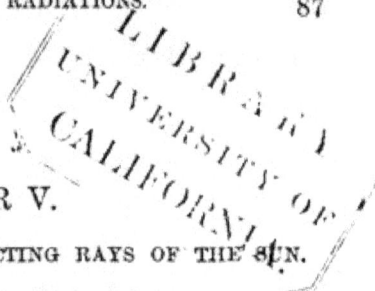

MEMOIR V.

ON THE NEGATIVE OR PROTECTING RAYS OF THE SUN.

From the Philosophical Magazine, Feb., 1847.

CONTENTS :—*Original discovery of protecting radiations.*—*Case of a daguerreotype plate.*—*Spectrum-photographs made in Virginia.*—*Protecting action of the less refrangible rays.*—*Protecting action of the extreme violet.*—*Variations of the protecting action.*—*Spectrum of darkness and spectrum of daylight.*—*Interference of rays of different colors.*—*Action of waves of red, yellow, and violet light with wavelengths* 2, 1½, 1.—*Use of the diffraction spectrum.*

IN a letter published in the *Philosophical Magazine* Nov., 1842, I had occasion to make some incidental remarks respecting a class of rays existing in the sunlight which have the quality of exerting a negative or antagonizing action upon those engaged in producing daguerreotype results.

In October last, MM. Foucault and Fizeau having made a communication to the French Academy of Sciences to a similar effect, and M. Edmond Becquerel, in criticising their results, having referred to me as the original author of the fact, I may on this occasion be excused for offering a few observations on this, which perhaps is destined to become one of the most important phenomena in relation to the chemical action of the sunlight.

That the opposite ends of the solar spectrum possess opposite qualities is an idea which has been floating among chemists for many years. The first distinct statement in relation to it with which I am acquainted occurs in a work published by Mr. B. Wilson, the second edition of which dates as early as 1776. It is entitled "A Series of Experiments on Phosphori." He shows that it is the more refrangible ráys which excite the phosphorescence of sulphide of lime, *but the less refrangible ones extinguish it when shining.*

In 1801 Ritter found that chloride of silver which had been blackened in the violet rays had its color partially restored when placed in the red. He states also that phosphorus, which is oxidized with the production of fumes in the invisible red, is instantly extinguished in the violet.

The well-known experiments of Wollaston with guaiacum served to show the opposite relations of the red and violet rays. It is remarkable that he subsequently abandoned this interpretation of the phenomenon, on discovering that green guaiacum changed its color by the application of a hot silver spoon.

In 1839 Sir J. Herschel encountered the same action in the case of some of the preparations of silver. His first idea was that of a positive and negative polarity of the spectrum; but this was subsequently modified for the reasons set forth in his memoir (*Phil. Trans.,* 1840, § 60, etc.).

In 1842 I had obtained some very fine daguerreotype impressions of the solar spectrum in Virginia, and since that time have never doubted the actual existence of these negative or protecting rays; and on this occasion, when that existence is reasserted by Lerebours, Fizeau, and Foucault, I will make known certain new facts, premising that I do not think the views taken by M. Becquerel are correct. They are founded on what seems to me to be a misapprehension of the phenomenon of the daguerreotype.

A daguerreotype plate can exhibit three different varieties of surface: 1st, a black aspect on those re-

gions where it has been unaffected by light; 2d, various
shades of white; 3d, a colored blackness, the tint of
which may be of a deep watch-spring lustre, or sometimes of an olive shade. Persons familiar with the process will understand completely what I mean. The first
of these conditions is represented in the deep shadows
of such a photograph, the places where the light never
acted; the second is exhibited in the various intensities
of whiteness, which constitute the figures of the picture,
the whiteness varying in intensity according to the intensity of the light; the third is the solarized or overdone
condition, which arises from too long an exposure to the
rays. Like the first, this may be spoken of as a blackness, but in reality it is a dark green or blue or tawny
tint. It is this solarized condition of surface which M.
Becquerel confounds with the first, the blackness arising
from the unchanged state; and it is precisely on this
point that the whole argument turns. For the sake of
having distinctive words to mark out these three conditions, I will call the first the unaffected state, the second the white state, and the third the solarized state.

The observations I made in Virginia were as follows:
That if a solar spectrum be received on a daguerreotype
plate on which a weak daylight was simultaneously
acting, the red, orange, yellow, green, and part of the
blue rays arrested the action of the daylight on that
portion of the plate on which they fell, and maintained
it in the unaffected state; while the residue of the blue,
the indigo and violet, carried their part of the plate to
a completely solarized condition. This therefore seemed
to justify the assertion that the less refrangible rays
protect Daguerre's preparation from the action of a diffused daylight.

It was also found that if the plate were exposed to
the daylight for a few seconds, so that had it been then

mercurialized it would have whitened uniformly all
over, on being made to receive the spectrum the less
refrangible rays actually carried it back to the unaffected condition, reversing what had been already done.
While the more refrangible rays were forcing it on to
the solarized state, these were returning it into the condition of shadow: they therefore not only *protect*, but
seem even to exert a *negative* or antagonistic action.

Sir J. Herschel has critically examined one of these
specimens, and has suggested an explanation of their
appearance on the Newtonian principle of the tints exhibited by thin films (*Phil. Mag.*, Feb., 1843). But I
found that it was immaterial whether the exposure to
the spectrum was for thirty seconds or one hour—the
result was the same. The final action had been produced, the less refrangible rays had carried their region
to the unaffected state, while the more refrangible had
solarized theirs. Now if the phenomenon were due, as
M. Becquerel supposes, to an unequal action of the same
kind in the different rays, it is obvious that the final
result ought to depend on the time of exposure; the
red ray, aided by the daylight, should carry its portion
through the various shades of white, and solarize it at
last. But this in the longest exposure never takes place;
that part of the plate remains as though a ray of light
had never fallen upon it.

Such are the facts I observed, and they seem to have
been reproduced by MM. Foucault and Fizeau; but
there are also others of a much more singular nature.
In these Virginia specimens *the same protecting action
reappears beyond the violet.*

The only impressions in which I have ever seen this
protecting action beyond the violet are those made in
Virginia in 1842; they were made in the month of July.
Struck with this peculiarity, on my return to New York

the following August I made many attempts to obtain similar specimens, but in no instance could the extra-violet protecting action be traced, though the analogous action of the red, orange, yellow, green, and blue was perfectly given.   Supposing, therefore, that the difference must be due either to impurities in the iodine or to differences in the method of conducting the experiment, I tried it again and again in every possible way. *To my surprise I soon found that the negative effect was gradually disappearing;* and on Sept. 29 it could no longer be traced, except at the highest part corresponding to the yellow and green rays.   In December it had become still more imperfect, but on the 19th of the following March the red and orange rays had recovered their original protective power.   It seemed, therefore, that in the early part of the year a protective action had made its appearance in the red ray, and about July extended over all the less refrangible regions, and as the year went on it had retreated upwards.

Are there, then, periodic changes in the nature of the sun's light?   The absorptive action of the earth's atmosphere is out of the question: if that were the cause, the character of these spectrum impressions should vary with the hour of the day.   Or is it not more probable that these singular phenomena rather depend on incidental changes in the experiment, such as external temperature, variations of moisture, the color of the sky, etc.?

Under proper circumstances there is no difficulty in exhibiting the power which the less refrangible rays exert in arresting the action of the daylight: under such circumstances a daguerreotype impression of the sun's spectrum yields all three of the varieties of surface before alluded to.   The plate in the less refrangible and extreme violet region is unaffected; a narrow space of white separates these unaffected portions from the indigo and violet spaces, which are in a highly solarized condition.

But a totally different result is obtained when the daylight is *not* allowed to fall on the plate, either before or during its exposure to the spectrum. Under these circumstances the rays which would otherwise protect now act on the plate and slowly whiten it. A daguerreotype spectrum formed in *darkness* and without previous exposure to the light exhibits a white stain over all the less refrangible regions, and bears a marked contrast to one formed under the simultaneous action of a weak daylight. For brevity I will call the former the spectrum of darkness, and the latter the spectrum of daylight. The following are some additional observations:

In the spectrum of darkness there is in the white stain a point of maximum action. This corresponds with the maximum of protection in the spectrum of daylight.

The white stain of the spectrum of darkness is apparently narrower than the protected space in the spectrum of daylight.

Rays of luminous or of non-luminous heat projected on the darkness or daylight spectra during their formation appear to exert no kind of special influence on the result.

The white fringe which borders the solarized portion is not due to anything analogous to conduction. These chemical changes, unlike thermal changes, cannot be conducted.

By interposing between the prism and the daguerreotype plate a small convex lens of short focus, so as to intercept in succession each of the colored rays, I threw all over the plate, while the spectrum was in the act of being impressed upon it, red, orange, yellow, and other lights in succession; the object being to ascertain how far the impressed spectrum would change when these monochromatic rays were used along with daylight, Herschel having previously shown in similar experiments that new phenomena arise during the conjoint action of rays (*Phil. Trans.*, 1840, § 64). The following are some of the observations I made; their date is Sept. 24, 1842.

The red ray when projected increases the length of the solarized portion, and also of its white extremities.

The yellow ray shortens the solarized portion.

The green ray exerts a greater action of the same kind.

The indigo ray gives a most remarkable result. It totally inverts the action of the less refrangible rays, and they solarize the plate, acting in the same way that the more refrangible rays commonly do, causing it to exhibit a watch-spring lustre.

I further found that when different rays are brought to act upon each other, the result does not alone depend upon their intrinsic differences, but also on their relative intensities. Thus the green and lower half of the blue rays, when of a certain intensity, protect the plate from the action of the daylight; but if of a less intensity, they aid the daylight.

The red and orange rays, when of a certain intensity, increase the action of daylight on the plate; but if of a less intensity, they restrain it.

These facts seem to be connected with the circumstance that there is often to be traced on daguerreotype plates a remarkable difference between the central and lateral parts of a spectrum. Thus if a line be drawn through the centre of such a spectrum and a parallel to it on one of the edges, the action at any point on the central line is the reverse of that at the corresponding point on the edge. A similar remark, as respects impressions on paper, has been previously made by Herschel.

Such are the chief facts I have observed in relation to

the daguerreotype spectrum.   It would seem at first
sight that their diversity is so great that we can have
but little hope of reducing them to a common system of
results originating in the same cause.   I have, however,
been long led to believe that the explanation is to be
met with in the great and fertile principle of interference.
From this point of view I regard the action of rays of
every kind as being essentially positive, and that action
mainly consists in impressing a vibratory movement on
the atoms of the decomposing substance.   It is to my
mind a fact of no common significance, that in those Virginia specimens the places of maximum protection in the
less and more refrangible regions fall where the lengths
of the luminous waves have the extraordinary relation
of 2 : 1.   Then, when we also see that, before a perfect
neutralization of action between two rays ensues, those
rays must be adjusted in intensity to each other, does it
not show that interference of some kind is going on?
Again, the yellow ray is in numerous instances the ray
which most completely antagonizes those at the red and
violet extremes of the spectrum: to use the language of
Herschel, "This ray may be considered as marking a
sort of chemical centre, a point of equilibrium, or rather
a change of action in the spectrum."   I cannot avoid seeing that these phenomena are connected with the remarkable fact that the waves of red, yellow, and violet light
are of lengths which correspond to 2, $1\frac{1}{2}$, 1.

If, then, a powerful yellow ray can hold in check a feeble violet one, and prevent it from decomposing iodide
of silver merely because their relation of length is in
the ratio of $1\frac{1}{2}$: 1, it should follow on the same principles that a red ray acting conjointly with a violet should
give rise to an increased effect, because the lengths have
now become 2 : 1.   And that this is in reality the case I
found by direct experiment; for on projecting the red

upon the violet, so that the colors should half overlap each other, I found that at the point of concourse the plate instantly solarized, and assumed a splendid green metallic color.

I have now explained the acceptation in which I receive the *negative ray* as a synonym (in this instance of iodide of silver) for the yellow ray, and alluded to the mechanism which seems to be the cause of protecting action generally. Perhaps on a review of his own experiments M. Becquerel may find reason to believe that there are in reality antagonizing actions in different parts of the spectrum; actions not limited to the daguerreotype, but occurring in all kinds of cases. They have been met with by every one who has examined the spectrum with sensitive papers, and, in a different series of phenomena, M. Becquerel has himself furnished a conclusive illustration. He shows that when sulphide of lime and other phosphorescent bodies in a shining state are exposed to the spectrum, the more refrangible rays increase the glow, but the less extinguish it.

It is proper to observe that some of the phenomena recorded in this communication which seem to be in opposition to the principle set forth are not so in reality. All reasonings founded on the decomposition of light by the prism, and the action of the prismatic spectrum on changeable surfaces, are liable to error. The only method free from these difficulties is to employ the diffraction spectrum formed by a ruled surface or a grating, a method which was proposed eight years ago by Herschel with a view of getting rid of the disturbing agencies arising from the ideal coloration of glass, and which I first carried into effect in 1844 with so much success that the resulting daguerreotype impressions contained Fraunhofer's lines, even with microscopic minuteness. With this spectrum we avoid a far more serious diffi-

culty than that of the ideal coloration of glass, a difficulty
arising from the magnitude of the refracting faces of the
prism.   It is this which makes a prismatic spectrum
blacken paper, made sensitive with the bromide of silver,
from the red to the violet end; whereas the diffraction
spectrum shows that the true action is confined to the
more refrangible side, and stops short of the centre of
the yellow space.

University of New York, *Dec.* 24, 1846.

MEMOIR VI.

ON THE DIFFRACTION SPECTRUM.

From the Philosophical Magazine, March, 1857.

CONTENTS :—*Mode of obtaining the diffraction spectrum.—The yellow is in its middle ; it is a centre of chemical action.—It is also the hottest ray.—Diffraction spectra by reflection.—Cold lines.—Dilatation of the more refrangible rays in the prismatic spectrum, compression of the less refrangible.—Action of the diffraction spectrum on salts of silver.—Use of wave-lengths for spectrum division.*

M. EISENLOHR having published in Poggendorff's *Annalen* for June and August, 1856, some researches on the diffraction spectrum, from which it appeared that he was not aware of the results I had obtained and published in 1844, I thought it well to call attention to them, and this I did in a letter to the editors of the *Philosophical Magazine*, which they published in their journal for March, 1857, and from which the following extracts are made :

At first I obtained the diffraction spectrum very much in the same manner that M. Eisenlohr has done, by passing a beam of light directed through a vertical slit by a heliostat, and through a piece of ruled glass at twelve feet distance. It was then received on an achromatic lens of about four-feet focus, which gave a sharply defined spectrum on a ground-glass or photographic surface. Subsequently I found that it was much better to silver the ruled surface with tin amalgam, in the manner of a looking-glass, and thus employ a reflected spectrum. This is far more brilliant than the transmitted one, and the silvering acts so perfectly that the most minute fixed lines may be seen.

G

In a work published in 1844 "On the Forces which Produce the Organization of Plants," the results I obtained were illustrated by steel engravings, one of them colored, with a view of comparing the prismatic and diffraction spectra, and of showing the photographic action on iodide, chloride, and bromide of silver. These results are substantially the same as those of M. Eisenlohr. They give the wave-lengths at which action begins and ceases on each of those substances. Some of the details may be found in a former number of the *Philosophical Magazine* (June, 1845), and in that article it is particularly recommended to use wave-lengths for designating rays, instead of titles of color, as red, yellow, etc.

The chemical action of the diffraction spectrum is, however, only given in part in M. Eisenlohr's publication. He speaks of what occurs in the more refrangible regions, and takes no notice of the action in the centre of the spectrum, and in its less refrangible space.

I will state in detail what I here mean. It is well known that in such a spectrum the yellow occupies the middle space, and that the light grades off in one direction to the red, and in the other to the violet. These terminate at equal distances from the yellow, so that the wave-lengths for the extreme violet, the centre of the yellow, and the extreme red, are as 1, $1\frac{1}{2}$, 2.

Now from the extreme violet to the yellow the spectrum blackens silver preparations, and this is what is commonly understood by its chemical action by those who have written on photographic subjects. But from the yellow to the extreme red the spectrum is also active, though in a different way. This half is in antagonism with the other half. It can suspend or arrest the blackening which would be caused by contemporaneously acting diffused light; nay, even more, it can undo what such light may have done some time before. Some remarks on this topic may be seen in the *Philosophical Magazine*, February, 1847 (and in Memoir V. of the present work).

The centre of the yellow space is therefore the point of a change in photographic action. From the commencement of the red to that point the action is negative; from that point to the extreme violet it is positive. The phase of action changes in the centre of the yellow.

M. Eisenlohr has made an allusion to the heating power of the spectrum, and to that point my attention has also been directed. This is the result at which I eventually arrived—that the centre of the yellow is the hottest space, and that the heat declines equally to the two ends of the spectrum. I attempted to form a diffraction spectrum without the use of any dioptric media, endeavoring to get rid of all the disturbances which arise through the absorptive action of glasses by using as the grating a polished surface of steel, on which parallel lines had been drawn with a diamond, and employing a concave mirror instead of an achromatic lens; and though my results were imperfect and incomplete, I saw enough to convince me that this is the right course to be taken in investigating the problem. It is absolutely necessary to employ a spectrum which has been formed by reflection alone.

I will also add that, in the experiments here referred to, a method was employed for determining the temperatures of narrow spaces, which may be recommended to those who are disposed to resume such inquiries. It is to use a blackened platinum wire, $\frac{1}{4}$ of an inch long, and $\frac{1}{30}$ of an inch in diameter, one end of which is fastened to the end of a bar of bismuth, and the other end to the end of a bar of antimony, each of these bars being $\frac{1}{4}$ of an inch square on the end and 4 inches long, their distant extremities communicating with a galvanometer. By carrying the platinum wire transversely along the spectrum, I expected not only to determine the distribution of heat, but also to ascertain whether the heat spectrum has fixed lines, like the luminous and chemical ones.

This was at a time when I first began to suspect that the essential difference between light and heat is this, viz., that while the vibrations of light are transverse, those of heat are normal, and its waves in that respect analogous to the waves of sound; but so feeble is the intensity of these spectra that I could do no more than satisfy myself that in the diffraction spectrum the centre of the yellow is really the hottest space, as well as the most luminous. I believe that there is a cold line in the spectrum answering in position to the dark line H, but I could not absolutely demonstrate it.

Nevertheless, so certain does it appear that the distribution of heat corresponds to the distribution of light, that I have not hesitated since that time to look upon the centre of the yellow space as the point of maximum heat, from which there is a decline to each end of the spectrum. And I accordingly made this the basis of a theory of vision, published in my Treatise on "Human Physiology." Such a view has the advantage of being sustained by many facts of comparative anatomy. It obliges us, it is true, to return to the opinions entertained of the functions of the black pigment more than a century ago; but then it gives a very elegant explanation of the uses of the different parts of the retina, examined in its radial section after the manner of M. H. Müller, and of those of the choroid coat—structures which are without meaning on the ordinary theory of vision.

UNIVERSITY OF NEW YORK, *Jan.* 26, 1857.

NOTE.—To the foregoing paragraphs I may add the following as contributions to the theory of vision.

I may also refer to my Treatise on "Human Physiology" (1856).

There are many rays emitted by the sun and other shining bodies to which our eyes are entirely blind.

Two different reasons may be alleged for our inability to perceive such rays: first, they may not be able to reach the retina, the media of the eye not transmitting them; second, the retina may be so constituted as to be unable to receive their impressions.

It has long been known that rays which come from sheet-iron heated by a lamp cannot pass either through the cornea or through the crystalline lens. Even of those that are furnished by an Argand flame, used as a luminous source of heat, less than one fifth pass through the cornea alone, and scarcely one fiftieth when the crystalline lens is interposed. Cima showed that of the heat-rays emitted by a flame, less than one tenth pass through the cornea, lens, and vitreous humor conjointly. Janssen, using a flame, compared the heat transparency of the separate media of the eye with that of water included between glass plates, showing that there is a perfect accordance between them if taken of equal thickness. From this it is to be concluded that invisible rays to a certain extent reach the retina. Franz, by carefully conducted experiments with a thermo-electric pile, came to the conclusion that a quantity of obscure rays detectable by the thermometer can reach the retina, which therefore must be so constituted as not to be able to perceive them.

This settles the question so far as the less refrangible or ultra-red rays are concerned. We have then to determine how it is with those at the opposite or more refrangible end of the spectrum. Do these pass through the media of the eye, or are they arrested and never reach the retina?

I made a series of experiments on these rays, and found that they passed through the different media of the eye, examined separately, and what is more to the point, through them all collectively, with but little loss. There was no difficulty in obtaining a dark stain on paper made sensitive with chloride of silver, and placed at the back of the eye of an ox, from which the sclerotic and pigment had been suitably removed. In a general manner the media of the eye act like water on the transmissibility of these rays.

Admitting from these experiments that invisible as well as visible rays reach the retina, we may next consider the nature of the impression made upon it, and are thus brought directly to an investigation of the act of vision.

There are three hypotheses to be considered:

1. That rays falling on the retina or black pigment impart to those structures a rise of temperature. This may be termed the calorific hypothesis.

2. That rays falling on the retina occasion a chemical change or metamorphosis in its structure, implying the occurrence of waste in it, and therefore the necessity of repair. This may be termed the chemical hypothesis.

3. That rays falling on the retina throw its parts into a vibratory movement, not necessarily attended by any metamorphosis of tissue, as waves of sound occasion consentaneous pulsations in the auditory apparatus of the ear. This may be termed the mechanical hypothesis.

First, *of the calorific hypothesis of vision.* Comparative anatomy offers certain facts which lend plausibility to this hypothesis. Some of the most remarkable of these relate to the construction of the eye in lower animals. The ocelli, which consist of dark-colored or black spots, or black cup-shaped membranes containing within them the rudiment of an optic nerve, are the beginning of an organ of vision. There being no optical apparatus for the production of images, the luminous impression must be felt as heat. For this the dark pigment is well designed. It is an old physical experiment to lay upon the snow on a sunshiny winter day pieces of differently colored cloth. They will melt their way to a greater depth in proportion as their tint is deeper; the black, becoming the warmest, sinks deepest; the white, reflecting most of the heat, scarcely melts the snow at all. Now an animal destitute of any visual organ can only be affected by the impressions of light in a very doubtful manner; but if there be upon its exterior a black spot, not only is there a much higher sensitiveness because of the increased absorptive power for heat, but the sphere of its consciousness is greatly extended, from the possibility of acquiring a knowledge of directions in space—a knowledge that becomes more and more exact with the increasing number and symmetrical arrangement of these ocelli.

If we apply these principles to a more perfect form of eye, as that of man, we are led to a new interpretation of the function of some of its parts. The black pigment becomes the receiving surface for images of external things, and rays falling upon it, in their diversity of color, brightness, and shade, in the act of becoming extinguished, engender heat. As with the tip of the finger passing over an object we can discover, even in the dark, spaces that are warm and those that are cool, so the rods and cones of Jacob's membrane, acting as tactile organs, convey to the brain a knowledge of the momentary distribution of heat on the dark concave of the eye. The pigment has therefore a far more important office to discharge than that of merely extinguishing stray light and darkening the inside of the globe.

But this calorific hypothesis is not without great dif-

ficulties. Heat suffers conduction. If this black pigment officiated as a transformer of light rays into heat
by producing extinction, there must unavoidably be a
lateral spread from the boundaries of warm to cooler
spaces, the edges of images must be nebulous and without sharpness of contour. Moreover, there is reason to
believe that the visual apparatus cannot take cognizance
of heat merely as such. Calorific rays reach the black
pigment and raise its temperature without the retina
being affected.

Such considerations seem, therefore, to exclude the calorific hypothesis, and prepare us for an examination of
the chemical.

SECOND, *of the chemical hypothesis of vision.* Numerous discoveries made of late years in relation to the
chemical action of light put us in possession of many
facts having a bearing on this hypothesis. A majority
of compound substances, both inorganic and organic, suffer chemical modifications when exposed to the access of
light, and, what is very significant, these changes are occasioned by definite classes of rays. One substance finds
its maximum of action in the violet region, another in
the yellow, another in the red. The effect in every instance grades off towards the less and more refrangible
spaces respectively.

In these actions of decomposition there is nothing like
lateral spreading, nothing answering to conduction. No
better proof of this is necessary than the exquisite sharpness of photographic pictures—a sharpness only limited
by the optical imperfections of the lens with which they
are made. The molecules on which the light falls are
the only ones that experience change; there is no propagation of effect from part to part—an important particular, because it is what we observe in the case of sight.

The retina, the nervous expansion of the eye, is so constituted that a maximum effect upon it is occasioned by the yellow ray, the action declining on one side to the red, and on the other towards the violet, and ceasing at the extremes of those rays. For this reason, when a solar spectrum is examined by the eye, the yellow is the most brilliant space, there being a decline in intensity from it to the two extremes.

In my experiments on the decomposition of carbonic acid by plants in the sunlight, to be described hereafter in these memoirs, the maximum of action was found to be in the yellow, with a gradation of effect towards the red and violet ends of the spectrum respectively. From this it would appear that a relation exists between light and compounds having a carbon nucleus, answering to that observed in the case of the retina of the eye. Such a relation is very well illustrated in the case of other chemical elements, as silver, a metal which is the basis of all ordinary photographic preparations. The ray of maximum action is in the indigo space. Objects viewed by a retina having a silver sensitive nucleus would present an appearance altogether unlike that they would offer to a carbon nucleus. The order of brilliancy in the lights would be no longer the same. The red and yellow parts of objects would be black, that is to say, invisible, and other rays beyond the violet would come into view.

Among experiments I have made on this subject, there is one of much physiological interest. The element phosphorus finds its maximum impression in the more refrangible portion of the spectrum, in that respect resembling silver. Upon a portion of translucent phosphorus, enclosed out of contact of air in a flattened glass tube, into which it had been drawn while melted, and then suffered to solidify, a solar spectrum was cast. The effect of light upon this kind of phosphorus is to turn it

eventually to a deep mahogany red, and chemically to
throw it from an active into an inactive state. As amor-
phous phosphorus, otherwise prepared, it ceases to shine
in the dark. In the experiments now alluded to, it appeared that this reddening takes place in the indigo and
violet spaces, so that the fixed lines known by spectroscopists as those about H were beautifully depicted.
Now some physiologists have supposed that nerve ves-
icle tissue owes its property to the presence of unox-
idized phosphorus, but if the principles we are contemplating be correct, and this were the case, the most brilliant ray in the spectrum should be the indigo, and not
the yellow. Therefore, if vision be performed by chemical change in the substance of the retina, it is carbon
and not phosphorus that is concerned.

If we admit that during the act of vision the retina,
as a structure with a carbon nucleus, undergoes metamorphosis, the principles of photo-chemistry would lead
us to expect that the yellow must be the brightest ray,
and a harmony is thus established between this and
other functional changes in the body. We also perceive
the significance of certain structures of the eye which
otherwise would appear to be without meaning. The
rapid retrograde metamorphosis which must be taking
place in the retina involves the provision of some means
for moving away the wasted products and of supplying
nutrition with the utmost quickness. And this is the
office discharged by the choroid.

But such removals and supplies require time. Time,
therefore, enters as an element in the visual operation.
Sight commences instantaneously, but the image of an
object may be seen long after the reality has disappeared. This instantaneous commencement of a retinal
impression may be very strikingly illustrated. The
spark of a Leyden-jar, though it does not last, as is af-

firmed, the millionth of a second, can without any difficulty be photographed even on so sluggish a compound
as silver iodide. On the far more sensitive retina the
chemical impression must be practically contemporaneous with the impinging of the light.

If after the eyelids have been closed for some time,
we suddenly and steadfastly gaze at a bright object,
and then quickly close the lids again, a phantom image
is perceived existing in the indefinite darkness before
us. By degrees the image becomes less and less distinct;
in a minute or two it has disappeared.

The chemical hypothesis renders a very clear explanation of this effect—an explanation that commends itself
to our attention as casting light in many cases on the
curious phenomena of apparitions: phenomena that have
been not without influence on the history of mankind.

The duration and gradual extinction of the retinal
phantoms correspond to the destruction and renovation
taking place in the retina itself. The blood supply is
very ample, as are likewise the channels for the removal
of waste, but the operations require time to be accomplished. As in machines contrived by man, so in natural
organs, the practical working does not always come up
to the theoretical standard. Theoretically, as the retina
suffers change under the incident light, the removal of
waste and nutrition should go on in an equal manner
both as to time and quantity. A marvellous approach
to the ideal perfection is attained, for though the action
of light must necessarily be cumulative, that is, increasing with the continuance of exposure, objects do not
become brighter and brighter as we look at them, but
they attain their predestined distinctness at once. The
action of the light, the removal of the waste it is occa-
sioning, and the supply for renovation are all contemporaneously going on with an equal step, or so nearly

so that such may be considered to be the practical effect.

THIRD, *of the mechanical hypothesis of vision.* There is a growing belief among those who are cultivating photo-chemistry that the mode of operation of a ray of light in accomplishing chemical changes is by establishing vibratory movements among the molecules of the substance affected. As has been affirmed, perhaps fancifully, of certain singers, that they could cause a glass goblet to fly to pieces by a proper intonation of their voice, through the attempt of the glass by resonance to execute incompatible vibrations, so it is thought that an incident ray may break asunder a group of molecules by establishing among them discordant agitations. Chemical decompositions by radiations become thus connected theoretically with vibratory movements.

But these are vibrations not necessarily attended by any destruction of tissue. Waves of sound occasion such pulsations in the apparatus of the ear without producing any chemical change in the auditory nerve.

If we consider the retina as an elastic shell, of which the parts are put into a purely mechanical movement by the pulsations of light, we abandon without explanation some of the most interesting portions of the structure of the eye. Of what use is that wonderful net-work of vessels constituting the choroid? It is a principle in physiology that the supply of blood to a part is proportional to its functional activity. The elaborate vascular mechanism in juxtaposition with the retina will bear no other interpretation than that that tissue is the seat of incessant chemical changes.

Moreover, physical science in its present state is not sufficiently advanced to furnish the means of clearly comprehending such purely mechanical motions executed by the ultimate particles of things. We may

conceive of the comparatively slow swaying of groups
of molecules under the influence of normal pulsations in
the air, but not of the dance of atoms disturbed by transverse vibrations in the ether. If, therefore, there were
no arguments of an anatomical kind to be presented
against the admission of this hypothesis, we should be
compelled to turn aside from it because of the inade-
quacy of our knowledge in tracing its conditions to their
applications.

This, therefore, is the conclusion at which we finally
arrive—that vision depends on chemical changes, especially of oxidation, in the retina, and that they approach in
their nature those that we speak of as photographic.
There is no difficulty in understanding how such changes
may give rise to an influence transmitted along the optic
nerve to the brain, when we reflect that the oxidation
of a few particles of zinc may accomplish specific mechanical results through many miles of intervening tele-
graphic wire, producing mechanical motions as in the
telegraph of Morse, or chemical changes as in that of
Bain.

We have remarked that a critical study of the function of vision cannot fail to lead to interesting results
respecting the nervous system generally. Guided by
that remark, we may perhaps profitably consider further
the vestiges of visual impressions, and the physical conditions under which they disturb us or spontaneously
obtrude themselves on our attention.

The perception of external objects depends on the
rays of light entering the eye, and converging so as to
produce images, which make an impression on the retina,
and through the optic nerve are delivered to the brain.
The direction of these influences, so far as the observer
is concerned, is from without to within, from the object
to the brain.

But the inverse of this is possible. Impressions existing in the brain may take, as it were, an outward direction, and be projected or localized among external forms; or if the eyes be closed, as in sleep, or the observer be in darkness, they will fill up the empty space before him with scenery of their own.

Inverse vision depends primarily on the condition that former impressions, enclosed in the optic thalami, or registering ganglia at the base of the brain, assume such a degree of relative intensity that they can arrest the attention of the mind. The moment that an equality is established between the intensity of these vestiges and sensations contemporaneously received from the outer world, or that the latter are wholly extinguished, as in sleep, inverse sight occurs, presenting, as the occasions may vary, apparitions, visions, dreams.

From the moral effect that arises, we are very liable to connect these with the supernatural. In truth, however, they are the natural results of the action of the nervous mechanism, which of necessity produces them whenever it is placed, either by normal or morbid or artificial causes, in the proper condition. It confounds the subjective and the objective together. It can act either directly, as in ordinary vision, or inversely, as in cerebral sight, and in this respect resembles those instruments which equally yield a musical note whether the air is blown through them or drawn in.

The hours of sleep continually present us, in a state of perfect health, illusions that address themselves to the eye rather than to any other organ of sense, and these commonly combine into moving and acting scenes, a dream being truly a drama of the night. In certain states of health appearances of a like nature intrude themselves before us even in the open day, but these, being corrected by the realities by which they are sur-

rounded, impress us very differently. The want of uni-
son between such images and the things among which
they have intruded themselves, the anachronism of their
advent, or other obvious incongruities, restrain the mind
from delivering itself up to that absolute belief in their
reality which so completely possesses us in our dreams.
Yet, nevertheless, such is the constitution of man, the
bravest and the wisest encounter these fictions of their
own organization with awe.

The visions of an Arab merchant have ended in tinct-
uring the daily life of half the people of Asia and Africa
for a thousand years. A spectre that came into the
camp at Sardis the night before the battle of Philippi
unnerved the heart of Brutus, and thereby put an end
to the political system that had made the Roman republic the arbiter of the world. A phantom that appeared
to Constantine strengthened his hand to that most difficult of all the tasks of a statesman, the destruction of an
ancient faith.

Hallucinations are of two kinds—those seen when the
eyes are open and those perceived when they are closed.
To the former the designation of apparitions, to the latter that of visions, may be given.

In a physiological sense, simple apparitions may be
considered as arising from disturbances or diseases of
the retina; visions, from the traces of impressions enclosed at a former time in the corpora quadrigemina and
optic thalami.

From flying specks floating before us, the first rudi-
ments of apparitions, it is but a step to the intercalation
of simple or even grotesque images among the real objects at which we are looking; and indeed this is the
manner in which they always offer themselves, as resting
or moving among the actually existing things.

The method by which the first photograph of the diffraction spectrum was obtained is described as follows in the *Philosophical Magazine*, June, 1845.

The prismatic spectrum, even when every precaution has been used to obtain it in a state of purity, its fixed lines being visible, is liable to lead us into many errors. As respects its luminous or photic properties, we cannot determine the distribution or intensity of the light because the violet extremity is unduly dilated. As respects the chemical effects, the same difficulty occurs, for these are necessarily controlled and disturbed by the law of distribution. All chemical actions occurring in the more refrangible regions, by being spread over a great space appear to be more feeble than what they actually are.

In a perfect spectrum the most luminous portion of the yellow should be in the centre; and from this the intensity of the light should gradually decline, fading away on one side in the red, and on the other in the violet. At equal distances from the middle yellow point on either hand the intensity of the light should be equal. These beautiful results are due to Mosotti, who also states that the length of the extreme red wave is to that of the extreme violet in the simple ratio of 2:1.

The prismatic spectrum does not exhibit these facts. The yellow is not in the centre; the blue, indigo, violet are abnormally spread out, the spectrum having its own law of distribution. But the diffraction spectrum enables us to observe them.

By the aid of a heliostat, I arranged horizontally in a dark room a narrow ribbon of light, coming through a slit $\frac{1}{30}$ of an inch wide, set vertically. At the distance of twelve feet it fell perpendicularly on a piece of flat glass, the surface of which was ruled with equidistant parallel lines by a diamond; and having been silvered with tin foil, after the manner of a mirror, it served the purpose of a grating. The reflected beam went out through the slit at which it entered, and on either side of it to the right and left the well-known double series of diffraction spectra made their appearance. I selected, for the obvious reason that it was not overlapped by its successor, the first of the series, and intercepting it by an achromatic object-glass, placed in the focus a frame capable of holding a ground-glass or sensitive surface. This frame was adjusted until the fixed lines were distinctly depicted upon it.

For a further description of the reflected diffraction spectrum I may refer to any of the elementary works on optics. It is sufficient for my purpose here to recall that the angular deviations of any two colors from the primitive incident ray are to one another as the lengths of their respective undulations.

On the ground-glass we see the fixed lines, and the length of waves corresponding to those lines has been carefully determined by Fraunhofer. The following table is extracted from Herschel's treatise on Light, the Paris inch being divided into one hundred millions of equal parts:

Length of wave corresponding to the fixed line B, 2541 parts.
"   "   "   "   "   " C, 2422 "
"   "   "   "   "   " D, 2175 "
"   "   "   "   "   " E, 1945 "
"   "   "   "   "   " F, 1794 "
"   "   "   "   "   " G, 1587 "
"   "   "   "   "   " H, 1464 "

When, therefore, we have any chemical, luminous, thermic, phosphorogenic, or any other effect under discussion, in relation to the spectrum, we have only to determine its place among the fixed lines, remembering in the diffraction spectrum the simple law that connects the deviations and wave-lengths. In the mode of operation here described absolute exactitude is not reached because our

H

measures are obtained from a flat surface, and not upon a circular arc.

To the diffraction spectrum thus formed I exposed for half an hour a daguerreotype plate, rendered sensitive by iodine and then by bromine. It resulted that the bromide of silver is decomposed at a maximum by a wave which is 0.00001538 of a Paris inch in length. The action does not extend equally, as we might have supposed, towards the more and less refrangible regions.

I exposed a silver plate which had been prepared by iodine, bromine, and chloride of iodine successively. The point of maximum fell as before at 0.00001538. The time of exposure one hour. The decomposition commenced by a wave in the green space, the length of which was 0.00002007, and was terminated by one in the violet, the length of which was 0.00001257. The point of maximum action, therefore, inclined to the violet, and was not midway between the extremities of the photograph. The absolute length of the stain depends, however, on the time of exposure.

I need not multiply these results. It is sufficient to add that in several trials I obtained in these delicate experiments photographs of the diffraction spectrum on different surfaces in great perfection. The fixed lines which are crowded close together were beautifully distinct.

I would suggest, therefore, that when we wish to indicate spectrum regions with precision, we should use wave-lengths. By doing this we shall connect the various actinic phenomena with a great many of the numerical results of optics, and have fixed points of comparison.

UNIVERSITY OF NEW YORK, *June*, 1845.

MEMOIR VII.

STUDIES IN THE DIFFRACTION SPECTRUM.

A popular exposition condensed from Harper's New Monthly Magazine, Vol. LV.

CONTENTS : — *Elementary description of the diffraction spectrum. — Young's discovery of interference. — Fresnel's discovery of transverse vibrations.—Gratings and the spectra they yield.—Gratings on reflecting surfaces. — Optical action of a grating. — Its spectra of different orders.—Interpretation of wave-lengths by the mind.—Mental appreciation of multiple wave-lengths. — Refutation of the principle that to every color there belongs an invariable wave-length.—Increase in the range of perception in the eye.—Extension to photographic impressions. —Encroachment on the first dark space.—Photographs of the diffraction spectrum.—Proposal to use wave-lengths for spectrum divisions.— Replacement of imaginary imponderable principles by wave-lengths.*

WHAT is a diffraction spectrum? Every person who has read a book on light is familiar with the prismatic spectrum, in the study of which Newton displayed his transcendent philosophical powers. The diffraction I have had occasion to refer to several times, and since it is less known, will now describe it. Some very curious phenomena connected with it I have personally examined. It carries us to a true interpretation of the relations of heat, light, and actinism; it offers some important suggestions respecting the mode of action of that most wonderful of all organs, the brain, and therefore commends itself to our most earnest attention.

If we look at a candle flame placed ten or a dozen feet distant, the eyelids being so nearly closed that the eyelashes intercept the incoming rays, we see on either side of the true image of the flame a succession of colored ones—rainbow streaks or fringes, as it were. Examining these particularly, we find that each of them is blue on the side nearest to the true image, and red on the more distant. Our investigation will be simplified if we consider the action of a single eyelash. We can then reason from that to the conjoint action of all.

It is necessary, however, in the first place, to recall some facts connected with the wave theory of light. The foundations of this theory were laid by Huyghens, the great Dutch philosopher, contemporary with Newton, but its construction advanced very slowly, being opposed by the great authority of Newton, who favored the corpuscular or emission theory, and regarded light as consisting of particles emitted with excessive velocity from shining bodies. Although there were facts, such as those connected with double refraction, easily accounted for by the system of undulations, but inexplicable on the emission theory, these were put aside, in the expectation that they would in the course of time be successfully dealt with. It was not until the publication of a course of lectures on natural and experimental philosophy by Dr. Young in 1802, in which he announced the great discovery of the interference of light, that the undulatory theory could no longer be overlooked. This discovery was, however, still ridiculed by the *Edinburgh Review*, and Young's explanations so bitterly attacked that he was constrained to publish a pamphlet in reply. Of this it is said that only a single copy was sold.

In 1819, a memoir by Fresnel was crowned by the French Academy of Sciences. He discovered that the vibratory movements in the ether constituting light are transverse to the course of the ray. His views are embodied in what is now known as the theory of transverse vibrations.

The conflict between the rival theories was eventually settled by the experiments of Fizeau and Foucault. On

Newton's principles the particles of light should move faster through water than through air; on the theory of Huyghens, waves of light must move slower in water than in air. The experiments of the French physicists proved that the latter is the case. This may, then, be considered as the successful establishment of the undulatory theory. It has, moreover, given that striking proof of its truth which may be considered as the criterion of any theory—the ability to foretell results. This it did in the case of the discovery of conical refraction.

Light, therefore, consists in the transference of energy or force, not in the transference of matter.

The grating I employed in the experiments hereinafter related was made for me by Mr. Saxton, at the United States Mint in Philadelphia, more than thirty years ago. Though from the work it did for me I cannot but speak of it with admiration—it enabled me to make the first photograph that was ever executed of the diffraction spectrum—yet it was far from being equal to the magnificent ones of Mr. Rutherfurd. This grating was five eighths of an inch long and one third of an inch in breadth. Mr. Rutherfurd's gratings have in some specimens 17,240 lines to the inch. I had found previously to 1843 that it is more advantageous practically to use a reflecting than a transparent grating, and accordingly I silvered mine with mercury-tin amalgam, such as is used in ordinary looking-glasses. Mr. Rutherfurd's reflecting gratings are coated with pure silver, by an operation more recently discovered.

I will now relate the use of these gratings, and describe some of the important discoveries made by them.

Let a beam of light, S A', Fig. 11, pass through a narrow slit, S, and fall perpendicularly on the ruled grating,

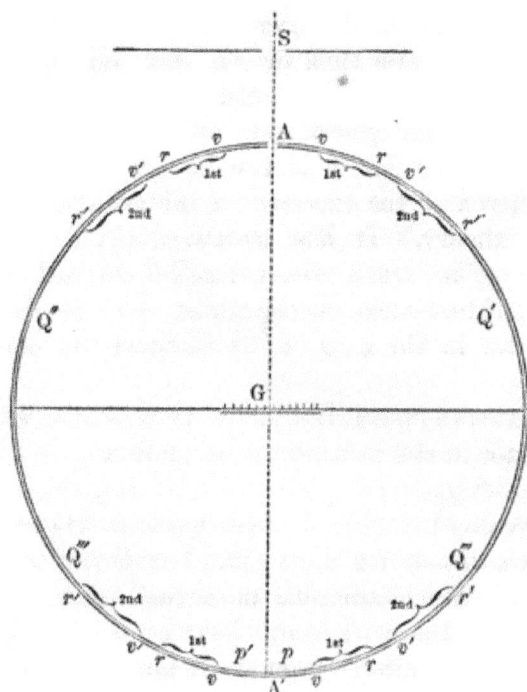

Fig 11.

G, the lines of which are parallel to the sides of the slit. Concentric with the middle line of the grating let there be placed a circular zone or screen, Q', Q'', Q''', Q'''', of white paper, through which there is an opening at A, to admit the intromitted beam.

A beam of parallel rays passing along S G will give a bright image of the slit S when it impinges on the screen at A'. This is the image by transmission. It would also give another similar image at A, were it not for the opening arranged there. This is the image by reflection. Also from G, as from a central axis, there falls upon the cylindrical paper zone, covering its surface all over, an infinite number of radiations.

These effects are seen with much more precision if

there be placed behind the grating a convex lens, or, still better, if the lens be the objective of a telescope.

Now the eye can only be impressed by special radiations consisting of waves of a determinate length. Its vision is limited to those that impart to it a sensation of red on one hand, and of violet on the other. To all others it is blind. Then, though the whole paper zone is receiving radiations of every kind, the eye selects out only those that it can perceive, and, as a result, sees in the four quadrants, Q', Q'', Q''', Q'''', those only for which it is fitted.

It follows, therefore, that at A' there is a white image of the slit S, and to the right and left of this there are equal spaces, p, p', completely dark. Beyond, and symmetrically on each side, there is a series of spectra, $v\ r$, $v'\ r'$, $v''\ r''$, etc., of which the violet ends are nearest A', and the red ends most distant. These spectra are designated respectively as being of the 1st, 2d, 3d, etc., order. On each side the 1st spectrum is separated from the 2d by an obscure space, $r\ v'$, which is shorter than the first dark spaces, p, p', and the red end of the 2d spectrum is overlapped by the violet of the 3d. In like manner the 3d is overlapped by the 4th, etc. If the intromitted ray be of sunlight, and a convex lens or small telescope be used, the dark Fraunhofer lines are seen in these spectra.

Such are the results seen in the quadrants Q''', Q'''', from the light transmitted through the grating. In the quadrants Q', Q'', exactly the same train of phenomena will be discovered—dark spaces and spectra, the latter having their violet ends nearest to A, and the overlapping of successive ones taking place in the manner above described.

Since the results are thus symmetrical in all the four quadrants, it is sufficient to select one of them for detailed examination. Let it be the quadrant Q''''.

Selecting one of the fixed lines, that in the yellow space, the sodium line D, for example, in the successive spectra, it will be found that the distance which intervenes between it and the middle of the white image A′ is in the second double, in the third triple, etc., the distance it is in the first. These angular distances are designated as the deviations of the ray under examination. Fraunhofer proved that

(1) The deviation of the same ray, *e. g.*, D, depends on the sum of the width of a groove in the grating and of a transparent interval, being in the inverse ratio of that sum.

(2) The deviation of any one of the colors of the spectrum of the first order, multiplied by the sum of a transparent interval and a groove, gives the length of a wave of light of that color.

(3) The deviations of the same color in the successive spectra increase as the whole numbers 1, 2, 3, 4, etc.

(4) The deviations of two colors in the same spectrum are to each other as the length of their undulations. Hence in all the violet is nearest to A′, and the red the most distant.

The undulatory theory gives a rigorous explanation of all these facts. The lengths of the waves of light have hence been most critically and accurately determined.

We may now examine more closely the spectrum that is nearest to A′—the spectrum of the first order. Being completely separated from the others, it presents the special facts most distinctly. At the point where the light first becomes visible, the violet or inner end of this spectrum, the wave-length of the incident radiation is, as Angström has proved, 3933, and the wave-length of the last visible radiation at the outer or red end is 7604, ten millionths of a millimeter. If we accept the velocity of light as determined by the experiments of Foucault, the number of vibrations made by the ether in the former of these radiations is 754 millions of millions in one second, and the number in the latter case is 392 millions of millions in one second.

Or, to quote measures which are perhaps more familiar, and numbers as given by Herschel, though not so exact as those of Angström, the number of undulations contained in one English inch at the extreme violet end is 59,750, and the number of vibrations executed in one second is 727,000,000,000,000. The number of undulations in one English inch at the extreme red end is 37,640, the number of vibrations executed there in one second being 458,000,000,000,000. The velocity of light used in these computations is 192,000 miles per second, that used in the preceding paragraph, 186,000.

Knowing the rate at which light moves in a second, and the wave-length of any particular color, it is easy to compute the number of vibrations made by the ether in one second for the production of that color. This is obtained by dividing the distance that light passes over in one second by the wave-length of the color in question.

The numbers we thus obtain give us an idea of the scale of space and time upon which Nature carries forward her works among the particles of matter. They also indicate to us the amazing activity of those portions of the brain which execute motions in accordance with those scales.

The distribution of the colored spaces in the diffraction spectrum is not the same as in the prismatic. In the former the yellow space, which is the most luminous radiation, is in the middle of the spectrum, and is not crowded down or compressed towards the red end, as in the latter. So the maximum intensity or illuminating power is, as Mosotti first observed, in the centre, the intensity of the light declining symmetrically on each side to the end.

The Italians have a clear perception, a quick appreciation of the symmetrical and beautiful. When Mosotti first stated this peculiarity of the diffraction spectrum, at a meeting of one of the Italian scientific societies, the announcement was received by the audience with loud acclamations of joy.

I may now describe some of my own studies of these beautiful spectra.

Recalling, then, the principle that the wave-length of an incident radiation is proportional to its deviation, let us select upon the paper zone previously described the point where a ray is falling having a wave-length 7866. It is, of course, twice as far from A' as was the violet end of the first spectrum, for the selected deviation is double. If we inquire what interpretation the mind will give of a radiation having such a wave-length, an inspection of the zone shows that not only is it visible, but that it is regarded as being of a violet color.

This is an important fact. We find that a radiation consisting of waves of a given length which is visible will also be visible when the constituent waves are twice that length. And in like manner it might be shown that the same will hold good when they are three, four, five, etc., times that length. Moreover, in all these cases the color impression imparted to the mind will be the same.

Again, let us select upon the paper zone another point where the wave-length is 15,208. It will have double the deviation of the red end of the first spectrum. Now, agreeably to the foregoing remarks, this point should be visible to the eye, and, for anything that has thus far been said, it should be interpreted by the mind as red light, its wave-length being twice that of the red of the first spectrum. But it is obvious that here a new consideration must enter into account. If this radiation has double the wave-length of the first red, it has triple the wave-length of the first yellow-green. On the principle just laid down, the mind may interpret it as red light or as yellow-green. Which will it do?

Examination of the paper zone, or, better still, through a telescope, shows that the mind adopts both these interpretations, and the same principle applying to other wave-lengths, this constitutes what we have spoken of as the overlapping of the second spectrum by the third, etc. At the point here specially considered, both red and yellow-green light are seen.

From what has here been presented, it follows that the principle considered as established in optics, that to every color there belongs a determinate wave-length, must be modified, since the same color impression will be given to the mind by waves that have twice, thrice, etc., that determinate wave-length. But should the wave-lengths under consideration answer to multiples of that of some other color, the mind will interpret them as being of that color too.

Moreover, these observations lead us to extend the range of perception of the eye. The prism would lead us to infer that it can only be affected by waves the length of which is between 3933 and 7604. Comparisons have hence been drawn between the organ of vision and the organ of hearing, to the disparagement of the former. The ear, it is said, can embrace a range of several octaves, but the eye is influenced by less than one. The grating, however, leads us to reject the restriction, and to place the eye more nearly on an equality with the ear.

It is also to be borne in mind that by using very condensed sunlight, or by resorting to fluorescent or other
optical contrivances, as several experimenters have done,
the range of vision may be carried beyond the proper
violet limit.

The principles here indicated must not be restricted
to the luminous radiations; they apply to all others too.
Thus if a photographic sensitive surface be made to receive the first spectrum, it will be impressed by certain
of its radiations, chiefly by those above the line G.   If
it be exposed in the second, third, etc., spectra, it will
again be impressed by the corresponding undulations,
having two, three, etc., times the former length.   From
this it may, therefore, be inferred that a chemical decomposition of a given substance, brought about by undulations of a certain length, will also be accomplished by
radiations that are octaves of the first.

It has been stated that a dark space, p, intervenes between the violet end of the first spectrum and the bright
streak A′.   This dark space is at present an attractive
and wonderful field of optical investigation.

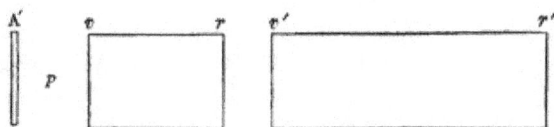

Fig. 12.

In Fig. 12, let A′ represent the white streak in the
position of A′ in Fig. 11; then from A′ to v is the first
dark space, p; from v to r, the spectrum of the first order,
its violet end, v, nearest to A′, its red end, r, more distant; from r to v' the second dark space; and from v'
to r' the spectrum of the second order.   The third spectrum overlaps this second, and the fourth the third, etc.;

but these it is not necessary to consider. The two dark spaces, and especially the first, are the objects mainly to be examined.

Previously to 1844 I had attempted to obtain diffraction photographs with the grating that Mr. Saxton gave me, and had met with great success. In that year I published engravings of them, the originals having been made on silver daguerreotype tablets, in use at that time. By these I carried spectrum impressions as far as the wave-length 3800, and therefore encroached considerably on the dark space p, towards A'. But collodion, since introduced, is a much more sensitive preparation. It has enabled Henry Draper, who has produced superb photographs of the more refrangible regions, to carry the impressions as far as 3032.

According to M. Mascart, waves are emitted by incandescent cadmium having a length not exceeding 2200. These stand still further in the dark space p.

In these excursions into the dark space the experiments of Professor Stokes on the long spectrum of electric light become not only interesting, but very important; for as we gradually approach A', the wave-length of the incident radiation is continually diminishing, and at A' it becomes zero. That point is the supreme limit, beyond which no radiant manifestation of any kind is possible.

The goal towards which experimental investigation is tending is therefore obvious. We are gradually groping the way across the dark space, and expect one day to reach the bright streak that lies at its terminus. At every step of advance the ether waves are becoming shorter and shorter, and the vibrations more and more rapid. When the journey is accomplished, a region will have been gained in which the waves are infinitely short, and the vibrations infinitely rapid.

Several years before the announcement of the discovery of photography by Daguerre and Talbot (1839) I had made use of that process for the purpose of ascertaining whether the so-called chemical rays exhibited interference, and in 1837 published the results in the Journal of the Franklin Institute, Philadelphia (July, 1837, p. 45). In this, as will be seen by consulting that publication, I was successful.

Encouraged by this result, I some years subsequently attempted to photograph the diffraction spectrum itself.

The following is an extract from the publication I made of this experiment in 1844: "Through a narrow fissure or slit, *a*, Fig. 13, I direct a beam of light horizontally, and at a distance of twelve feet receive it on a grating, *b c*, the lines of which are parallel to the slit. Having found that there are advantages in using a reflecting grating, I silvered this with tin amalgam, which copies the ruling perfectly. There is no difficulty in placing *b c* so that the ray coming from *a* falls perpendicularly on it, for all that is required is to move the grating into such a position that the light, after reflection from it, goes back through the fissure *a*. At a distance from *b c* of six inches I place an achromatic object-glass, *d*, in such a position that it

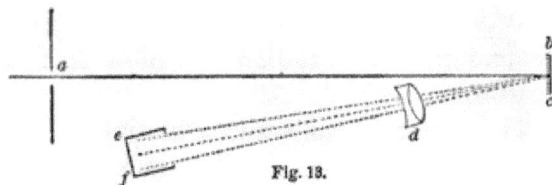

Fig. 13.

shall receive perpendicularly the reflected rays of the spectrum of the first order. The lens is brought as near to the grating as possible without its edge intercepting the ray coming from *a*. In the focus of this lens, at *e f*, a ground glass is placed. This portion of the apparatus is, however, nothing more than the sliding part of a common photographic camera, which contains the ground glass and shields for sensitive preparations."

In the publication above referred to I gave engravings of the results thus obtained; the fixed lines were marked by their wave-lengths. The photographs were very clear and beautiful; they bore magnifying six or eight times without injury to their sharpness.

I may here be permitted to add that it was on the publication of these researches in 1844 that I first made the suggestion to describe spectrum effects by wave-lengths, or what, perhaps, is still better, by ether vibrations—a method now generally adopted. I may give the following extracts:

"In the earlier discussions of the chemical effects of light, the different regions of the spectrum were marked out by the designations of the different colored rays, and effects were described as taking place in the red or yellow or violet regions respectively. An improved plan was proposed by Herschel, and followed by him in his various writings. It consists essentially in dividing the space which exists between the red and yellow ray as insulated by cobalt blue glass into 13.30 parts, taking the centre of the yellow ray as the zero point, and continuing the divisions equally into the more and less refrangible regions.

"Over these methods the use of the fixed lines possesses very great advantages, inasmuch as we make reference to actually visible points existing in the spectrum.

"It has been stated that the deviations of the different fixed lines in the diffraction spectrum are proportional to the lengths of the undulations which they respectively represent. By designating the different points of the spectrum by their wave-lengths, the subdivision may be carried to any degree of minuteness, the measures of one author will compare with those of another, and the dif-

ferent phenomena of chemical changes occurring through
the agency of light become at once allied to a multitude
of other optical results. If it were necessary, by a very
simple arithmetical process we could determine the number of vibrations executed by a ray bringing about a
given decomposition in billionths of a second. The
fixed lines used in this way enable us at once to divide
the diffraction spectrum into any number of parts, and,
by comparing wave-lengths and the velocity of light, to
indicate effects either in space or in time."

The diffraction spectrum, as we have seen, differs
strikingly from the prismatic in the arrangement of its
colored spaces. In the latter, the less refrangible parts
are compressed more and more in proportion as their
refrangibility is less. Now there is reason to believe
that in the former the colored spaces are equally warm,
though so feeble is the calorific effect that all attempts
at the direct measurement of the heat have proved
unsatisfactory. I first made such attempts with very
delicate thermo-electric apparatus, but could not obtain
sufficiently striking results. Admitting, however, that
every ray, irrespective of its color, in the act of extinction will generate the same amount of heat, it necessarily
follows that in the prismatic spectrum the heat should
appear to increase steadily from the more to the less
refrangible end, because in it the compression of the
colored spaces is becoming greater and greater, and this
is what is actually observed.

These considerations respecting the distribution of
heat in the spectrum lead naturally to the examination
of a much more comprehensive problem—indeed, one of
the most important problems that science presents—viz.,
the constitution of the sunbeam.

Until the time of Newton it was universally admitted

that light is a pure undecomposable elementary principle. He showed that this conclusion must be modified. No one, except Leibnitz, in those days, and no one for a long time subsequently, discerned the ominous import of the discoveries of this Prince of geometers. Of his detection of the origin of Kepler's laws, and its necessary consequence of the mode in which the government of the universe is conducted, I have nothing here to say. Let us see how it was with his discoveries concerning light.

His interpretation of the experiment he made in the "dark chamber" was this, that light is not an undecomposable element, as was at that time supposed, but that in reality it consists of not fewer than seven different constituents, recognizable by their color. These, if mixed in any manner together, whether by grinding tinted pigments or revolving parti-colored sectors, or converging the spectrum through a convex lens, would, by their union, produce white light. His felicitous experiment with the two reversed prisms silenced the carping critics of that day, who had declared that the colored tints with which he was working had no such origin as he affirmed—in difference of refrangibility—but were analogous to the iridescent play of light on a pigeon's breast, or the more gorgeous lustre of a peacock's tail. It cost only a short struggle, and the theory of the composite nature of light made good its ground.

When, therefore, Herschel, in his examination of the sun's surface through colored glasses, came to the conclusion that the heat emitted by the sun is essentially and intrinsically distinct from the light, and that these elements may be parted from each other by refraction, he did no more than develop the principle that had been announced by Newton; and when, at a later period, Melloni extended these researches, and it was universal-

I

ly admitted that there are heat rays which, like light rays, have various refrangibilities, this conclusion was quite accordant with Newton's results. Heat was considered as existing in the solar beam independent and irrespective of light. In fact, the one might be easily separated from the other.

When, again, the Swedish chemist Scheele, investigating the chemical action of light, showed that there are rays invisible to the eye, and of greater refrangibility than the violet, which can produce the decomposition of certain compounds of silver, these were considered to be an additional element, and passed under the designation of chemical rays, deoxidizing rays, etc. Treated of in the works of physics of those times as imponderable bodies, there seemed to be no necessary limit to their number. More than half a hundred ponderable substances were known. Why, then, should there not be as many of these imponderable ones? This was the view universally entertained at the time I began the experimental study of radiations. For such as are concerned in producing chemical changes I suggested a special designation, which, however, did not find acceptance: the inappropriate and unmeaning appellation, *actinic* rays, was preferred.

Meantime, however, the undulatory theory of light had been steadily making its way. It was exhibiting all the aspect of a great physical truth, in not only rendering an explanation of known facts, but also in predicting the occurrence of other facts previously unknown. Persons who were in the front of the scientific movement in this direction had thus their attention forcibly drawn to a contemplation of the whole subject from this new point of view. They very soon perceived that from it bonds of interconnection between facts hitherto supposed to be isolated might be discerned;

things that were fragmentary and confused spontaneously fell into an orderly arrangement.

While the theory of optics was making this great
advance, another important science, physiology, was presenting a similar development. It was casting off the
Vital Force of the older medical authors, and acknowl-
edging the dominion of chemical and physical forces.
It had become plain that the interpretation of many
phenomena, as hitherto received, must be changed.

We may apparently have heat without light, and
light without heat. In the darkest room we cannot
perceive vessels filled with boiling water, yet the warmth
we experience on approaching them assures us that they
are emitting radiations. Is not this heat without light?
If we stand in the rays of the full moon, we cannot detect any increase of temperature. Is this not light without heat? It is true that in this latter instance we are
mistaken as to the fact; but overlooking that—for the
heat to be detected in the moonbeams requires the most
sensitive apparatus—do not such observations assure us
that heat and light are independent of each other, physical principles having an existence separate from each
other?

Such were some of the arguments on which was sustained the hypothesis of the intrinsic difference of light
and heat. In this, no account was taken of the optical
functions of the eye. Qualities were incorrectly attributed to radiations which, in truth, were due to peculiarities in the organ of vision.

The great service which the diffraction spectrum has
rendered to science is the abolishment of all these imaginary independent existences — heat, light, actinism,
etc.—and the substitution for them of the simpler conception of vibratory motions in the ether. The only
difference existing among the radiations that issue from

a grating, in the manner we have been describing, is in their wave-lengths, or what comes to the same thing, in their times of vibration. The diversity of effects produced depends on the quality of the surface on which they fall. If on a dark surface, and the more so in proportion to its blackness, they engender heat; if on the retina, they are interpreted by the mind as light; if on photographic preparations, they produce decompositions, designated *actinic effects*.

Heat, light, actinism, are, then, not natural principles existing independently of each other, but effects arising in bodies from the reception of motions in the ether, motions which differ from each other ,in their rapidity. Of those that the eye can take cognizance of, the most rapid impart to the mind the sensation of violet light, the slowest the sensation of red, and intermediate ones the intermediate optical tints. Colors, like light itself, are nothing existing exteriorly. They are merely mental interpretations of modes of motion in the ether, and in this they represent musical sounds, which exist only as interpretations by the mind of waves in the air.

MEMOIR VIII.

ON THE PHOSPHORESCENCE OF BODIES.

From the Philosophical Magazine, Feb., 1851; Harper's New Monthly Magazine, No. 325.

CONTENTS : — *Early observations on phosphorescence. — The diamond. — Duration of shadows. — Lémery's theory. — Du Fay's theory. — Qualities of diamond and fluor-spar. — The volume of a phosphorescent body does not change during its glow. — A structural change accompanies the phosphorescence of bodies ; there is a minute disengagement of heat. — Phosphorescence is not communicable. — Absolute quantity of light emitted.*

THERE are some surfaces on which if a shadow falls, it can be brought into view a long time subsequently.

A belief in the existence of the carbuncle, a stone supposed to have the property of shining in the dark, appears to have been current from the very infancy of chemistry. It gave rise to many legends among the alchemists, and early travellers relate marvellous stories respecting self-shining mountains and gems. Thus it was said that the King of Pegu wore a carbuncle so brilliant that if any of his subjects looked upon him in the dark, his countenance seemed as though it was irradiated by the sun, and that in a certain part of North America there was a mountain which illuminated the country for many miles, and served by its rays to guide the Indians at night. The story seems to indicate that the locality of this wonder was somewhere in the western part of Pennsylvania. Mr. Boyle relates that a governor of one of the American colonies imparted this fact to him at a time when he was charged with the superin-

tendence of those important settlements, and that an expedition had been despatched to ascertain the facts correctly. It saw the shining wonder from afar, but the
light diminished as the place was approached, and becoming at length invisible, the locality could not be
determined with certainty.

These legends had for some time been passing into
discredit, when Vincenzio Cascariola, a cobbler of Bolog-
na in Italy, who had abandoned the mending of shoes
for the purpose of finding the philosopher's stone, discovered his celebrated phosphorus, the Bolognian stone, or,
as it was then designated, sun-stone (*lapis solaris*). He
had seduced himself into the expectation that a heavy
mineral he had met with—barium sulphate—contained
silver, and in an attempt to melt out that precious metal
was astonished to see that the burned substance shone
like an ignited coal in the dark. This was in the year
1602.

Some time afterwards a Saxon of the name of Baldwin
conceived the idea of obtaining the soul of the world by
distilling in a retort chalk which had been dissolved in
aqua fortis. In this extraordinary pursuit accident led
him to observe that the substance he was working with
possessed the quality of shining in the dark after it had
been exposed to the light of the sun. The alchemist
Kunckel, who relates the incident, tells us with gravity
how he stole a piece of this substance on the occasion of
a visit he made to Baldwin one night when that adept
was trying to make his phosphorus shine by the light
absorbed from a candle, and also from its image reflected by a concave mirror. In consequence of this theft,
Kunckel succeeded in discovering what the substance
was, and made known the method of its preparation.

The special condition under which these preparations
shine in the dark was very quickly detected. Isidore,

of Seville, speaking of the "lightning-stone," says, "Si sub divo positus fuerit fulgorem rapit sidereum." That condition is previous exposure to light.

The discovery of the elementary substance now known as phosphorus drew the attention of the cultivators of natural science to this singular property, and under the names of sun-stones, light magnets, noctilucas, etc., various shining bodies were introduced. But the first truly scientific examination of the subject was made by Boyle, on the occasion of observing that a certain diamond belonging to Mr. Clayton, and subsequently purchased by Charles II., emitted light in the dark. Though he does not seem to have been aware of it, the fact itself was not new, for the alchemist Albertus Magnus says in the thirteenth century that he had seen a diamond which glowed when it was put into warm water. A diamond rubbed upon gold becomes beautifully luminous; as Bernouilli remarks, it shines like a burning coal excited by the bellows.

A diamond rubbed upon gold emitting light! the imaginary or intrinsic value of the substances employed adds to the glory of the phenomenon. A light, too, that cannot be extinguished by water, and yet so ethereal and pure that it can set nothing on fire. Here certainly were facts of interest enough to excite the philosophers of the last century.

The chief points ascertained by Boyle respecting the diamond were that it shone by friction with various bodies, and at the same time displayed electrical development; that it also glowed when warmed by a candle, the fire, a hot iron, or even when placed on the skin. Under the latter circumstances it exhibited no electricity, being unable to attract a hair held near to it. He also found that it would shine under water, various acid or alkaline liquids, or when covered with saliva, and that the glow was increased when the gem was put into hot water.

These results led to the conclusion that though under certain circumstances the light was accompanied by electrical development, as when friction had been used, there was no necessary connection between the two properties. The gem would shine without the least trace of attractive power.

(Among substances endowed with this property, one of the best was discovered about a century ago by Canton. Still known as Canton's phosphorus, it is easily made by burning oyster-shells in an open fire until they have become white; then, having pulverized them with about a quarter of their weight of flowers of sulphur, they are once more brought to a dull red heat in a crucible. This completes the preparation. A convenient mode of using the substance is to provide a piece of tin plate two or three inches square, brush over one side of it with gum or glue water, then dust upon it from a fine sieve some of the powdered phosphorus. In this manner a uniform white surface is procured, well adapted for experiments.

If on such a surface a key or other opaque object be laid, and it then be exposed for a moment to daylight, on carrying it into a dark room and removing the key, a spectral shadow will be seen, depicted in black, and its contour marked out by the brilliantly glowing phosphorus surrounding it. After continuing to shine for some minutes, the light gradually fades, and finally becomes extinct. If, this having been accomplished, the phosphorized plate be put away in a box or drawer where not a ray of light can reach it, and kept therein for days or even weeks, on exposing it in a dark room, on a plate of warm metal, the phantom shadow will emerge, perhaps even more strongly than at first.

A wonderful experiment, truly. Shadows, then, are not such fleeting, such fugitive things as poets say. They may bury themselves in stony substances, and be made to come forth at our pleasure.

The persistence of such surface phantoms may be strikingly illustrated by a simple experiment in which light is not concerned. If on a cold polished metal, as a new razor, an object such as a small coin be laid, and the metal be then breathed upon, and when the moisture has had time to disappear, the coin be thrown off, though now the most critical inspection of the polished surface can discover no trace of any form, if we breathe once more upon it a spectral image of the coin comes plainly into view. And this may be done again and again. Nay, more, if the razor be put carefully aside where nothing can deteriorate its surface, and be so kept for many months, on breathing again upon it the shadowy form emerges.)

Early in the last century two hypotheses were introduced for the explanation of the various cases of phosphorescence:

1. That phosphorescent bodies act like sponges to light, absorbing it, and retaining it by so feeble a power that very trivial causes suffice for its extrication. This was the view of Lémery, and was published in 1709.

2. That phosphorescence arises from an actual combustion taking place in the sulphureous parts of the glowing body. It is to be remembered that sulphur figured largely in the chemistry of those days. This was the hypothesis of Du Fay.

To this celebrated electrician we owe a very able investigation of the phosphorescence of various bodies, and especially of the diamond. He recognized the fact, overlooked by Boyle, that the gem must first be exposed to

the light; and then, when taken into a dark place, it
shines for a time, the light gradually fading away.   But
the glow can be re-established by raising the temperature, and an exposure of a single second to the sun is
quite enough to commence the process.

To recognize feeble degrees of luminosity, it is necessary for the observer to remain in the dark until the
pupil of the eye is quite dilated, and the impression of
light to which the retina has been exposed is worn off.
Du Fay gives a singular but very serviceable practical
process.   He recommends the experimenter to keep one
eye bound up or closed for the purpose of observing in
the dark, and to use the other in conducting his processes in the light.   He remarks the curious fact that the
eye which has been shut will not have the delicacy of
its indications affected by that which has been exposed
to the light.

In this manner Du Fay found that of four hundred
yellow diamonds all were phosphorescent; but some
that were white or rose-colored or blue or green were
not.   Nor was there any external indication by which
it could be told whether any given one of these kinds
would shine.   He discovered, too, that the glow took
place under various-colored media, as stained-glass,
water, milk, but not under ink.   He also made attempts
to compel the gem to preserve its light by enveloping it
in opaque media, such as ink, black wax, etc., under the
idea that the light could not get out, and concluded that
he had partially succeeded, because in some instances the
diamonds would shine after being so shut up for six or
twelve hours.   He verified Boyle's fact on the effects of
hot water and heating generally, and carried his temperatures to far higher degrees, even above a white heat,
finding that the stone had lost none of its qualities, for
it would take light again when it was cold on a momen-

tary exposure to the sun. He also investigated how far
the glow was connected with electrical relations, and
showed its perfect independence. He also greatly increased the list of phosphori, asserting that, so far from
the quality being a peculiarity of the Bolognian stone,
Baldwin's compound, the diamond, all solid substances
except the metals, are phosphorescent when rightly treated, and even these he believed would eventually be
found to have the same property.

There is one point upon which Du Fay dwells that
deserves more than a passing remark—the connection
between phosphorescence and temperature. He proved
that phosphori cannot absorb light so well when they
are warm as when they are cold, and that a rise of temperature always makes them disengage their light.

It is obvious that these early investigators labored
under great difficulties arising from the imperfect chemical science of those times. They confounded together
things that were essentially different, such as the shining
of urine-phosphorus with the glow of the diamond, and
this again with the electrical light arising when friction
has been employed. Then, again, their erroneous views
of the composition of bodies were constantly leading
them astray. Thus Du Fay, finding that the Bolognian
stone (barium sulphate) emitted a sulphury smell, and
thinking that it shone because of the burning of the
sulphur, transferred the same explanation to the case of
yellow diamonds, and asserted that they also glowed
through the combustion of the sulphur that discolored
them.

I do not intend now to give a review of the subsequent discoveries and hypotheses brought forward by
the numerous experimenters of the last century, or by
Heinrich, the Becquerels, Biot, Poggendorff, Pearsall, and
many others in this. I may, however, recall attention to

a very elaborate memoir by Osann (Poggendorff's *Annalen*, 1834, vol. xxxiii., p. 405), in which he discusses the various theories of combustion, absorption, excitation, etc., and gives many new facts.

(Respecting the phosphorescence of diamonds, I have recently had an opportunity of making a curious observation. A lady, a relative of mine, has a pair of earrings in which are set two large and beautiful gems, both of which phosphoresce after exposure to an electrical spark; she has also another pair in which both the diamonds in like manner phosphoresce. Judging from these four instances, one might regard this property as very common. Curiously enough, the necklace belonging to this set, containing thirty-eight stones of very fine water, has only one that will phosphoresce. This necklace would, therefore, lead us to reverse the conclusion to which the ear-rings had led us, and to infer that phosphorescing diamonds are comparatively rare.)

All solid substances, except the metals, possess the phosphorescent quality. We may, however, by making a judicious selection of the bodies which are to serve as our means of experiment, disembarrass the inquiry of many of its complications. If we employ the Bolognian stone (barium sulphate) or Canton's phosphorus (calcium sulphide), or, indeed, any other substance liable to undergo chemical changes in the air, we introduce unnecessary phenomena, and cannot distinctly prove whether the shining is due to a direct combustion of the parts or to other causes.

Among selections that might be made, diamond and fluor-spar possess qualities rendering them very eligible for these purposes — unchangeability in the air and under water. Even between these there is a choice, for fluor-spar possesses all the good qualities of diamond. It might be said, considering the chemical relationships

of diamond, that when it glows it undergoes a kind of
surface combustion, which is the cause of the light; but
though direct experiments prove that this is not the
case, it is much better to resort to fluor-spar, which is
free from such an objection. It is absolutely incombustible. Besides, it can be obtained perfectly transparent
or nearly opaque; it occurs of many tints of color; can
be easily cut and polished to any figure, and obtained in
pieces of any required size. Its phosphorescent powers
are very high; indeed, it yields, when properly treated,
to no other substance, not even to Canton's phosphorus,
in that respect, and greatly exceeds the potash sulphate,
a substance which, however, possesses many eligible
qualities. It will therefore be understood that in selecting fluor-spar and its varieties as the subjects for experiment, it has been done with a view of bringing the results to their simplest conditions. In such inquiries
Canton's phosphorus and bodies chemically changeable
are wholly inadmissible.

The specimens of fluor-spar employed by me were derived from many different sources, American and European. The color of the light they emitted was in some
cases blue, in some green, in some yellow. Among them
was an American variety of chlorophane of a pale flesh-
colored aspect, translucent on the edges, and excelling
all the others in the splendor of its light. It equalled
the best Canton's phosphorus in power, yielding a superb
emerald green light when it received the rays of the sun
or of an electric spark, or had its temperature raised.
The warmth of the hand in a dark place made it shine.
Considering the facility with which we can regulate the
intensity of an electric spark, measuring out the quantities of light used in a given experiment, it is clear that
there are great advantages in resorting to it in prefer-
ence to the variable rays of the sun. Our choice of a

substance should be controlled by these conditions, and fluor-spar completely fulfils all the indications.

It will be seen, however, that I have not restricted myself to the use of this body, but whenever other substances could be compared with it, have resorted to them also. The general principles here set forth as applicable to fluor-spar may be likewise extended to them.

———

To what cause are we to attribute phosphorescence? What are the changes taking place in the glowing body?

We have already seen that a century ago two different answers had been given to these questions. Lémery supposed that all bodies act towards light as they do to heat, absorbing it and then giving it out: Du Fay that all phosphorescences are cases of combustion.

Before we can reach a decision there are evidently many preliminary points to be settled. If chemical changes between the glowing body and the air are disposed of, and the action is recognized to be of a purely physical or molecular kind, it is necessary to determine (1) whether there is any expansion or contraction of the shining body during its glow; (2) whether there is any structural change; (3) whether there is any evolution of heat along with the light; or (4) any development of electricity. These inquiries will now be taken up in succession.

1. *Is there any change of volume in a phosphorescent body during its glow?*

I attempted to ascertain this by causing various bodies to shine brightly when enclosed in glass vessels filled with water, so that if there were any expansion the water might be pressed out into a slender tube, and the amount of dilatation thereby determined. The arrangement was as follows:

A glass tube about two inches long and three quarters of an inch in diameter was closed at its upper end by means of a plate of polished quartz, cemented airtight. Immediately beneath the quartz the phosphorescent body was supported. Through a cork which closed the other end of the tube there passed a piece of thermometer tube bent on one side, and to it was affixed a scale. The arrangement was supported on a suitable stand, so that the quartz was uppermost, and at a little distance above it the spark from a Leyden-jar could be passed between a pair of stout iron wires maintained at an invariable distance, and thus produced phosphorescence in the body. It may be remarked that these effects of an electric spark do not take place well through glass, and hence a plate of quartz, which readily transmits them, must be used.

In Fig. 14, $a\ a$ is the glass tube, $b\ b$ the plate of polished quartz, c the phosphorescent body, $d\ d$ the cork closing the lower end of the tube, $e\ e$ the bent tube, f its scale, $g\ g$ the iron wires connected with a Leyden-jar, and giving a spark. The index drop at h refers not to this, but to a subsequent experiment.

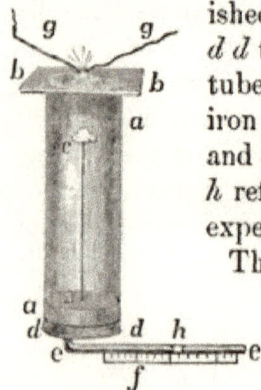

Fig. 14.

The large tube containing the phosphorescent body must be filled quite full of water, free from air, as also must be the thermometer tube to a given mark on its scale. If an electric spark be now passed between the wires to make the phosphorus shine, it is clear that if there be any expansion or contraction of its volume, there will be a corresponding movement in the water of the thermometer tube.

On making the trial, and using in succession a crystal of violet-colored fluor-spar, a piece of flesh-colored chlorophane, and a mass of Canton's phosphorus, the result in all cases was negative; for, though these different substances glowed very brilliantly as soon as the spark passed, there was not the smallest movement perceptible in the index liquid of the thermometer tube.

With a view of estimating the delicacy of the means thus used for determining any change in the volume of the spar, the solid content of a piece of chlorophane was determined by weighing in water; also the value of each division of the scale was ascertained. The value of each such division was equal to $\frac{1}{1200}$ of the volume of the spar, and a movement equal to one tenth of that value could have been detected.

It may therefore be concluded that *a phosphorescent body, when at its maximum of glow, has not changed its volume perceptibly.*

The conclusion thus arrived at is strengthened by another mode of experiment. If change of volume be connected with this evolution of light, it might reasonably be expected that a sudden, severe, but equable compression, exerted on a piece of spar, the light of which is just fading out, would compel it to regain a portion of its brilliancy. A piece of chlorophane in that condition was placed in water contained in the apparatus known as Oersted's instrument for measuring the compressibility of water, and which is described in most of the treatises on physics; but though, by suitably turning the screw, pressures varying from one to four atmospheres were suddenly put on the spar and as suddenly removed, no change whatever was seen in the glowing mass, the light of which continued steadily to die away.

In Fig. 15, Oersted's instrument for proving the compressibility of water, *a a* is the glass cylinder filled with

water, b the pressure screw, c the phosphorescent spar or substance.

These experiments have a bearing on Lémery's theory. A mass of iron suddenly compressed grows hot; so, too, does atmospheric
air. It would, therefore, not be unreasonable
to expect that if a phosphorus acted like a
sponge to light, and were thus pressed upon,
it would yield up its light. But conceptions
derived from the old theories of specific heat
are perhaps scarcely applicable here.

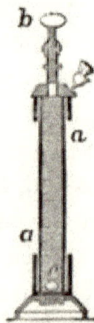
Fig. 15.

When unequal pressure is applied, the result is different. A piece of chlorophane pressed by a forceps glows
brightly; if crushed, the fragments sparkle like little
fire-works as they fly through the air. If the spar be
previously powdered, a shining is still produced, and
when the pulverization is conducted in an agate mortar
in the dark, bright eddies of light follow the track of
the pestle. In these cases, however, the separation of
the laminæ of the crystal and the heat produced by
friction probably determine the result. Canton's phosphorus did not shine when compressed or submitted to
friction.

2. *Does any structural change accompany the phosphorescence of bodies?*

The foregoing experiments appearing to prove that
if there be any expansion of a phosphorescing body, it
is to a very small amount, I next endeavored to determine whether there is any molecular change, or new
structural arrangement, which can be detected by polarized light.

A flat piece of fluor-spar, polished on both sides, was
placed in a polariscope, and a pair of blunt iron wires
connected with a Leyden-jar were adjusted near the

K

front of it, so that when the spark passed, a brilliant
glow arose in the spar, which was at once viewed
through the analyzer of the instrument. But though
the experiments were made both by daylight and lamplight, no kind of effect could be detected. Had any
molecular change occurred, it could not have escaped
notice.

In Fig. 16, $a\,b$ is the polariscope, c the flat piece of
fluor-spar, b the analyzer, $d\,d$
the wires giving the electric
spark.

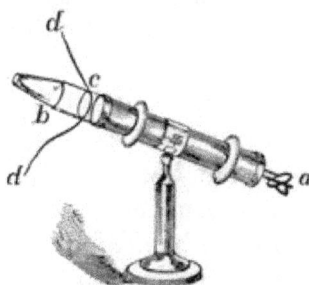

Fig. 16.

These experiments were first
made by using as the analyzer
a doubly refracting achromatic
prism; they were, however, repeated with a Nicol, in which
the eye is not disturbed by a
bright image as in the other
case. Having fixed the plate of polished fluor in the
polariscope, it was readily perceived that it possessed
naturally a structural arrangement, for there were cloudy
spaces or lines in it which contrasted with the faint
white light passing in the adjacent parts. It was also
seen that this structural arrangement could be deranged
in a transient manner, either by pressure or an unequal
warming, as is well known of other bodies; but when a
powerful electric discharge was passed near the spar,
and a brilliant phosphorescence took place, no impression could be detected. Even when the iron wires rested on the spar, and the explosion passed over its surface,
nothing was perceptible except along the line between
the ends of the wires, where the surface was roughened
or abraded by the force of the discharge.

But though these experiments with polarized light
give a negative result, or, at all events, prove that a

phosphorus when shining has its molecular condition so little disturbed that the change cannot be detected in this way, there can be no doubt that if the means of testing were more delicate, such a change would be discovered, for many years ago Mr. Pearsall found that specimens of fluor, not possessing phosphorescence naturally, might have that quality communicated to them by repeated exposure to many powerful electric discharges, which also gave rise to a change in their natural color. Now there can be no doubt that such an alteration of tint implies an alteration of structure.

Besides the test by polarized light, there is another which may be resorted to for the detection of structural changes when they are merely superficial; it is the mode in which various vapors will condense. I described several such cases in the *Philosophical Magazine* for September, 1840, some time previously to the publications on the subject that were made by M. Moser. They were brought forward at that time as an illustration of the manner in which mercurial vapors condense on a daguerreotype plate and develop images which it has received. Proceeding on this principle, a large plate of fluor-spar, the surface of which was finely polished, was made to phosphoresce brightly along a given line determined by the ends of two iron wires, which served as a discharger for a Leyden spark, and were placed near to the polished surface. The spar was forthwith suspended in the mercurial box of a daguerreotype apparatus and kept there an hour. The mercury condensed upon it faintly in the manner it would have done on a daguerreotype plate, especially on and in the vicinity of those parts that were more immediately exposed to the spark. This, therefore, seems to prove that there is in these cases a molecular modification of the shining surface.

3. *When a phosphorescent body glows, does it likewise emit heat?*

A very thin bulb, half an inch in diameter, was blown on a piece of thermometer tube, and after being washed over with gum-water, finely powdered chlorophane was dusted on until it was neatly coated all over. A drop of water was then introduced into the tube to serve as an index. Although the instrument was very sensitive to heat, when the chlorophane was made to shine and emit a gorgeous emerald green light by the passage of a powerful electric spark near it, no movement whatever of the index ensued. From this it would appear that the quantity of heat developed by phosphorescence must be very small.

In Fig. 17, *a a* is the glass bulb covered with a coat-

Fig. 17.

ing of powdered chlorophane, *b* a drop of water serving as an index.

A modification of this exper- iment, which appeared to offer several advantages, was tried. The instrument repre- sented in Fig. 14 was emptied of its water, and a single drop, *h*, put into the index tube. It was supposed that when the rays of the electric spark passed through the quartz and made the phosphorus shine, the air contained in the tube, warmed thereby, would expand, and a move- ment in the index liquid of the thermometer tube take place. But in several trials, in which different bodies— chlorophane, Canton's phosphorus, etc.—were employed, the results were uniformly negative; for though these different substances glowed splendidly as soon as the spark passed, there was not the slightest rise of tem- perature perceptible.

A further attempt was made as follows: The disk of quartz being removed and replaced by a cork, through

which a pair of iron wires to serve as a discharger passed
air-tight, and descended to within a short distance of the
phosphorus, sufficient time was allowed in various repetitions for the index liquid to come to rest.   It was hoped
that this form of experiment would have advantages over
the preceding, because the discharging wires could be
brought nearer to the phosphorus, and the effect take
place without the intervention of the quartz.   When
the spark was made to pass, there was a great movement in the index tube, as in the instrument known as
Kinnersley's electrometer, but the liquid immediately returned to within a short distance of its first place; then
a slow dilatation occurred, as though the air was gradually warming.   Thus in one experiment the liquid stood
at 24°, after the explosion it returned to 26°, and then
there was a gradual dilatation to 32°.

To eliminate the various disturbing causes in this experiment, it was repeated many times, the spar being alternately introduced into the glass tube, and alternately removed.   It was found that whenever the spar was
present the gradual dilatation alluded to took place;
but when the spar was not in the tube, instead of a
dilatation, there was a gradual contraction until the index liquid recovered its original position.

From this it appears that *with the evolution of light
there is a feeble extrication of heat.*

The quantities of heat thus liberated are so small, and
the causes of error are so numerous, that I endeavored
by other methods to obtain more trustworthy results.
Thus I attempted to determine the surface temperature
of a flat piece of chlorophane while phosphorescing by
means of the thermo-electric multiplier.   The thermo-
pile was placed in a vertical position, and the spar
having been attached to a piece of wood, which served
as a handle, intense phosphorescence was communicated

by a Leyden spark, and the flat and shining surface instantly put on the upper face of the pile. But there was no movement of the astatic needles.

Then, taking the stone by its handle, it was touched with the tip of the finger for one second, and quickly placed on the pile. A prompt movement of the needles, amounting to four degrees, ensued. These experiments were repeatedly tried, and the results were uniformly the same.

In Fig. 18, *a a* is the thermo-electric pile, *b b* the plate of chlorophane, *c* the handle.

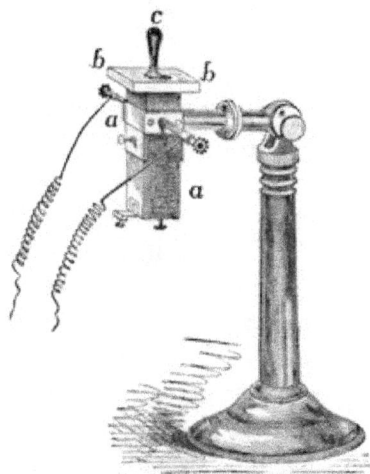
Fig. 18.

On considering these results, it appears that as the temperature of the air near the multiplier in one of the experiments was 53°, and the estimated temperature of the skin 94°, the amount of heat which the stone received from the touch of the finger must have been very small. I made a comparative trial by touching the bulb of a thermometer for the same space of time, in the same way, and found that there was a rise of about $1\frac{1}{2}°$. But the conductibility of quicksilver is much greater than that of chlorophane.

It is to be inferred, therefore, that the quantity of heat set free during phosphorescence is very small, and that the surface of the chlorophane does not change its temperature by one fourth of a degree; for had it done so, the multiplier would have instantly detected it.

4. *Is phosphorescence accompanied with a development of electricity?*

It has been stated already that the experimenters of the last century paid a good deal of attention to this point. Du Fay established the fact that though in many cases of phosphorescence there is a development of electricity, there are many others in which the light seems to be wholly unattended by any disturbance of that kind.

I have repeated some of these experiments, and with the same result, proper care being taken to avoid friction and other obvious causes of electrical excitement. Thus a flat piece of chlorophane, phosphorescing powerfully, was put on the cap of a very delicate gold-leaf electroscope, but no disturbance whatever was perceptible.

A large crystal of fluor-spar was made to phosphoresce brilliantly along a line about half an inch in length by passing the spark of a Leyden-jar between two blunt iron wires, the ends of which were that distance apart, and resting on the face of the crystal. Over this line of blue light, which was pretty sharply marked, and which lasted for several minutes, a fine hair was held. This would have been readily attracted and repelled by the feeblest excitation of sealing-wax, but in this case it wholly failed to yield any indication whatsoever.

In connection with the foregoing experiments, I may mention some miscellaneous facts. Some attempts were made to determine whether phosphorescent bodies in the field of a powerful electro-magnet would exhibit any change of property. Six Grove's pairs were caused to magnetize a good electro-magnet; the power they could give to it would enable the keeper to support about half a ton. Between its polar pieces chlorophane, Can-

ton's phosphorus, etc., which were made to glow by exposure to a Leyden spark, were placed. But it made no
difference in the light whether the magnetism was on
or not.

It was also found that the electric spark from a contact-breaker would communicate phosphorescence to all
the various bodies in use in these experiments, and that
up to a certain point the intensity of the light increased
with the number of sparks received.

Phosphorescence is not communicable from one body
to another. Having provided two polished plates of
fluor-spar, one of them was made to glow by an electric
spark, and the other was immediately put upon it. No
communication of phosphorescence took place; the second piece remained perfectly dark.

Some authors state that fluor-spar does not become
phosphorescent by exposure to the sun; but this remark does not apply to all varieties of it. Thus some
chlorophane, which had been ignited in a glass tube till
it had ceased to shine, was pulverized and again ignited in a platinum crucible. It emitted an emerald light.
A slip of wood was now put on it to screen a part of
its surface, and it was exposed to the sun for a few
minutes. On ignition, it shone again finely, with a green
light, the shadow of the wood being beautifully depicted. The same having been repeated a great many
times, it appeared that the phosphorescence at last began to decrease, perhaps by frequent ignition causing a
change.

A screen of yellow glass intervening between the sun
and some powdered chlorophane prevented phosphorescence, but it took place through a plate of polished fluor-
spar. When the light of an electric spark was used instead of the sunshine in this experiment, the fluor-spar
prevented phosphorescence.

GENERAL CONCLUSIONS.

The results to which the foregoing experiments bring us are therefore—

1st. That the methods employed in these experiments are not sufficiently delicate to detect any increase in the dimensions of a phosphorus while it is in a glowing state.

2d. No structural change can be discovered by resorting to polarized light; but there is reason to believe from the change of color which certain bodies exhibit when the quality of shining is communicated to them, and from the manner in which vapors condense on their surfaces, that such has actually taken place.

3d. That phosphorescence is attended with a minute rise of temperature.

4th. That it is not necessarily connected with any electrical disturbance.

On comparing these conclusions, it is obvious that if the third be correct there must necessarily be a change of volume, and that the reason the dilatation is not discovered by direct experiment is owing to the insufficiency of the means employed.

The general definition given of phosphorescence is that it is the extrication of light without heat (Gmelin). But these results show that that definition is essentially incorrect; for if the experiment be made with due care, a rise of temperature can be detected, though its absolute amount may be very small.

Determination of the absolute quantity of light emitted by phosphori.

And now we may inquire how it is with the light itself? do we not deceive ourselves respecting it? We ought to recollect that it is barely perceptible in the open day, and that these experiments require to be made in the dark. We should also recollect the great sensitiveness of the eye, and how feeble a luminous impression it can detect. Impressed with these facts, I have endeavored to compare the absolute quantity of light given by the most brilliant phosphori with some well-known standards. The result of these experiments puts a new view on the whole subject.

The first attempts I made for this purpose were conducted on the principle of comparing the stains formed on a daguerreotype plate by the phosphorus under trial, and by an oil-lamp, receiving the rays from each on a concave metallic mirror eight inches in diameter and fifteen in focus, arranged as a reflecting camera obscura. There were set, side by side, a small oil-lamp, a piece of white paper illuminated by the lamp, and a fragment of chlorophane, arranging things in such a manner that the chlorophane might be illuminated by rays coming from a contact-breaker worked by two Grove's pairs. The contact-breaker was kept in action fifteen minutes, and then, to prove the sensitiveness of the plate, the lamp was moved for one minute to a new position, and the experiment closed.

On developing, it was found that the impressions of the lamp had solarized, both that of fifteen minutes and that of one, proving that such a light in one minute is amply sufficient to change the plate to its maximum. Also the electric spark of the contact-breaker was solarized, and the image of the piece of white paper beautifully given of a clear white; but the phosphorescing spar had made no impression, except from one portion where it had reflected the rays of the spark.

Suspecting that the spark from the contact-breaker might not have been powerful enough, I repeated the experiment, using sparks from a Leyden-jar. The oil-lamp was exposed in front of the mirror one minute, and then removed; then ten strong sparks were passed over the spar, each of which made it emit an emerald light; but during the moment of the passage of each spark a screen was interposed, that no direct or reflected light, and, indeed, none but that of the phosphorus, could reach the mirror and sensitive plate.

On mercurializing, it was found as before that the lamp was beautifully depicted but the spar was invisible.

In Fig. 19, a is the oil-lamp, b the white paper, c the chlorophane, cut and polished, d the contact-breaker with its wires, f f the concave mirror. Its concavity faces the above-named objects, and reflects their images inverted and reversed on a sensitive plate, e. The mirror and sensitive plate are enclosed in a darkened box not shown in the figure.

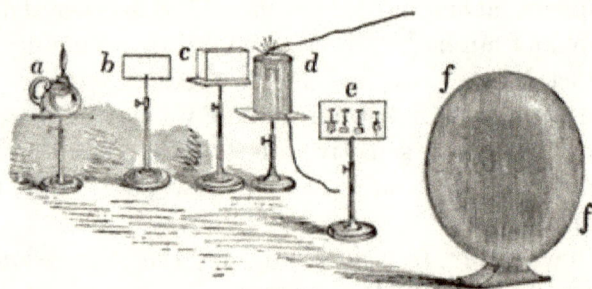

Fig. 19.

Estimated, therefore, by the chemical effects they can produce, the light from chlorophane is incomparably less intense than that from a common lamp. For there can be no doubt that each of the ten Leyden sparks gave a light which made the spar phosphoresce brilliantly for six seconds, and the whole phosphorescence was equal in duration to that produced by the light of the lamp; yet the latter had changed the plate to a maximum, while the former had not made the smallest perceptible impression.

As the foregoing attempt to obtain photographic effects had failed, I varied the experiment as follows: In a Bohemian glass tube a quantity of chlorophane in coarse fragments, sufficient to occupy about three inches in length of the tube, was placed. The reflecting camera with its sensitive silver plate was set in a proper position. When everything was arranged, a spirit-lamp was applied to the chlorophane, which soon emitted a superb emerald light, and continued to do so for about two minutes. As the light began to decline, the spar splintered by decrepitation. The process went on in a very satisfactory way. An oil-lamp was then placed in front of the camera for five seconds. On developing, the image of the lamp-flame came out, but no trace whatever of the chlorophane could be detected. Thus it appears that the splendid green light emitted when the spar is heated is at least twenty-four times less intense than the light emitted by a small oil flame. It should be remembered that this is a measure of absolute intensity, and not of illuminating power.

But as it is known that green light is not very efficient in changing a sensitive surface, I tried to determine the intensity of the light emitted by chlorophane by the optical method of Bouguer, described in the first of these Memoirs, p. 39.

The spar being heated by a current of hot air arising from the flame of a spirit-lamp, the light of which was carefully screened by a chimney and other contrivances of sheet-iron, a comparison was made with a very small oil-lamp, the flame of which was about six tenths of an inch high and the wick one sixth of an inch thick. It was covered with a glass shade.

The spar, when it began to glow, cast a reddish shadow on the paper, which shadow was extinguished when at its maximum by the lamp at about 25 inches, the spar being at 5 inches.

The distances of the chlorophane and lamp from the paper were, therefore, as $1:5$. The illuminating effect is as the squares of those numbers, and therefore $1:25$. But for extinction it requires that one light should be sixty times as intense as the other; it follows, therefore, that at those distances the illuminating power of the lamp is fifteen hundred times as intense as the illuminating effect of the spar.

But the quantity of spar used in this experiment exposed a surface much greater than that of the flame; it was estimated to be at least twice as great. This, therefore, would bring us to the conclusion that the intrinsic brilliancy of the chlorophane is not $\frac{1}{3000}$ part that of the lamp.

This experiment was several times repeated. Thus it was found that the lamp extinguished the shadow from the spar when the relative distances were $1:4$. The lamp at 4 was therefore sixty times as luminous as the spar at 1; that is, their illuminating power was as $1:960$. But it was estimated that the surface of the spar was $3\frac{1}{2}$ times that of the flame of the lamp, so this would make the intrinsic brilliancy $\frac{1}{3360}$, a result of the same order as the preceding.

From this we conclude that *the intrinsic brilliancy of phosphori is very small; a fine specimen of chlorophane, at its maximum of brightness, yielding a light three thousand times less intense than the flame of a very small oil-lamp.*

It was stated above that these photometric experiments put a new view on the whole subject; in fact, they explain all the difficulties of the foregoing inquiries. How could we expect to be able to measure the heat of phosphorescence? The radiant heat of the little oil-lamp here employed would require at such distances a very delicate thermometer to measure it. Is it likely, then, that we could detect that of a source three thousand times less intense?

I conclude, therefore, that all phosphoric bodies emit radiant heat as well as light; but that its quantity is so small that we have no means delicate enough to measure it, though the eye is so sensitive that it can detect the light, the absolute intensity of which has, however, hitherto been greatly overrated. I believe that the quantities of both are of the same order, and if this be true we should scarcely expect to discover any dilatation of the glowing body unless means much more refined than those here resorted to are employed.

MEMOIR IX.

ON THE EFFECTS OF HEAT ON PHOSPHORESCENCE.

From the Philosophical Magazine, Feb., 1851.

Contents:— *Experiment of Albertus Magnus.* — *Degree of phosphorescence at different temperatures.* — *The quantity of light a substance can retain is inversely as its temperature ; the quantity it can receive is directly as the intensity and quantity of light to which it has been exposed.* — *Phosphorescent images of the moon.* — *Action of ether waves.* — *Effects of cohesion.* — *Reason that gases, liquids, and metals are non-phosphorescent.*

It has been already observed that the effect of heat in promoting the disengagement of light is an old discovery. Albertus Magnus remarked it in the case of a diamond plunged into hot water.

It is customary in later works which treat systematically on phosphorescence to group the different facts under two heads—1st, phosphorescence produced by insolation ; 2d, by heat. An example of this is offered in the standard work on chemistry by Gmelin.

A division of this kind brings the whole subject into confusion. It assigns different causes for things that are essentially allied. It leads to the inference that as under certain circumstances the sunlight or an electric spark can make bodies glow, so under other circumstances heat will produce the same effect, and this wholly independent of incandescence.

But what are the facts? If a yellow diamond placed upon ice be submitted to the sun, and then brought into a dark room the temperature of which is 60°, for a time there is a glow, but presently the light dies out. If the

diamond be now put into water at 100°, it shines again, and again its light dies away. If next it be removed from that water and suffered to cool, and then be re-immersed, it will not shine again; but if the water be heated to 200°, and the diamond be dropped into it, again it glows, and again its light dies away.

There is, therefore, a correspondence between the light disengaged and the temperature. We are not to conclude from the foregoing illustration that when the diamond has its temperature raised from 100° to 200° the light is due to the heat. On the contrary, the light is unquestionably due to the primitive exposure to the sun; just as in Lémery's illustration of the sponge, if we exert a little pressure a portion of the water flows out; if a stronger pressure, still more; and for each degree of pressure there will be a corresponding quantity of water expelled.

The connection between phosphorescence and temperature may be instructively illustrated as follows:

Suppose that three yellow diamonds, a, b, c, have been simultaneously exposed to the sun, a being kept at 32°, b at 60°, c at 100°, and that they are then simultaneously removed to a bath of water at 100° in a dark room; it will be found that a emits a bright light, b shines more feebly, and c scarcely at all. This is what ought to be expected from the principle laid down above; for if at a particular temperature a certain quantity of light is set free, it is clear that a has the advantage of b, so that it will disengage all the light to be set free between 32° and 60°.

From such experiments and considerations it is to be inferred that there is an intimate connection between temperature and phosphorescence which may be conveniently expressed in the following terms. *The quantity of light a substance can retain is inversely as its temperature.*

This principle furnishes the explanation of a multitude of facts. Thus Du Fay discovered that the Bolognian stone shines brighter when exposed to the sky than to the sun. In the latter case the temperature rises, and the quantity of light retained is less. Under violet and other glasses, stained with such colors as impede the warming effect, phosphorescence is even more vivid than when no glass has intervened. On the same principle we have an explanation of Du Fay's apparently successful attempt to prevent the escape of light from diamonds by putting them in ink or covering them with black wax. When removed from the ink and brought into the air, they became somewhat warmer—perhaps the touch of the finger aided the effect—and a corresponding quantity of light was set free.

But though temperature is a controlling, it is not the only condition involved. If it were, phosphorescence after insolation should occur only after a rise of temperature. The fundamental fact of the whole inquiry proves that a glowing body can retain more light in presence of a lucid surface than it can in the dark.

Is not this fact analogous to what we meet with in the exchanges of heat? A substance can retain more heat in presence of a hot body than a cold one. The brilliancy and quantity of light to which a phosphorus is exposed goes very far to determine the intensity of the subsequent glow. Thus I found that a piece of chlorophane exposed to one spark of a contact-breaker shone feebly, but if it had received one hundred sparks, its light was very vivid; and it has long been known that in delicate phosphori a certain degree of luminosity can be communicated by the moonbeams, a more intense one by lamp-light, and one still more brilliant by the sunshine or a Leyden spark. This, therefore, leads to the conclusion that *the quantity of light a phosphorus*

L

can receive is directly as the intensity and quantity of light to which it has been exposed.

(With respect to the light of the moon, I have succeeded in obtaining an image of that satellite on Canton's phosphorus, by the aid of a concave metallic mirror.)

The various facts herein cited indicate that when a ray of light falls on a surface, it throws the particles thereof into vibration. An examination of the action of the differently colored rays dispersed by a prism shows that in general the greater the frequency of vibration of the impinging ray, the more brilliant is the phosphorescence. But in such a prismatic examination we have constantly to bear in mind the disturbing agencies which are present, and especially the antagonizing effects of heat; that this determines the amount of light that a phosphorus can receive, and also the rate of its subsequent extrication. In Memoir V., p. 87, I have shown how the photographic action of light betrays the general principle of an interference of vibratory movements, and the production of antagonizing results in different parts of the solar spectrum. An argument is there brought forward to the effect that as the violet end produces phosphorescence and the red extinguishes it, this is a proof of opposition of action. In explaining this fact, M. E. Becquerel supposes the darkening power of the red rays to be due to the more rapid disengagement of the phosphorescence by reason of the heat produced by those rays, and the apparent antagonization is not attributable to the supposition of vibratory movements of light rays of different frequency, but to the relations of caloric and light. The force of this explanation, however, disappears when it is understood that light and heat, the chemical and phosphorogenic rays, are, according to the principles of the able experimenter, all manifestations of the same agent. It avails us nothing to say that a want of phosphorescence at the less refrangible end of the spectrum is due to the heat-giving powers of those rays, when that very heat-giving power is under the hypothesis dependent on their comparative rapidity of vibration.

We may, therefore, in the explanation of phosphorescence, abandon expressions derived from the material theory of light, and assume that whenever a radiation falls upon any surface, it throws the particles thereof into a state of vibration, just as in the experiment of Fracaster, in which a stretched string is made to vibrate in sympathy with a distant sound, and yield harmonics, and form nodes. Such a view includes at once the facts of the radiation of heat and the theory of calorific exchanges; it also offers an explanation of the connection of the atomic weights of bodies and their specific heats. It suggests that all cases of decomposition of compound molecules under the influence of a radiation are owing to a want of consentaneousness in the vibrations of the impinging ray and those of the molecular group, which, unable to maintain itself, is broken down under the periodic impulses it is receiving into other groups, which can vibrate along with the ray.

If a hot body, a, be placed in presence of a cold body, b, the theory of the exchanges of heat teaches that the temperature of the latter will steadily rise until equilibrium between the two takes place. The molecules of a communicate their vibratory movement to the ether, and this in its turn imparts an analogous movement to the molecules of b. For, as the ethereal medium is of vastly less density than the vibrating molecules, each of these oscillations will produce in it a determinate wave, which is propagated through it according to the ordinary laws of undulations, in such a way that the ether would be in repose after the wave had passed, were it not for the recurrence of the continuing vibration of the molecules. At each vibration the molecules of *a* lose a part of their *vis viva*, by the quantity they have communicated to the ethereal wave, the intensity or amplitude of the wave becoming less and less as this abstraction of force is going on. But the ether being of uniform density and elasticity throughout, each of its particles communicates the whole *vis viva* it has received to the next adjacent, and would instantly come to rest were it not again disturbed by the vibrations of the material molecules. These elementary considerations show how it is that a wave of sound passes through the air, or of light through the ether, and the particles of those media come instantly to rest; but a hot body or a vibrating string persists in its motions, which only undergo a gradual decline. If the vibrating molecule were in a medium of the same density, it would impart to it all its motion at once, and in the same way that a heavy molecule gradually communicates its motion to the ether, so in its turn does the ether to other systems of molecules.

Upon these principles we may explain the phenomena of phosphorescence. From a shining body undulations are propagated in the ether, and these, impinging on a phosphorescent surface, throw its molecules into a vibratory movement. These in their turn impress on the ether undulations; but by reason of the difference of its density compared with that of the molecules, they do not lose their motion at once; it continues for a time, gradually declining away and ceasing when the *vis viva* of the molecules is exhausted.

When a phosphorescent surface is exposed to the luminous source, it necessarily undergoes a rise of temperature, and the cohesion of its parts is diminished; but after its removal from that source, as the temperature declines and radiation goes on, the cohesion increases, and a restraint is put on those motions.

Now let the phosphorus have its temperature raised, and the cohesion of its molecules be thereby weakened, and the restraint on their motions abated. At once they resume their oscillations, and continue them to an extent that belongs to the temperature used. When this has passed away, a still higher temperature will release them once more, and the glowing will again be resumed.

What would be the result if we could cause the surface of a mass of water on which circular waves are rising and falling to be instantaneously congealed? It might be kept in that condition for a thousand years, and then, if instantaneously thawed, the waves would resume their ancient motion from the point at which it was arrested, and it would now go on to its completion.

So with these phosphori. Exposed to light of a suitable intensity, their parts begin to vibrate; but the freedom of those motions is interfered with by their cohesion. Amplitude of vibration must always be affected by cohesion, and if the ray be removed and the temperature be permitted to decline, the restraint becomes greater and greater, and they pass into a condition somewhat like that which has just been illustrated. It matters not how long a time may intervene, rise of temperature will enable them to resume their motions.

These principles give an explanation of all the facts we observe. We see how it is that as we advance from one temperature to another the phosphorus will resume its glow, and that there is, as it were, for every degree a certain amount of vibratory movement that can be accomplished, or, to use a different phrase, a certain amount of light that can be set free. It also necessarily follows that different solids will display these motions with different degrees of facility, and hence shine for a longer or shorter time, and with lights of different intensities.

But in liquids and gases, which want that particular

condition of cohesion characteristic of the solid state, and the parts of which move freely among each other, phosphorescence cannot take place, for it depends on the influence that cohesion has had in restraining the vibratory movements.

Further, the condition of opacity does not permit phosphorescence to be established. The exciting ray cannot find access to disturb the interior layers of the mass, and even if it did, and phosphorescence ensued, how could we expect to be able to discover it through the impervious veil of the superficial layers? The light of the most brilliant phosphorus cannot be seen through the thinnest gold-leaf. Its intensity is vastly too small. These are the reasons that no one has ever yet succeeded in detecting phosphorescence in metals and black bodies.

It will be gathered from this explanation that I am led to believe that all the facts of phosphorescence can be fully explained on the principles of the communication of vibratory motion through the ether; that as upon that theory an incandescent body maintained at incandescence would eventually compel a cold body in its presence to come up to its own temperature by making its particles execute movements like those of its own, so the sunshine or the flash of an electric spark compels a vibratory movement in the bodies on which its rays fall; that these vibrations are interfered with by cohesion in the case of solids, but that they are instantly established and almost as instantly cease in the case of liquids and gases; that reducing the cohesion of a solid by raising its temperature permits a resumption of the movement; and that the condition of opacity, whether metallic or otherwise, is a bar to the whole phenomenon.

MEMOIR X.

ON THE DECOMPOSITION OF CARBONIC-ACID GAS BY PLANTS IN THE PRISMATIC SPECTRUM.

From the Proceedings of the American Philosophical Society, May, 1843; Philosophical Magazine, Sept., 1843; American Journal of Science and Arts, Vol. XXVI., 1843; Philosophical Magazine, Sept., 1844.

CONTENTS:—*The decomposition of carbonic acid by light formerly attributed to the violet ray.—It can be successfully accomplished in the prismatic spectrum.—It takes place not in the violet but in the yellow ray.—Decomposition by yellow absorbent media.—Analysis of the gas evolved; it always contains nitrogen.—Decomposition of alkaline carbonates and bicarbonates.—Analysis of the gas evolved in different rays.*

FOR many years it has been known that the green parts of plants, under the influence of sunlight, possess the power of decomposing carbonic acid, and setting free its oxygen. It is remarkable that this, which is a fundamental fact in vegetable physiology, should not have been investigated in an accurate manner. The statements met with in the books are often far from being correct. It is sometimes said that pure oxygen gas is evolved, that the decomposition is brought about by the so-called "chemical rays;" these and a multitude of such errors pass current. So far as my reading goes, no one has yet attempted an examination of the phenomena by the aid of the prism, the only way in which it can be correctly discussed.

In a paper by Dr. Daubeny, inserted in the *Philosophical Transactions* for 1836, two facts, which I shall verify in this communication, are fully established. These are, 1st, the occurrence of nitrogen gas in mixture with the

oxygen, an observation originally due to Saussure, or
some earlier writer; and, 2d, that the act of decomposition is due to the LIGHT of the sun. This latter effect, obtained by employing colored glasses or absorbent media, has not been generally received. Doubt will
always hang about results obtained in that way, and
nothing but an examination by the prism can be satisfactory.

In its connection with organic chemistry and physiology the experiment of the decomposition of carbonic
acid by leaves assumes extraordinary interest. When
we remember that this decomposition is the starting-
point for organization out of dead matter, that commencing with this action of the leaf the series of organized atoms goes forward in increasing complexity,
and blood and flesh and cerebral matter are at its terminus, it is clear that unusual importance belongs to
precise views of this the commencing change. The rays
of the sun are the authors of all organization.

There is but one way by which the question can be
finally settled — it is by conducting the experiment in
the prismatic spectrum itself. When we consider the
feebleness of effect which takes place by reason of the
dispersion of the incident beam through the action of
the prism, and the great loss of light through reflection
from its surface, it might appear to be a difficult operation to effect a determination in that way. Encouraged,
however, by the purity of the skies in America, I made
the trial, and met with complete success.

When the leaves of plants are placed in water from
which all air has been expelled by boiling, and exposed
to the sun's rays, no gas whatever is evolved from them.
When they are placed in common spring or pump water, bubbles quickly form, which, when collected and
analyzed, prove to be a mixture of oxygen and nitro-

gen gases; from a given quantity of water a definite
quantity of air is produced. When they are exposed
in water which has been boiled, and then impregnated
with carbonic acid, the decomposition goes on with rapidity, and large quantities of gas are evolved.

The obvious inference seeming to arise from these
facts is that all the oxygen collected is derived from
the direct decomposition of carbonic acid.

Having by long boiling and subsequent cooling obtained water free from dissolved air, I saturated it with
carbonic-acid gas. Some grass leaves, the surfaces of
which were carefully freed from adhering bubbles or
films of air by having been kept in carbonated water
for three or four days, were provided. Seven glass
tubes, each half an inch in diameter and six inches
long, were filled with carbonated water, and into the
upper part of each the same number of blades of grass
were placed, care being taken to have all as near as could
be alike. The tubes were placed side by side in a small
pneumatic trough. It is to be particularly remarked that
the leaves were of a pure green aspect as seen in the water; no glistening air-film such as is always on freshly
gathered leaves nor any air-bubbles were attached to
them. Great care was taken to secure this perfect freedom from air at the outset of the experiments.

The little trough was now placed in such a position
that a solar spectrum, kept motionless by a heliostat,
and dispersed in a horizontal direction by a flint-glass
prism, fell upon the tubes. By bringing the trough
nearer to the prism or moving it farther off, the different
colored spaces could be made to fall at pleasure on the
inverted tubes. The beam of light was about three
fourths of an inch in width.

In a few minutes after the beginning of the experiment, the tubes on which the orange, yellow, and green

rays fell commenced giving off minute gas-bubbles, and
in about an hour and a half a quantity was collected
sufficient for accurate measurement.

In Fig. 20, *a a* represents the trough, and R, O, Y, etc.,
the tubes containing the leaves and carbonated water
placed so as to receive the spectrum, *b b.*

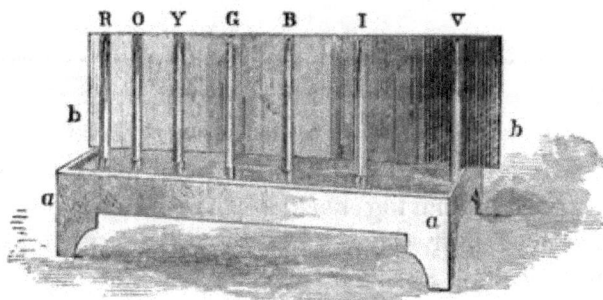

Fig. 20.

The gas thus collected in each tube having been transferred to another vessel and its quantity determined, the
little trough with all its tubes was freely exposed to the
sunshine. All the tubes now commenced actively evolving gas, which, when collected and measured, served to
show the capacity of each tube for carrying on the process. If the leaves in one were more sluggish or exposed
a smaller surface than the others, the quantity of gas
evolved in that tube was correspondingly less. And
though I could never get the tubes to act precisely
alike, after a little practice I brought them sufficiently
near for my purpose. In no instance was this testing
process of the power of each tube for evolving gas omitted after the experiment in the spectrum was over.

From the following table it appears that the rays which
cause the decomposition of carbonic-acid gas are the orange, the yellow, the green ; the extreme red, the blue, the
indigo, and the violet exerting no perceptible effect. We

Table of the Decomposition of Carbonic Acid by Light of Different Colors.

Experiment I.		Experiment II.	
Name of ray.	Volume of gas.	Name of ray.	Volume of gas.
Extreme red......	.33	Extreme red and red.	.00
Red and orange...	20.00	Red and orange......	24.75
Yellow and green.	36.00	Yellow and green.....	43.75
Green and blue...	.10	Green and blue.......	4.10
Blue	.00	Blue	1.00
Indigo............	.00	Indigo	.00
Violet...........	.00	Violet,...............	.00

should therefore expect that in a beam, passing through absorbent media of such a nature that the extreme red, the blue, the indigo, and violet are absorbed, this decomposition should nevertheless go on. A solution of bichromate of potash nearly fulfils these conditions. It transmits the luminous rays in question, except a trace of those which correspond to the more refrangible yellow and less refrangible green.

A remarkable proof of the correctness of the foregoing prismatic analysis comes out when leaves are made to act on carbonated water in light which has passed through a solution of bichromate of potash. I took a wooden box of about a cubic foot in dimensions, and having removed its bottom, adjusted to it a trough made of pieces of plate-glass. The box being set on one side, its lid served as a door, and the trough being filled with a solution of bichromate of potash, the sun's beams came through it, and in the interior of the box leaves and carbonated water could be exposed to the rays that had escaped absorption. The thickness of the liquid stratum was about half an inch. I had several such boxes made, so that I might compare the simultaneous effect of light that had undergone absorption by different media. They formed, as it were, little closets, in which bodies could be exposed to parti-colored light—blue, yellow, red, etc.

Fig. 21 represents one of these closets; *a a* is its side

Fig. 21.

of stained-glass, or the trough containing the absorbing solution, bichromate of potash, etc.; *b b*, the door.

Whenever an experiment was commenced in these closets, a similar one was simultaneously commenced in the unobstructed sunshine. It is needless to state that in all these care was taken to have the different arrangements for decomposition as nearly alike as possible.

On comparing the amount of gas evolved in unabsorbed light and in light that had undergone absorption by bichromate of potash, in three out of five trials the gas collected under the latter circumstances exceeded in volume that collected under the former; this was probably due to a higher temperature existing in the box.

On comparing the volume of gas collected under bichromate of potash and under litmus-water, the latter was not equal to half the former.

I compared the gas evolved in unobstructed light, under bichromate of potash, and under ammonia sulphate of copper; the results were as follows:

Unobstructed light.................... 4.75
Bichromate of potash................. 4.25
Ammonia sulphate of copper.......... .75

It therefore appears that light which has passed through bichromate of potash, by which the chemical rays have been absorbed, can accomplish this decomposition; but light which has passed through ammonia sulphate of copper, which transmits the chemical rays, fails to produce that effect.

For these reasons I conclude that the decomposition of carbonic acid by the leaves of plants is brought about by the rays of light, and that the calorific and so-called chemical rays do not participate in the phenomenon. The rays of light are therefore as much entitled to the appellation of chemical rays as those which have heretofore passed under that name.

Next I examined the constitution of the gaseous mixture given off during these decompositions. It proved to be not pure oxygen, but a variable mixture of oxygen, nitrogen, and carbonic acid. Omitting the carbonic acid, which diffused from the solution in variable quantities, the following table represents the composition of the gaseous mixture:

Analyses of Gas evolved from Carbonated Water.

Exp.	Name of plant.	Oxygen.	Nitrogen.
1	Pinus tæda.	16.16	8.34
2	"	27.16	13.84
3	"	22.33	21.67
4	Poa annua.	90.00	10.00
5	"	77.90	22.10

This table contains a few out of a great number of experiments, all of which might have been quoted as examples of the conclusions which I wish to deduce. 1st. They all coincide in this, that the oxygen is never evolved without the simultaneous appearance of nitrogen; 2d. That when certain leaves are employed, as those of the Pinus tæda, there seems to be a very simple relation between the volumes of oxygen and nitrogen. In the first and second of these experiments, the volume of oxygen is to that of nitrogen as two to one; in the third, as one to one. In certain cases this apparent simplicity of proportion is departed from; but from its frequent occurrence in many analyses I have made it seems to demand attentive consideration. Moreover, in other plants, as in experiments 4 and 5, the amount of oxygen is relatively greater, and between it and the nitrogen there does not appear any exact proportion.

In order to ascertain whether decompositions taking place under absorbent media, as bichromate of potash, yield the same results as indicated in the foregoing table, I made several analyses of gas collected under those circumstances. The presence of the absorbent medium did not seem to exert any influence whatever, the general results coming out as though it had not been employed.

It was also found that the alkaline carbonates and bicarbonates could be decomposed by leaves in yellow light. The alkaline bicarbonates, as is well known, undergo decomposition by a slight elevation of temperature. When boiled in water they gradually give off their second atom of carbonic acid, and slowly pass into the condition of neutral carbonate. In the experiments I made with them the boiling was not continued long enough to affect to any extent the constitution of the salt, and in each case any portion of carbonic acid extricated during cooling was removed by the air-pump. A few leaves placed in this solution showed no effect if kept in the dark, but if brought into the sunshine there was a copious evolution of gas-bubbles, which, on detonation with hydrogen, proved to be rich in oxygen gas. One experiment gave 88 of oxygen, 12 of nitrogen.

In a subsequent communication to the *Philosophical Magazine* (Sept., 1844), I gave additional analyses of the gas emitted in yellow light as follows:

Five tubes, each three eighths of an inch in diameter and six inches long, were inverted in a small trough of water containing carbonic acid, with which the tubes were also filled. Some blades of grass nearly of the

same size and volume were placed in each tube. This
grass had been kept for two days in the dark in a bottle
filled with carbonated water. During this time the film
of air which envelops all new leaves was removed, the
grass became perfectly free from all adhering gaseous
matter, and when in the carbonated water exhibited a
dark-green aspect.

I have previously found that leaves thus soaked emit
under the influence of light a larger amount of nitrogen
than usual; this comes from the incipient decay of some
of their nitrogenized constituents. When under these
circumstances they are placed in the sunshine, this nitrogen comes off along with the gas liberated from the carbonic acid.

In the experiment I am now relating, a tube arranged
like one of the foregoing five evolved *in the open sunshine* a certain volume of gas which was composed of—

$$
\left.
\begin{array}{l}
\text{Oxygen.................. 41} \\
\text{Nitrogen................. 59} \\
\text{Carbonic acid........... 00}
\end{array}
\right\} 100
$$

The five tubes were placed in the spectrum in the following colors, and emitted the quantities of gas represented in the following table:

1	Extreme red and red.	0.0
2	Orange and yellow....	19.8
3	Yellow and green.....	27.4
4	Blue " "	0.5
5	Indigo and violet.....	0.0

The gas in tube 2, which had been in the orange and
yellow ray, was then washed with a solution of caustic
potassa. After this it still measured 19.8. It contained,
therefore, no perceptible quantity of carbonic acid. It
was next examined for oxygen, and with the following
result:

Constitution of Gas emitted in Orange and Yellow Light.

Oxygen........	8.0		Oxygen........	40.4
Nitrogen.......	11.8	or	Nitrogen.......	59.6
Carbonic acid..	0.0		Carbonic acid..	00.0
	19.8			100.0

The gas evolved by the yellow and green rays was next analyzed. Like the former, it underwent no diminution by washing with caustic potash. After this treatment it therefore measured 27.4, and on being examined for oxygen yielded as follows:

Constitution of Gas emitted in Yellow and Green Light.

Oxygen........	12.5		Oxygen........	45.6
Nitrogen.......	14.9	or	Nitrogen.......	54.4
Carbonic acid..	00.0		Carbonic acid..	00.0
	27.4			100.0

In explanation of the large and variable amount of nitrogen occurring in these analyses, it will scarcely be necessary to remind the vegetable physiologist that it arises from the mode of conducting the experiment. In order to be absolutely certain that no atmospheric air infilmed the leaves, they were soaked in water, and then when brought into the sunlight the nitrogen which had accumulated in their tissues from incipient decay diffused out with the first portions of oxygen. As, therefore, more and more gas was evolved, the relative amount of the nitrogen diminished. Thus the reason that the third tube appeared to be richer in oxygen than the second was owing to its containing more gas. Any person, however, who is familiar with the physiological action of leaves will understand these things without any further explanation.

University of New York, *June* 15, 1844.

MEMOIR XI.

OF THE FORCE INCLUDED IN PLANTS.

Collected and condensed from Memoirs in the Journal of the Franklin Institute and
Harper's Monthly Magazine.

CONTENTS :— *Growth of a seed in darkness and in light.—Action of
plants and animals respectively on the atmosphere.—Examination of
Rumford's experiments.—Dark heat rays cannot decompose carbonic
acid.—Germination in colored rays.—Greening of leaves takes place in
the yellow and adjacent rays.—The essential condition of all chemical
changes by radiation is absorption.—Plants absorb force from the sun;
it is associated with their combustible parts, and is disengaged by oxidation.*

I HAVE given in Memoir II. a description of the structure of an ordinary flame. Its light is derived from particles of solid carbon issuing from combustible matters
with which the wick or the gas jet is fed; these solid
particles, passing from a low temperature to a white
heat, and undergoing eventually complete oxidation, escape into the atmosphere as carbonic-acid gas.

We now encounter a question of imposing interest:
Whence has the force which thus manifests itself as heat
and light been derived? Force cannot be created; it
cannot spring forth spontaneously out of nothing.

It may be said, without much error, of such flame-giv-
ing compounds as we are here considering, that they are
for the most part compounds of carbon with hydrogen.
To regard them as such will very much simplify the
facts we have now to present.

Under the form of oils and fats these combustible
substances are derived directly or indirectly from the
vegetable world; directly, as, for instance, in the case of

M

olive-oil; indirectly, as in the case of animal oils and fats. These have been collected by the animals from which we obtain them out of their vegetable food. Even fats derived from the carnivora have been procured from the herbivora, and came originally from plants.

This brings us therefore to a consideration of the chemical facts connected with the life of plants.

If a seed be planted in moist earth, the air having access and the temperature that of a pleasant spring day, germination in the course of a few hours will take place. Should the process be conducted in total darkness, as in a closet, the young plant shoots upward, pale, or at most of a faint tawny tint. We can easily verify this statement by placing a few turnip seeds in a flower-pot containing earth, put into a closet or drawer from which light has been carefully excluded.

A sickly-looking plant thus springs from a seed in the dark. It is etiolated, as botanists say. If we examine it carefully, making allowance for the water it contains, we shall find that no matter how tall it may be, its weight has not increased beyond the original weight of the seed from which it came. It has been developing at the expense of the seed, the substance of which has been suffering exhaustion for its supply of nourishment. We cannot continue this development in the dark indefinitely, for the seed-supply is soon exhausted, and then the shoot dies.

But if instead of exposing the seed which is the subject of our experiment to darkness, we cause the germination to take place in the open day, a very different train of consequences ensues. There is no longer that immoderate extension of a sickly etiolated stem upward, but the parts emerging into the light turn green. Very soon, to use a significant expression, they are weaned from the seed; they no longer use the material collected for them in the preceding year by the parent plant, and stored up for their use, but their leaves, expanding, turn green, and expose themselves to receive the rays of the sun. If they be now examined as in the previous instance, making allowance for the water they contain, it will be found that from day to day their weight is increasing; they are living independently of the seed. They are obtaining carbon and hydrogen, the former from carbonic acid, and the latter from water and ammonia—compounds existing in the air or furnished from the ground.

If a seedling, germinated in darkness and permitted to grow to a certain extent, be then exposed to light, provided its dark-life has not continued too long, its etiolated aspect will soon disappear; it turns green, and assumes all the characters of a healthy plant. This is in effect the natural process. For we bury seeds a little under the surface, covering them lightly with earth, the opacity of which secures the necessary darkness; but the mould being moist, the air having a ready access, and the temperature of the season suitable, all the conditions needful for germination—water, air, warmth, darkness—are present. The plumule, or shoot, makes its way out of the obscurity into the light; its reliance for nutrition on the seed ends; its independent life begins. It obtains carbon, hydrogen, oxygen, nitrogen from the air, and saline substances and water from the soil.

The facts which we thus bring into relief, as necessary for the further exposition of the subject, are these: In the first stage of the life of a plant, its dark-life, there is, excluding water, a diminution in the weight; in the second, or light-life, there is an increase, due very largely to the appropriation of carbon from the air. The atmospheric carbonic acid has been decomposed, its oxygen set free and, for the most part, permitted to escape, its other constituent, carbon, now ministering to the growth of the plant.

A stone trough standing in a garden received the waste water from a pump. There had accumulated on its sides a green slimy growth (conferva). From this growth, on the west side of the trough, which was receiving the morning rays of the sun, bubbles of gas were continually forming; and these, as they attained a sufficient size, rose through the water and escaped into the air. This effect on the west side diminished as the sun passed towards the meridian, but at mid-day the north side of the trough was in full activity. As evening came on, that in its turn gave forth fewer bubbles, and was succeeded in activity by the east side. At first it was thought that these bubbles were nothing more than the gas which is dissolved in all water, and analogous in composition to atmospheric air, but closer examination showed that it was oxygen, nearly pure. During the night no gas whatever was disengaged.

Priestley, Ingenhousz, Rumford, and other experimenters of the last century investigated these facts carefully. The conclusions to which they came may be thus summarized: All ordinary natural waters contain carbonic acid in solution; leaves or other green parts of plants placed in such water and kept in darkness exert no action upon it, but in the sunshine they decompose the carbonic acid, appropriating its carbon and setting its oxygen free, as gas. Soon, however, the supply in the sample of water is exhausted, and the action even in the sunlight ceases. It is again resumed if more carbonic acid be artificially dissolved in the water; and since the air expired from the lungs in the act of breathing contains much of that gas, it is sufficient, by the aid of a tube, or in any other suitable manner, to conduct such expired air into the water for the disengagement of oxygen to go on.

The experiments of these earlier chemists had thus established the important fact that from carbonic acid, which is extensively diffused through the atmosphere and in water, and even in the soil, through the influence of sunlight, oxygen is obtained. The sunlight, then, is the force which carries into effect the decomposition.

There is thus a perpetual drain on the supply of carbonic acid, a perpetual tendency to its diminution, and hence, for the order of nature to continue, there must be an incessant supply. The source of that supply was very strikingly indicated by some of Priestley's experiments. "Having rendered a quantity of air thoroughly noxious by mice breathing and dying in it, he divided it into two receivers inverted in water, introducing a few green leaves into one, and keeping the other receiver unaltered. The former was placed in light, the latter in darkness. After a certain time he found that the air in the former had become respirable, for a mouse lived very well in it; but that in the latter was still noxious, for a mouse died the moment it was put into it."

To Priestley chiefly, though he was aided by other investigators, we must refer the honor of one of the greatest discoveries of the last century. It was this, that the two great kingdoms of nature, the animal and the vegetable, stand at once in antagonism and alliance. What is done by the one is undone by the other. Each is absolutely essential to the existence of the other. There is a never-ending cycle through which material atoms run. Now they are in the atmosphere, then they are parts of plants, then they are transferred to animals, and by them they are conducted back to the atmosphere, to run through the same cycle of changes again. The sunlight supplies the force that carries them through these revolutions.

Previously to 1834 I had turned my attention to this

interesting subject. It had been asserted by Rumford that many other substances besides the leaves of plants would evolve oxygen gas. He specified raw silk and cotton fibres. My investigation commenced by an examination of this assertion. I soon found that two totally distinct things had been confounded. Ordinary water contains, as has been said, carbonic acid in solution, but it also necessarily contains the ingredients of atmospheric air. To this, for the sake of distinctness, the perhaps incorrect designation of water-gas may be given. Since oxygen is very much more soluble than nitrogen, this dissolved gas differs in composition from atmospheric air. It is relatively richer in oxygen.

I very soon found, on exposing raw silk, spun glass, and other such fibres, immersed in water, to the sun, that Rumford's assertion was correct—gas-bubbles were set free; but his inference was incorrect—the gas did not come from decomposed carbonic acid; it was merely the water-gas of the water. I was thus able to separate the true from the false portion of the experiment. And though I thus dispose of the subject in a few words, it is perhaps due to the labor that was expended to say that it cost several weeks of uninterrupted work and many scores of analyses before I felt absolutely certain that this was the indisputable interpretation of Rumford's experiments.

Now as a guide to a correct exposition of the experiments that I have to relate, it must be borne in mind that the assumption of a green color by a germinating plant and the decomposition of carbonic acid by it are identical events. Or, perhaps, to speak more correctly, the latter is the cause, the former the effect.

At that time very incorrect views of the nature of the sun-rays were entertained. It was believed that they contained three distinct principles: (1) heat, (2) light, (3) chemical or deoxidizing radiations. In common with all other chemists I accepted this view, and proposed to myself to determine to which of these principles the decomposition of carbonic acid and the greening of plant leaves are due.

And first, to ascertain if it were the heat radiation, I converged, by the aid of a large metallic mirror, the dark or invisible radiations emitted by an iron stove on some leaves placed in water. The gas which was set free was nothing more than the water-gas; there was no decomposition of carbonic acid.

In Fig. 22, a is the concave mirror, b an inverted flask containing the leaves and water; it dips into a glass, c, also containing water, and receives the reflected radiations of the stove, d d.

Then I tried a similar experiment, using the radiations emitted by a brightly burning wood fire. The result was the same as the preceding, and I even pushed the experiment so far that the water became very

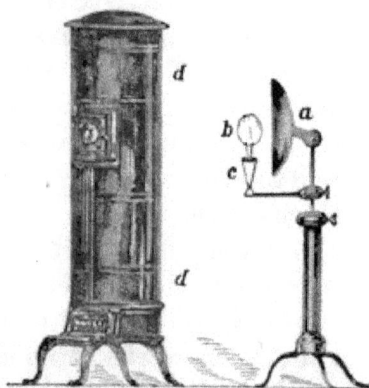

Fig. 22.

hot, a portion of its carbonic acid effervesced from it, and the leaves lost their bright green color. Still no decomposition of the carbonic acid could be detected.

At this time it was generally received that the essential characteristic of the more refrangible—the violet—rays is that they produce deoxidation. In accordance with this opinion a name—deoxidizing—had been given them. Now since the decomposition of carbonic acid is an effect of deoxidation, I was not surprised at the issue of the foregoing experiments, and expected to find that though the less refrangible radiations, those of heat, were inoperative, the more refrangible, the chemical or deoxidizing, would decompose carbonic acid readily.

But some collateral experiments had thrown a difficulty in the way of this conclusion. I had caused seeds to germinate in three little closets (Fig. 21, page 172), into which, by means of panes of colored glass, or through troughs filled with colored liquids—red, yellow, and violet—light respectively could be admitted I remarked with very great surprise that the seeds in the red-light closet and those in the violet one were just as much etiolated as they would have been had they grown in darkness; those in the yellow closet promptly assumed a green color, and developed themselves as well as if growing under natural circumstances.

But the light that comes through stained glass and colored solutions is far from being homogeneous; it contains rays of many refrangibilities. I therefore determined to attempt the greening of plants and the decomposition of carbonic acid by their leaves—phenomena which, as has been said, are equivalent—in the solar spectrum itself.

I arranged things so as to have a horizontal solar spectrum of several inches in length kept motionless by a heliostat. I had previously caused to germinate in a wooden box filled with earth, and of corresponding length, a crop of seeds. They were etiolated, or blanched, for the germination had taken place in the dark. These young plants I placed so as to receive the spectrum. Very soon those that were in the yellow space turned green, but those in the extreme red and extreme violet underwent no change, though the exposure might be kept up the whole day.

In Fig. 23, $a\ a$ is the box containing the germinating seeds, and placed so as to receive the colored spaces R,

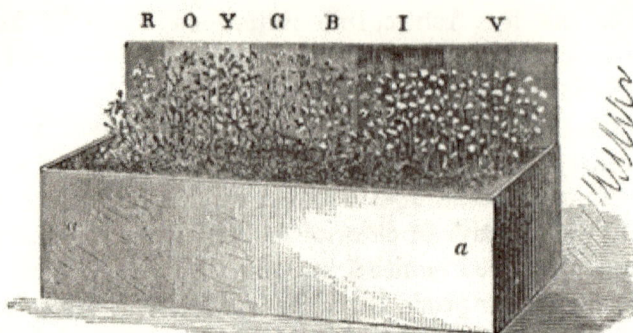

Fig. 23.

O, Y, G, B, I, V—red, orange, yellow, green, blue, indigo, violet—of the spectrum.

Many repetitions of this experiment satisfied me that it is the yellow and adjacent regions of the spectrum which occasion the greening of plants; the heat rays and the chemical rays have nothing whatever to do with it.

Next I attempted the decomposition of carbonic acid in the spectrum, and succeeded. I read before the American Philosophical Society in Philadelphia, at its centennial celebration in 1843, an account of this experiment as published in Memoir X.

In addition to the special interest of these experiments on plant life, they had a very important bearing on the general principles of actino-chemistry. They proved that it is altogether incorrect to suppose that chemical changes are brought about by the more refrangible rays only. They showed that every ray has its proper chemical function; for instance, the violet in the decomposition of compounds of silver, the yellow in the case of carbonic acid. And hence I proposed to abandon the conception of a tripartite division of the spectrum into heat, light, and chemical radiations, and to designate radiations by their wave-lengths, or, better still, by their number of vibrations—a method now universally adopted in spectrum analysis.

By other experiments—a narrative of which would be too long for the present occasion—I established this result: that for any ray to produce a chemical effect, it must be absorbed. For instance, when a ray has passed through a mixture of chlorine and hydrogen gases, and by causing them to unite has produced hydrochloric acid, it can no longer produce the same effect if made to pass through a second portion of the same mixture; its acting part has been detained or absorbed by the first. So, too, the radiations which have fallen on a daguerreotype plate, and impressed their image upon it, have lost the quality of producing a similar effect on a second plate that may be placed to receive them. Their active portion has been taken up or absorbed by the first. *The essential preliminary of all chemical changes by radiations is absorption.*

But it must not be supposed that the rays thus absorbed are annihilated or lost. They are simply held in reserve, ready to be surrendered again, undiminished and unimpaired, if the conditions under which they were absorbed are reversed. They may appear under some other form — as heat, electricity, motion — but their absolute energy remains unchanged. This is a necessary consequence of the theory of the Conservation and Correlation of Force.

From this point of view how interesting is that great discovery made by Angström, that an ignited gas emits the same rays it absorbs—a discovery that explained the Fraunhofer lines of the solar spectrum, and constituted an epoch in the history of spectrum analysis.

I have now presented the facts that are requisite for answering the question proposed on one of the foregoing pages: " Whence has the force which manifests itself as heat and light in a flame been derived? Force cannot be created; it cannot spring forth spontaneously out of nothing."

The answer is, it came from THE SUN.

Under the influence of his rays the growing plant decomposed carbonic acid obtained from the atmosphere, appropriating its carbon and setting its oxygen free. To accomplish this decomposition, this appropriation, it was necessary that a portion of the energy contained in those rays should be absorbed. Associated with this, the carbon could now form part of the plant, and, indeed, constituted the solid basis of which it was composed.

But the force thus associated with the carbon atoms was not annihilated; it was only concealed: through countless ages it might remain in this latent state, ready at any moment to come forth. All that is requisite is to oxidize the carbon, to turn it into carbonic acid, and the associated energy, under the form of heat and light, is set free.

When we read by gas or by the rays of a petroleum lamp, the light we use was derived from the sun perhaps millions of years ago. The plants of those ancient days, acting, as plants do now, under the influence of sunshine, separated carbon from the carbonic acid of the atmosphere by associating it with the radiant energy they had absorbed, and this remained for an indefinite time enclosed, as it were, in the now combustible material, ready to be disengaged as soon as the reverse action, oxidation, takes place, returning then to commingle as heat with the active forces of the world.

Much of what has here been said applies to hydrogen as well as to carbon. Hydrogen is derived, under similar conditions, from the decomposition of water or ammonia. When its oxidation recurs, it delivers up, under the form of heat, the energy it had absorbed. As, however, I am here speaking of the source of light in flames, in which carbon takes the leading and hydrogen only an indirect or subordinate part, it is not necessary to trace in further detail the action of the latter element.

A very interesting illustration of the principles here under consideration occurs in the case of the decomposition of water by an electric current. The constituents of the water, hydrogen and oxygen, are set free in the gaseous form. But for them to assume that form they must be furnished with caloric of elasticity. The current supplies them with this, and, indeed, the decomposition can only go on at the rate which is regulated by that supply. The heat they have thus assumed remains insensible in them, imparting to them their elastic or gaseous condition, until they are caused to reunite and reform water, when it is at once given up. The part that is played by that portion of the electric current which is thus transformed into heat—and furnishing their caloric of elasticity to the evolving gases is absolutely essential to the decomposition—has been hitherto too much overlooked by chemists.

Nature thus offers us in the instance we have been considering in this Memoir a striking illustration of the transmigration of matter and of force. Plants obtain carbon from the atmosphere; it constitutes the basis of their combustible portions. Sooner or later it suffers oxidation, turns back into the condition of carbonic acid, and is diffused again into the atmosphere. There is a never-ending series of cycles through which it runs: now it is in the air, now a part of a plant, now back again in the air. And the same is true as regards the energy with which it was associated. Derived from the sunbeam, it lay hidden in the plant, awaiting re-oxidation; then it was delivered, escaping under the form of heat or light, and remingling with the universal cosmic force from which it had been of old derived by the sun, or from which, perhaps more correctly speaking, the sun himself was derived, for he is the issue of nebular condensation.

I cannot close this Memoir without making reference to a point of surpassing interest. What goes on in the case of a flame, goes on in the case of an animal. Either from other animals or from plants, combustible material is obtained and used as food. Directly or indirectly it undergoes oxidation in the system, brought about by the air introduced through the process of respiration. Speaking in a general manner, though there are many intermediate products, the issue of this chemical action is the evolution of carbonic acid, ammonia, water, which pass into a common receptacle, the atmosphere. Thence their ingredients are taken by plants, and, under the agency of the sunlight, combustible material—food—is re-formed. The same particle is, therefore, now in the air, now in the plant, now in the animal, now back again in the air. It suffers a perpetual transmigration.

But in the case of an animal the oxidation may not be so sudden, so complete, as it is in the case of a flame. It may, and indeed generally does, go on stage by stage, step by step, partial oxidations occurring. It is thus that, from one original hydrocarbon, a long catalogue of fatty and oily substances may arise; the inevitable issue, however, is total oxidation. As the partial degradations go on, in corresponding degrees the latent energy or force is set free. It may assume any correlated form, as muscular motion; or, as heat, it may give warmth to the body; in certain fishes, as the gymnotus, it may turn into an electrical or nerve current; in certain insects, such as the fire-fly, into light.

As a cataract is only a form which any river may assume if it comes to a precipitous descent—a form which, though it may be outwardly unchanging, is interiorly never for two successive moments the same, for it is perpetually fed from above and is wasting away below—so the flame of a lamp is only a form, the aspect of which is determined by its environment. The changes it is undergoing issue in the liberation, the escape, of force, chiefly under the aspect of light and heat. Its life is very transitory. It dies out as soon as the oil that fed it is exhausted. We blow upon it, and it passes into nonentity.

And so, too, with an animal, the appearance of identity it presents is altogether deceptive. At no two successive moments are its parts the same. In a very short time all the old have been removed, and new ones have taken their places. The force that it derived from its food has been manifested in various ways, such as muscular motion or heat. But the material particles have not been destroyed; they have merely gone back into the atmosphere, and will be used by nature for the fabrication of other plant and animal forms over and over again. And so, too, the energy they have displayed—it has not ceased to exist; the heat, for instance, that once vivified them has merely mingled with that of the outer world, and is ready to discharge its special functions again and again. In the world there is thus an unceasing transmigration of matter, an unceasing transmigration of force.

MEMOIR XII.

EXPERIMENTS TO DETERMINE WHETHER LIGHT PRODUCES ANY MAGNETIC EFFECTS.

From the Journal of the Franklin Institute, Feb., 1835.

CONTENTS: — *Mrs. Somerville's experiments.* — *Christie's experiment.* — *Their results cannot be substantiated.* — *The violet ray has no effect.* — *Blue glasses and blue ribbons also ineffective.*

"THE more refrangible rays of light are said to possess the property of rendering iron and steel magnetic. The existence of this property was first asserted by Dr. Morichini of Rome. Other observers subsequently failed in obtaining the same results, but in the year 1825 the fact appeared to be decisively established by the learned and accomplished Mrs. Somerville in an Essay published in the *Transactions of the Royal Society.* In her experiments sewing-needles were rendered magnetic by exposure for two hours to the violet ray, and the magnetic virtue was communicated in still shorter time when the violet rays were concentrated by a lens. The indigo rays were found to possess a magnetizing power almost to the same extent as the violet; and it was observed, though in a less degree, in the blue and green rays. It is wanting in the yellow, orange, and red. Needles were likewise rendered magnetic by the sun's rays transmitted through green and blue glass. These results have been verified by M. Zantedeschi, of Pavia (*Bibl. Univ.*, May, 1829), but their accuracy has been doubted by Riess and Moser, who consider that the means employed by Mrs. Somerville for ascertaining the magnetic state of the needles were not sufficiently exact. They found the oscillation of the needles to be wholly unaffected by exposure to the prismatic colors (*Brewster's Journal*, Vol. II., p. 225, N. S.). This must still be regarded, therefore, as one of the disputed points in science" (*Turner's Chemistry*).

It has been supposed that these contradictory results arose entirely from local circumstances. A hazy atmosphere, such as is met with in the northern and middle countries of Europe, might perhaps influence in some manner this peculiar property of light when the clearer sky of Italy permitted the experiment to succeed. Some, indeed, have thought that the observers who were said to have verified the alleged results were deceived in not having previously ascertained the magnetic state of the needles they used.

During the past summer (1834) I have attempted to satisfy myself whether the more refrangible rays really exert any magnetic influence; and happening to reside in the south of Virginia, on the same parallel of latitude as Tunis and the more northerly African kingdoms, I thought the situation too favorable to suffer such an opportunity to pass without endeavoring to gain some information on this contested point.

In 1824 Mr. Christie found that a needle six inches long, contained in a brass compass-box with a glass cover, suspended by a hair and made to vibrate alternately shaded and exposed to the sun, came to rest much sooner in the latter than in the former case. That this was not occasioned by an increase of temperature was proved by the needle vibrating more rapidly when its temperature was raised by other means.

On repeating this experiment, I very quickly found that it depended in a great measure on the mode of suspension of the needle, and its position with respect to the incident light, what results would be obtained. If the needle was suspended on a point, or by a thread without torsion, the time and the number of vibrations were the same whether the needle was exposed to the sunbeam or not. But if the needle was suspended by a hair or other organic substance having torsion, the sunbeam would occasion a twist in the hair on its first exposure to light; and if the direction of that twist happened to coincide with the direction of the needle's motion, the momentum of the needle was increased and the vibrations continued longer. A needle which vibrated forty-four times in one minute would occasionally, owing to this cause, vibrate forty-six when suspended by a hair; but if by a silk fibre, its vibrations were always forty-four, the first arc of vibration being in every instance 40°.

Thinking to obtain more decisive effects, I concentrated the sunbeam with a lens on the south pole of the suspended needle, and found that the needle was thrown into a rapid, tremulous motion. But here the hot air ascending from the needle acts upon it as upon the sail of a windmill; and the same effect ought to take place to a certain extent on simple exposure of the half of a vibrating needle to direct light. But I found that a needle suspended in the vacuum of an air-pump by a thread without torsion is in no way affected by exposure to solar light.

It is said in the account of Christie's experiment that the needle was contained in a brass compass-box. It might have been that electrical currents were excited in that box which were the cause of the derangement in question. I therefore vibrated a needle under similar circumstances, with the result above stated. I should mention that this was done in a solid cylinder or ring of brass without any seam or soldered junction; but as

N

compass-boxes are generally made of sheet-brass with a soldered seam in the side, it was possible that the fine line of solder acted with the brass as a thermo-electric couple, capable of excitation by the warmth of the sunbeam. I therefore made a compound cylinder, half of

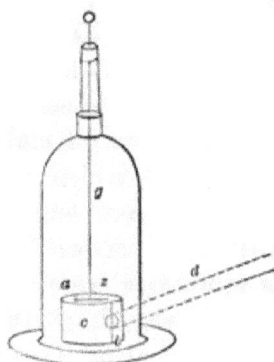
Fig. 24.

copper, the other half of zinc, c z, Fig. 24, the edges of which, at a and b respectively, are soldered together, one junction, b, being polished, the other, a, blackened. The needle was suspended in an exhausted receiver by a silk fibre, g, and a ray of light, d, coming from an aperture half an inch wide in the shutter fell upon the junction. The needle used in this experiment was of watch-spring. Its first vibration was performed in an arc of 40°, and when the compound cylinder was taken away it made thirty-two vibrations in sixty seconds in vacuo. On placing it concentrically with the compound cylinder, and suffering the ray to impinge on the polished junction, the moment that the arc of vibration had become 40° the number of oscillations in one minute was observed; six experiments gave severally thirty-two. On turning the blackened junction to the light the result was still thirty-two, and on substituting the solid brass cylinder three consecutive trials gave thirty-two. The thermometer stood in the sunshine at 103°.

By some this magnetic action of light has been attributed to the violet or more refrangible rays only. A needle of watch-spring about four inches long, which in an exhausted receiver suspended by a filament of silk exhibited no polarity, had one half exposed to the violet ray dispersed by a flint-glass prism. This ray was sepa-

rated from others of the spectrum by passing through a
slit in a metallic screen, and half the needle was screened
from its action by a piece of paper. After two hours'
exposure to the sun it was suspended again in the exhausted receiver, but still showed no polarity; it was
then exposed to the other rays successively, with the
same result. The needle was now slightly touched, and,
slowly vibrating, arranged itself in the magnetic meridian. The first vibration was performed in a semicircular
arc, the number of vibrations in one hundred seconds
was twenty-seven. But after four hours' exposure to
the violet ray, as before, no evidence of any change
either increasing or diminishing the number of oscillations could be obtained. A beam of violet light passing
through a disk of stained glass was concentrated on one
end of a sewing-needle by means of a lens without producing any change in the number of vibrations made in
one minute. This needle on some occasions would, however, give different results: when its first vibration was
performed in a semicircle, the number varied from forty-
one to forty-three in sixty seconds. On vibrating it in
vacuo, its results uniformly gave the latter number very
nearly.

The position of the needle to the incident ray is not
of any consequence, whether it receives it obliquely, in
the direction of the light, or across it. If soft iron be
substituted for steel, the results are still negative, even
if the needle be arranged in the magnetic meridian, the
line of dip, or any other position. I therefore came to
the conclusion that the violet ray exerts no influence on
the magnetic needle, and that all the other rays are
equally inert.

But as Mrs. Somerville found that a needle placed
under a piece of glass or blue ribbon, having half its
length protected by paper, became in a short time mag-

netic, I tried the same experiment, but in every instance failed to make the needle magnetic. When suspended by a silk fibre in vacuo needles showed no disposition to arrange themselves in any particular direction, and when they came to rest were found cutting the magnetic meridian at every angle, though the temperature of the sunbeam to which they had been exposed on one occasion was 124°. Great care was taken to ascertain the previous non-magnetic state of the needles, and they were suspended by a fibre without torsion.

MEMOIR XIII.

AN ACCOUNT OF SOME EXPERIMENTS ON THE LIGHT OF THE SUN, MADE IN THE SOUTH OF VIRGINIA.

Abstract from the Journal of the Franklin Institute of Philadelphia for June, July, August, and September, 1837; Philosophical Magazine, Feb., 1840.

CONTENTS:— *Absorption of luminous radiations.* — *The reference spectrum.*—*Absorption of heat radiations; the apparatus employed.*—*Absorption of chemical radiations.*—*Screen of bromide of silver.*—*Coloration of chloride and bromide of silver by radiations that have passed through correspondingly colored solutions.*—*Early application of photography to the investigation of physical problems.*—*Crystallization of camphor towards the light.*—*The side of vessels towards the sky is the colder.*

IF a beam of the sun's light be passed through a solution of chromate of potassa, it can no longer blacken a piece of sensitive paper—paper covered over with chloride or bromide of silver. If the light which has thus passed through a stratum of this liquid be converged by means of a lens, the chloride of silver will remain for a long time without change in the focus.

I made many such experiments with a view of determining the effect of absorbent media on the luminous, calorific, and chemical rays. Such, at that time, was the accepted subdivision of the solar radiations. A ray of the sun was caused to pass through a trough with parallel sides containing the absorbent solutions, and for the sake of exactness a reference spectrum was employed, such as is now used in spectroscopic experiments. In this manner the particular luminous rays absorbed by any given solution could be determined, the absorbed heat-rays could be ascertained by an air-thermometer, and the acting chemical rays by papers made sensitive by chloride of silver.

Into a darkened chamber, the shutter of which is represented in section at A A, Fig. 25, a beam of the sun's light is made to pass horizontally by means of a mirror of silvered glass, B. The mirror I use is one belonging to a solar microscope, and by turning the milled screws, $e\,e$, it can be brought into any position required to throw a beam horizontally into the room. A brass tube, f, two inches in diameter, can be screwed into the position figured. There is also a lens, a, Fig. 31, which may be introduced; its focus is nine inches, its diameter about two inches, and the diameter of the sun's image it gives nearly $\frac{1}{16}$ of an inch.

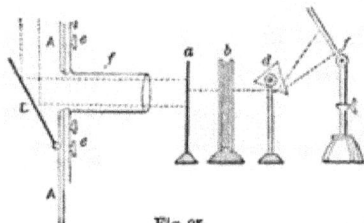

Fig. 25.

A piece of sheet-lead about a quarter of an inch thick is to be cut into the form of a horse-shoe of such size that a circle one inch in diameter may be inscribed in it. Upon this lead two pieces of glass are cemented, so as to form a trough for containing liquids. In Fig. 26, $a\,a$ is the wooden basis or foot of the trough; $c\,c\,c$, the leaden horse-shoe; $b\,b$, the glass plates.

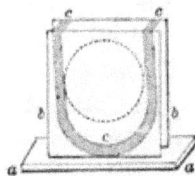

Fig. 26.

A thin metallic plate three or four inches square, having a longitudinal slit in it about an inch long and $\frac{1}{16}$ of an inch wide, is to be provided. In Fig. 27, $a\,a$ is the slit.

The lens, a, Fig. 31, having been removed, a beam of light is thrown horizontally into the room, as at Fig. 25; the slit, Fig. 27,

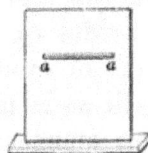

Fig. 27.

being placed as at *a*, Fig. 25, a narrow streak of light
passes through.   The trough, Fig. 26, is then placed
behind in such a position that half the light coming
through the slit may pass through the liquid in the
trough, and the other half pass by its side unintercepted.
Behind the trough is placed a flint-glass prism, as in Fig.
25, and further still a white pasteboard screen, *f*; *a* is
the slit plate, *b* the trough, *d* the prism.

The action of this arrangement is as follows: The
beam of light cast by the mirror into the room is intercepted, except the small portion that passes the slit.   Of
this a part passes through the trough and a part on one
side of it.   Two beams of light, therefore, fall on the
prism, one of which has passed through the trough
and suffered absorption, the other has not.   There are,
therefore, on the pasteboard screen two spectra side
by side.

Suppose, for example, that the trough is filled with
distilled water; the two spectra on the screen are alike,
as in Fig. 28.   If it be filled with a solution of chromate
of potassa, in that coming from the light that has passed
through the trough the blue, indigo, and violet rays are
missing, as in Fig. 29.   If the trough be filled with ammonia-sulphate of copper, the red, the orange, the yellow
have disappeared or been absorbed, as in Fig. 30.   The

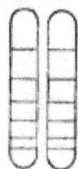

Fig. 28.

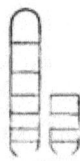

Fig. 29.

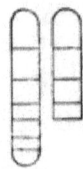

Fig. 30.

reference or undisturbed spectrum enables us to determine what rays have been absorbed with precision.

For the investigation of the absorption of heat the

following apparatus was used. The mirror being placed

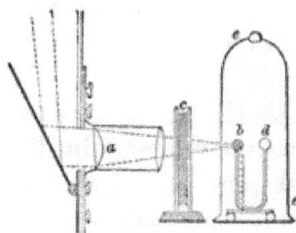

Fig. 31.

on the shutter, as in Fig. 31, a plano-convex lens, *a*, is screwed into the tube so as to bring the rays to a focus on one of the bulbs of a delicate differential thermometer; this gives the heat of the sunbeam as concentrated by the lens. To find the effect of any liquid medium in absorbing heat-rays, the trough filled with the substance under trial is placed as at *c*, Fig. 31. The cone of rays, converging from the lens, *a*, on the blackened bulb, *b*, forms an image upon it, and the differential thermometer yields a corresponding indication. In trying different solutions the same trough is always to be used, so that the solutions may always be of the same thickness. It is also requisite to cover the thermometer with a very thin shade of clear glass, *e e*, to prevent disturbance from air-currents.

I may select as examples the following: A solution of sulphate of copper and ammonia, absorbing the red and yellow light, transmitted twenty rays out of every hundred that fell on it.

A thin stratum of pitch enclosed between two plates of crown-glass, and transmitting a homogeneous red light, but absorbing all the other colors of the spectrum, allowed only 19 per cent. of the heat to pass through it.

In examinations of the transmissive powers of vapors and gases, a cubical bottle was used instead of the trough. The vapor of iodine was thus found to absorb two thirds of the heat falling on it; but the same bottle filled with nitrous acid, and which, therefore, was in a stratum of the same thickness, permitted much more heat to pass.

For the investigation of chemical radiations, there was placed upon the screen, *f*, Fig. 25, a paper covered with bromide of silver. It darkened in those parts on which the more refrangible rays fell.

Or, having removed the differential thermometer and its shade, Fig. 31, the cone of light converging from the lens passed through a solution of sulphate of copper and ammonia in the trough, and a piece of paper painted with chloride of silver was placed so as to receive the focus. Though but little heat was transmitted through the solution, a dark spot was at once produced characteristic of the blackening of the silver salts by the sun-rays. Though, therefore, this double salt transmits the rays of heat with difficulty, the rays of chemical action pass with facility. If in the trough there be placed a strong solution of bichromate of potassa, a far greater quantity of light will pass, and much more heat; but a paper painted with chloride of silver being placed in the focus, *no chemical change whatever goes on*, the chloride retaining its usual whiteness.

NOTE. — The foregoing experiments were made between 1834 and 1837. The reader of this volume must not regard them from the present elevated view of Radiations. At that time very little respecting these radiations was known. They present to me personally this point of interest — that they were the beginning of a series of researches to which I devoted many subsequent years.

The following is a list of solutions possessing an absorptive action on the more refrangible or chemical radiations:

Bichromate of potassa,
Chromate of potassa,
Yellow hydrosulphide of ammonia,
Hydrosulphide of lime,
Chloride of iron,
Chloride of gold,
Chloride of platinum.

It is to be remarked that all these solutions are yellow. I also found that a great many vegetable colored infusions, especially those which had a *yellow* tint, would in like manner absorb the chemical rays.

When pieces of paper covered with chloride or bromide of silver were exposed to a beam which had passed through a solution of red sulpho-cyanide of iron, the paper became of a brick-red color; if to a beam which had passed through a solution of ammonia-sulphate of copper, it became of a blue-brown; and on exposing a piece for five days to light acted on by bichromate of potassa, it became perceptibly of a faint yellowish green.

A beam which has passed through bichromate of potassa does not cause the union of a mixture of chlorine and hydrogen. I kept such a mixture for several hours in it, and could not perceive any change.

Ritter asserted that the opposite extremities of the spectrum possess opposite powers of chemical action: he states that phosphorus will emit fumes in the red ray, but if the violet be thrown on it, it ceases to smoke. This experiment I repeated often, and under favorable circumstances, but could not make it succeed.

I could succeed, however, in showing very beautifully the interference of that class of chemical rays which 'blacken chloride and bromide of silver, but failed for want of proper apparatus in trying to produce their polarization. An electric current circulating in a wire does not seem to have any influence on these chemical rays. I found that the same neat magnified image of the wire was obtained on chloride paper when it was placed in a beam diverging from a lens, whether the current was made to pass or was stopped.

Note.—These were instances of the application of photography to the solution of physical problems long before the announcement of the discoveries of Daguerre or Talbot.

The following are some mechanical results of solar light:

(*a.*) Having made a large air-pump jar clean and dry, a few pieces of camphor were placed on the plate of the pump, and the jar exhausted. The pump and its receiver were then set in the sunshine, and very soon all that side towards the sun was covered with crystals; but few or none were on the farther side.

(*b.*) A torricellian vacuum having been made in a tube half an inch or more in diameter, and upwards of thirty inches long, a fragment of camphor was passed up through the mercury into it. The tube might be kept for any length of time in the dark without anything happening, but on bringing it into the beams of the sun crystallization in a few minutes took place on that side of the tube next the luminary.

(*c.*) On the inside of an air-pump jar a circle of tinfoil an inch in diameter was pasted; and having operated as in experiment (*a*), that side was exposed to the sun. Crystals soon formed, but the tinfoil protected the glass in its vicinity, and none were found within a certain space round the metallic circle. Fig. 32.

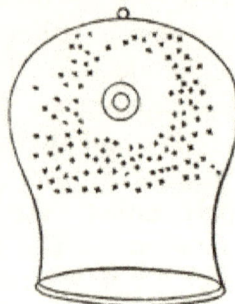

Fig. 32.

(*d.*) Crystallization is not necessarily connected with these results; the vapor of mercury in a torricellian void is condensed towards the light; so also the dew which settles on the inside of a jar containing water is always on the side nearest the window. The rays of the sun have also the power of decomposing chloride of gold: the metalline spangles are deposited on the side of the glass nearest to the light.

Artificial light gives none of these results.

(*e.*) Having removed the piece of tinfoil used in experiment (*c*), it was placed on a stand in front of the receiver; it hindered crystallization taking place on the parts on which its shadow was cast, and also for a certain space in the vicinity.

(*f.*) A piece of tinfoil was placed before a jar that had been already coated with crystals. It removed all those crystals that were within its shadow.

(*g.*) Instead of using a piece of tinfoil as in experiment (*c*), the receiver was made hot, and a piece of resin rubbed upon it, so as to leave a transparent circle of that substance on it. On exposure to the light it was found that the resin could not protect the glass.

(*h.*) Along the inside of a vessel about to be exposed to the sun a glass rod was rubbed. Rows of crystals were deposited on the lines described by the end of the rod. But the vessel must be very dry for this experiment to succeed. This curious fact was first observed in the case of an exhausted vessel having a small siphon gauge shut up in it, the extremity of which rested against the glass. By accident the gauge was moved half round the glass, and shortly afterwards a line of crystals was observed coinciding with the line of motion. The appearance was such as is represented in Fig. 33.

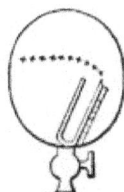
Fig. 33.

Now can we explain these singular results, excepting the last, on any other known principle than this: that the side of the jar nearest the sun radiates freely the heat it receives back again, while radiation is interfered with at the other side? that, in point of fact, the anterior side is the colder, and the other side the hotter?

MEMOIR XIV.

EXAMINATION OF THE PROCESS OF DAGUERREOTYPE.— NOTE ON LUNAR PHOTOGRAPHY.

From the Philosophical Magazine, Sept., 1840; Smithsonian Contributions to Knowledge, No. 180.

CONTENTS :—*Description of the process.—Cause of the deposition of mercury and production of the light parts of the picture.—Polishing of the plate.—The operation of iodizing.—Effect of keeping the iodide.— The achromatic lens.—Reduction of focal length in the non-achromatic.—The development.—Fixing by hyposulphite and galvanism.— Necessity of heating the plates.—Lunar impressions.—Artificial light. —Note on lunar photography—The first photographs of the moon.*

MORE than one hundred instances are recorded in Berzelius's Chemistry in which the agency of light brings about changes in bodies; these are of all kinds: formations of new compounds, re-arrangements of elements already in union, changes of crystallographic character, decompositions, and mechanical modifications.

The process of the daguerreotype is to expose a surface of pure silver to the action of the vapor of iodine, so as to give rise to a peculiar iodide of silver, which under certain circumstances is exceedingly sensitive to light. The different operations of polishing, washing with nitric acid, exposure to heat, etc., are only to secure a pure silver surface; the operation of hyposulphite of soda, and the process, which I shall presently describe, of galvanization, are to free the plate from its sensitive coating, and in nowise affect the depth of the shadows, as some of the French chemists at first supposed.

There is but one part of the daguerreotype which

does not yield to theory: on one point alone there is obscurity. Why does the vapor of mercury condense in a white form on those portions of the film of iodide which have been exposed to the influence of light?—condense to an amount almost proportional to the quantity of incident light?

Even on this point there are facts which appear to have a bearing.

(*a.*) It has long been known that if a piece of soap-stone or agalmatolite be made use of as a pencil to write with on glass, though the letters that may have been formed are invisible, and though the surface of the glass may subsequently have been well cleaned, yet they will come into view as soon as the glass is breathed on.

(*b.*) I have often noticed that if a piece of very clear and cool glass, or, what is better, a cold polished metallic reflector, has a little object such as a piece of metal laid upon it, and the surface be breathed over once, the object being then carefully removed, as often as you breathe again on the surface a spectral image of it may be seen, and this singular phenomenon may be exhibited for many days after the first trial was made.

(*c.*) Again, in the common experiment of engraving on glass by hydrofluoric acid, if the vapor has been very weak, no traces will be perceived on the glass after the wax has been removed; but on breathing over it, the moisture condenses in such a way as to bring the object into view.

(*d.*) In a former Memoir (XIH.) I described a phenomenon relating to the crystallization of camphor on surfaces of dry glass, on which invisible traces have been made by the pressure of a glass rod; this also appears to belong to the same class of effects.

Berzelius (*Traité*, Vol. II., p. 186) has attempted to explain (*a*) and (*c*) on this principle, that the changed and unchanged surfaces radiate heat unequally. There may be strong doubts as to the correctness of this, but is not the daguerreotype due to the same cause, whatever it may be?

We must separate carefully the chemical changes which iodide of silver undergoes in the sunbeam from the mechanical changes which happen to the sensitive film: iodide of silver turns black in the solar ray—the whole success of the daguerreotype artist depends on his checking the process before that change shall have supervened.

The coating of iodine is not *immediately* necessary to the production of images by the mercurial vapor. The condition seems to be traceable to the metallic surface. If you take a daguerreotype, clean off the mercury, polish the plate thoroughly with rottenstone, wash it with nitric acid, and bring it to a brilliant surface, yet if it has not been exposed to heat, the original picture will reappear on exposure to the mercurial vapor. Is not this a result of the same kind as those just referred to?

As a polishing material for the daguerreotype plate, common rottenstone and oil answer very well. The plate having been planished by the workman, is to be rubbed down to a good surface, and as high a polish given to it as possible; it is to be heated and washed with nitric acid, as indicated in the French account, and finished by being rubbed with whiting (*creta præparata*) in the state of a very fine powder, going over it for the last time with a piece of clean dry cotton; this gives an intensely black lustre, that cannot be obtained by rottenstone alone, and thoroughly removes any film which nitric acid may have left.

To coat with iodine, I make use of a box about two inches deep, in the bottom of which that substance in coarse flakes is deposited; no cloth intervenes, but the silvered plate, with a temporary handle attached to it, is brought within half an inch of the crystals, and it becomes perfectly coated in the course of from one to three minutes; no metallic strips are necessary to insure this effect; if the edges and corners are thoroughly clean, the golden hue will appear uniformly.

M. Daguerre recommends that the plate, after being iodized, shall be placed in the camera without loss of time. The longest interval, he says, ought not to exceed an hour. "Beyond this space the action of the iodine and silver no longer possesses the requisite photogenic properties."

There may be something peculiar in the preparation of the plate as I have described it, but it is certain that this observation must be received with some limitation. A plate which has been iodized does not appear so quickly to lose its sensitiveness. On the other hand, by keeping it in the dark for twelve or twenty-four hours, its sensitiveness is *often remarkably increased.* Other advantages also accrue. Those who have made many of these photogenic experiments will have had frequent occasion to remark that the film of iodine is not equally sensitive all over, that there are spots or cloudy places which do not evolve any impression, and often the whole is in that condition that the bright parts alone come out, while the parts that are in shadow do not evolve correspondingly, nor can they be well developed, except at the risk of solarizing the picture. Now a plate that has been kept for several hours is by no means so liable to these effects: I do not pretend to give any reason for this, but merely mention it as a fact, of considerable importance to the travelling daguerreotyper; he will find that the iodine does not lose its sensitiveness in many days.

In a paper read before the Royal Society, Herschel

O

states that there is an absolute necessity of a perfect achromaticity in the object-glass of a photographic camera. M. Daguerre appears to have been under the same impression, and recommends in his published account such an object-glass.

All the rays of light, with perhaps the exception of the yellow, leave an impression on the iodide of silver. The less refrangible rays, however, act much more slowly than those at the opposite end of the spectrum. In the common kinds of glass, the most energetic action takes place in the indigo, or on the boundary of the blue. Now the retina receives an impression with equal facility from each of the different rays, the yellow light acting as quickly upon it as the red or the blue. Vision is therefore performed independently of time, the eye catching all the colors of the spectrum with equal facility and with equal speed. But it is not so with these photogenic preparations. In the action of light upon them, time enters as an element; the blue ray may have effected its full change, while the red is yet only beginning slowly to act; and the red may have completed its change before the yellow has made any sensible impression. On these principles, it is plain that an achromatic object-glass is by no means essential for the production of fine photographs; for if the plate be withdrawn at a certain period, when the rays that have a maximum energy have just completed their action, those that are more dispersed but of slower effect will not have had time to leave any stain. We work, in fact, with a temporary monochromatic light.

Upon these principles I constructed the camera I am in the habit of using, with a double convex non-achromatic lens. Some of the finest proofs were procured with a common spectacle lens, of fourteen inches focus, arranged at the end of a cigar-box as a camera; a lens

of this diameter answers very well for plates four inches
by three, reproducing the objects with the most admirable finish, copper-plate engravings being represented in
the minutest particulars, and the marks of the tool becoming quite distinct under a magnifier.

In this instance, it is true, owing to the magnitude of
the focal length compared with the aperture, but little
difficulty ensues from chromatic aberration; but when
with the same focal length the aperture is increased to
three or four inches, the dispersion becomes very sensible, and yet good proofs can be procured by working
in the method here indicated, the chief difficulty then
arising from spherical aberration.

It has already been stated that the ray of maximum
action for the daguerreotype, when colorless French
plate-glass is used, lies probably within the indigo space:
it therefore follows that the length of the camera should
be diminished, after arranging it to the luminous focus.
The importance of this has been pointed out in a paper
by Mr. Towson; I was, however, in the habit of using
that adjustment before reading the suggestions contained in his communication. The amount of shortening to
be given to the camera, where the lens is fifteen inches
focus, does not commonly exceed three tenths of an inch.
If the luminous focus be used, the proof comes out indistinctly.

In the subsequent process of mercurializing, it is of little importance what the angular position may be. Several experimenters were for a time under the idea that
an angle of 45° or 48° is a necessary inclination, in order
that the plate should take the vapor; this arose from a
misinterpretation of the printed account. Plates mercurialize equally well in a horizontal as in any other position; perhaps a slight inclination may be of advantage,
in allowing the vapor to flow with uniformity over the

iodized surface, but the chief use of an angle of 45° is to allow the operator to inspect the process through the glass of the mercury-box.

Sometimes it is advantageous to heat the mercury a second time, when the proof is not distinctly evolved at first. Indeed, it occasionally happens that a proof which did not evolve at all at first will come out quite fairly on raising the temperature of the mercury again.

M. Daguerre recommends two methods of removing the sensitive coating from the plate—by washes of hyposulphite of soda, and by a solution of common salt. The former answers perfectly, the second only indifferently well. There is, however, another process, which is very simple, and has an advantage over the former of these in cheapness. It adds not a little to the magic of the whole operation, in the eyes of those who are unaccustomed to chemical results. The plate, having been dipped into cold water, is placed into a solution of common salt, of moderate strength ; it remains without being acted upon at all; but if it be now touched on one corner with a piece of zinc, which has been scraped bright, the yellow coat of iodide moves off like a wave and disappears. It is a very pretty process. The zinc and silver forming together a voltaic couple, with the salt water intervening, oxidation of the zinc takes place, and the silver surface commences to evolve hydrogen gas; while this is in a nascent condition it decomposes the film of iodide of silver, giving rise to the production of hydriodic acid, which is very soluble in water, and hence instantly removed.

This process, therefore, differs from that with hyposulphite. The latter acts by dissolving the iodide of silver, the former by decomposing it. It is necessary not to leave the zinc in contact too long, or it deposits stains, and in large plates the contact should be made at the four corners successively, to avoid this accident.

After the proof is washed, all the defects in the preparation of the plate become apparent. If a film of mercury has existed on it, due to its not having been burned sufficiently long, there will be found a want of distinctness in the shadows; or if the plate has not been burned at all, perhaps the former impressions will reappear. This accident frequently happened in my earlier trials, when care had not been taken to give a due exposure each time to the spirit-flame. Spectral appearances of former objects, on different parts of it, emerged—an interior with Paul Pry coming out, when the camera had been pointed at a church.

There is no difficulty in procuring impressions of the moon by the daguerreotype beyond that arising from her motion. By the aid of a lens and a heliostat, I caused the moonbeams to converge on a plate, the lens being three inches in diameter. In half an hour a very strong impression was obtained. With another arrangement of lenses I obtained a stain nearly an inch in diameter, and of the general figure of the moon, in which the places of the dark spots might be indistinctly traced.

An iodized plate, being exposed for fifteen seconds only close to the flame of a gas-light, was very distinctly stained; in one minute there was a very strong impression.

On receiving the image of a gas-light, eight feet distant, in the camera, for half an hour, a good representation was obtained.

The flame of a gas-lamp was arranged within a magic-lantern, and a portion of the image of a grotesque on one of the slides received on a plate; a very good representation was procured.

With Drummond's light, and the rays from a lime-pea in the oxy-hydrogen blowpipe, the same results were obtained.

Note on Lunar Photography. — To the foregoing paragraph respecting photographs of the moon, I may here add an extract from a publication by my son, Dr. Henry Draper, in the *Smithsonian Contributions to Knowledge*, No. 180, p. 33, entitled, " On the Construction of a Silvered Glass Telescope, and its Use in Celestial Photography :"

" The first photographic representations of the moon ever made were taken by my father, Professor John W. Draper, and a notice of them published in his quarto work ' On the Forces that Organize Plants,' and also in the September number (1840) of the *London, Edinburgh, and Dublin Philosophical Magazine*. He presented the specimens to the New York Lyceum of Natural History.   The Secretary of that association has sent me the following extract from their minutes:

" ' March 23d, 1840. Dr. Draper announced that he had succeeded in getting a representation of the moon's surface by the daguerreotype. . . . The time occupied was twenty minutes, and the size of the figure about one inch in diameter. Daguerre had attempted the same thing, but did not succeed. This is the first time that anything like a distinct representation of the moon's surface has been obtained.—Robert H. Brownne, *Secretary*.' "

MEMOIR XV.

ON THE TAKING OF PORTRAITS FROM LIFE BY PHOTOGRAPHY.

From the Philosophical Magazine, September, 1840.

CONTENTS :—*History of the invention.—First attempts by whitening the face.—Use of reflecting mirrors.—Use of a blue-colored trough.—Kind of camera necessary.—The seat or support.—The background.—Appropriate dresses.—Ladies' dresses.—Arrangement of the shadow.—Reflecting camera.*

HISTORICAL NOTE.—This Memoir contains the first published description of the process for taking daguerreotype portraits. Of late, since the introduction of collodion, this art has been much cultivated and improved. It now forms an important branch of industrial occupation. That it was possible by photogenic processes, such as the daguerreotype, to obtain likenesses from the life, was first announced by the author of this volume in a note to the editors of the *Philosophical Magazine*, dated March 31, 1840, as may be seen in that journal for June, 1840, p. 535. The first portraits to which allusion is made in the following Memoir were produced in 1839, almost immediately after Daguerre's discovery was known in America.

In the *Edinburgh Review* for January, 1843, there is an important article on Photography. In that the invention of the art of taking photographic portraits is attributed to its true source—the author of this book. It says: "He was the first, we believe, who, under the brilliant summer sun of New York, took portraits by the daguerreotype. This branch of photography seems

not to have been regarded as a possible application of
Daguerre's invention, and no notice is taken of it in the
reports made to the legislative bodies of France. We
have been told that Daguerre had not at that period
taken any portraits; and when we consider the period
of time—twenty or twenty-five minutes—which was
then deemed necessary to get a daguerreotype landscape,
we do not wonder at the observation of a French author,
who describes the taking of portraits as ' *Toujours un
terrain un peu fabuleux pour le Daguerreotype.*' "

Very soon after M. Daguerre's remarkable process for
photogenic drawing was known in America, I made attempts to accomplish its application to the taking of
portraits from the life. M. Arago had already stated in
his address to the Chamber of Deputies that M. Daguerre expected by a slight advance to meet with success, but as yet no account had reached us of that object
being attained.

In the first experiments I made for obtaining portraits
from the life, the face of the sitter was dusted with a
white powder, under an idea that otherwise no impression could be obtained. A very few trials showed the
error of this; for even when the sun was only dimly
shining there was no difficulty in delineating the features.

When the sun, the sitter, and the camera are situated
in the same vertical plane, if a double convex non-achromatic lens, four inches in diameter and fourteen in focus,
be employed, perfect miniatures can be obtained *in the
open air*, in a period varying with the character of the
light, from twenty to ninety seconds. The dress is admi-
rably given, even if it should be black; the slight differences of illumination are sufficient to characterize it, as
well as to show each button, button-hole, and every fold.

Partly owing to the intensity of such light, which cannot be endured without a distortion of the features, but
chiefly owing to the circumstance that the rays descend
at too great an angle, such pictures have the disadvan-
tage of not exhibiting the eyes with distinctness, the
shadow from the eyebrows and forehead encroaching on
them.

To procure fine proofs, the best position is to have the
line joining the head of the sitter and the camera so arranged as to make an angle with the incident rays of less
than ten degrees, so that all the space beneath the eyebrows shall be illuminated, and a slight shadow cast from
the nose. This involves obviously the use of reflecting
mirrors to direct the ray. A single mirror would answer,
and would economize time, but in practice it is often convenient to employ two; one placed, with a suitable mechanism, to direct the rays in vertical lines; and the second
above it, to direct them in an invariable course towards
the sitter.

On a bright day, and with a sensitive plate, portraits
can be obtained in the course of five or seven minutes, in
the diffused daylight. The advantages, however, which
might be supposed to accrue from the features being
more composed, and of a more natural aspect, are more
than counterbalanced by the difficulty of retaining them
so long in one constant mode of expression.

But in the reflected sunshine the eye cannot support
the effulgence of the rays. It is therefore absolutely
necessary to pass them through some blue medium, which
shall abstract from them their heat, and take away their
offensive brilliancy. I have used for this purpose blue
glass, and also ammonia-sulphate of copper, contained in
a large trough of plate glass, the interstice being about
an inch thick, and the fluid diluted to such a point as to
permit the eye to bear the light, and yet to intercept

no more than was necessary. It is not requisite, when colored glass is employed, to make use of a large surface; for if the camera operation be carried on until the proof *almost* solarizes, no traces can be seen in the portrait of its edges and boundaries; but if the process is stopped at an earlier interval, there will commonly be found a stain, corresponding to the figure of the glass.

The camera I have used, though much better ones might be constructed, has for its objective two double convex lenses, the united focus of which for parallel rays is only eight inches; they are four inches in diameter in the clear, and are mounted in a tube, in front of which the aperture is narrowed down to $3\frac{1}{2}$ inches, after the manner of Daguerre's.

The chair in which the sitter is placed has a staff at its back, terminating in an iron ring, which supports the head, so arranged as to have motion in directions to suit any stature and any attitude. By simply resting the back or side of the head against this ring, it may be kept sufficiently still to allow the minutest marks on the face to be copied. The hands should never rest upon the chest, for the motion of respiration disturbs them so much as to make them of a thick and clumsy appearance, destroying also the representation of the veins on the back, which, if they are held motionless, are copied with surprising beauty.

It has already been stated that certain pictorial advantages attend an arrangement in which the light is thrown upon the face at a small angle. This also allows us to get rid entirely of the shadow on the background, or to compose it more gracefully in the picture; for this it is well that the chair should be brought forward from the background from three to six feet.

Those who undertake daguerreotype portraitures will of course arrange the backgrounds of their pictures according to their own tastes. When one that is quite uniform is desired, a blanket, or a cloth of a drab color, properly suspended, will be found to answer very well. Attention must be paid to the tint: white, reflecting too much light, would solarize upon the proof before the face had had time to come out, and owing to its reflecting *all* the different rays, a blur or irradiation would appear on all edges, due to chromatic aberration. It will be readily understood, that if it be desired to introduce a vase, an urn, or other ornament, it must not be arranged against the background, but brought forward until it appears perfectly distinct on the ground-glass of the camera.

Different parts of the dress, for the same reason, require intervals differing considerably, to be fairly copied —the white parts of a costume passing on to solarization before the yellow or black parts have made any decisive representation. We have therefore to make use of temporary expedients. A person dressed in a black coat and open waistcoat of the same color must put on a temporary front of a drab or flesh color, or by the time that his face and the fine shadows of his woollen clothing are evolved, his shirt will be solarized, and be blue or even black, with a white halo around it. Where, however, the white parts of the dress do not expose much surface, or expose it obliquely, these precautions are not essential; the white shirt collar will scarcely solarize until the face is passing into the same condition.

Precautions of the same kind are necessary in ladies' dresses, which should not be selected of tints contrasting strongly.

It will now be readily understood that the whole art of taking daguerreotype miniatures, consists in directing an almost horizontal beam of light, through a blue-colored medium, upon the face of the sitter, who is retained in an unconstrained posture, by an appropriate

but simple mechanism, at such a distance from the background, or so arranged with respect to the camera, that
his shadow shall not be copied as a part of his body;
the aperture of the camera should be three and a half or
four inches at least, indeed the larger the better, if the
objective be aplanatic.

If two mirrors be made use of, the time actually occupied by the camera operation varies from forty seconds to
two minutes, according to the intensity of the light.   If
only one mirror is employed, the time is about one fourth
shorter.   In the direct sunshine, and out in the open air,
the time varies upwards from half a minute.

Looking - glasses which are used to direct the solar
rays after a short time undergo a serious deterioration,
the silvering assuming a dull granular aspect, and losing
its black brilliancy.   Hence the time, in copying, becomes
gradually prolonged.

The arrangement of the camera above indicated gives
reversed pictures, the right and left sides changing places.
Mr. Woolcott, an ingenious mechanician of this city, has
taken out a patent for the use of an elliptical mirror for
portraiture; it is about seven inches in aperture, and
allows him to work conveniently with plates two inches
square.   The concave mirror possesses this capital advantage over the convex lens, *that the proof is given in its
right position, that is to say, not reversed;* but it has the
serious inconveniences of limiting the size of the plate,
and representing parts that are at all distant from the
centre in a very confused manner.   With the lens, plates
might be worked a foot square, or even larger.

Miniatures procured in the manner here laid down are
in most cases striking likenesses, though not in all.   They
give, of course, all the individual peculiarities—a mole, a
freckle, a wart.   Owing to the circumstance that yellow
and yellowish browns are long before they impress the

substance of the daguerreotype, persons whose faces are freckled all over give rise to the most ludicrous results, a white portrait mottled with just as many black dots as the sitter had yellow ones. The eye appears beautifully: the iris with sharpness, and the white dot of light upon it with such strength and so much of reality and life as to surprise those who have never before seen it. Many are persuaded that the pencil of the painter has been secretly employed to give this finishing touch.

UNIVERSITY OF NEW YORK, *September*, 1840.

.

MEMOIR XVI.

ON THE CHEMICAL CONDITION OF A DAGUERREOTYPE SURFACE.

From the Philosophical Magazine, September, 1841.

Contents :—*Mercury exists all over a daguerreotype surface.—There is no superposition of the parts.—The shadows have metallic mercury ; the lights silver amalgam.—No iodine is ever evolved from the plate. —Action of a solution of gum and one of gelatine in tearing off the films.—The starch experiment.—The etching of daguerreotypes.*

As many of the results to be given in the next Memoir depend on the use of daguerreotype tablets, I shall in this examine the chemical and physical condition of those surfaces and offer proof of the following facts:

1st. That metallic mercury exists all over the surface of an ordinary daguerreotype—in the shadows as well as in the lights—in the shadows as metallic mercury, in the lights as silver amalgam.

2d. That in an iodized daguerreotype as taken from the mercury-bath there is no order of superposition of the parts, that is to say, the iodide is neither *upon* nor *beneath* the mercury, but both are, as it were, in the same plane.

3d. That when a ray of light falls upon the surface of this preparation, through all the intervening steps, and up to the point of maximum action, no iodine is evolved from the plate, but that in the common daguerreotype the light communicates a tendency to the atoms of the iodide to yield up to the mercurial vapor their silver, while the iodine retires and combines with the unaffected silver beneath. It follows that when such a plate is withdrawn from the mercurial vapor, there is all over it a uniform film of iodide of silver of the very same thickness as at first, and that this has happened through a *direct corrosion* of the silver by the iodine, while it was undergoing the mercurial operation.

I pass at once to the proofs of these several propositions, and, 1st, *That metallic mercury exists all over the surface of an ordinary daguerreotype, in the shadows as well as in the lights—in the shadows it is as metallic mercury, in the lights as silver amalgam.*

I took a plated copper three inches by four in surface, and having prepared it with care, I exposed half of it to the diffused light of the day, screening the other half; it was then mercurialized at 170° Fahr., the iodide removed by hyposulphite of soda, and washed. And now, a plate on which a gold-leaf was spread was placed over it, but separated, as shown in Fig. 34, in the points a, b, c, by three slips of glass. By means of a spirit-lamp

Fig. 34.

the photographic plate $d\ e$ was heated, and the gilded plate $g\ k$ kept cool by occasionally wetting it. On parting the plates, it was perceived that faint but distinct traces of whitening were visible all over the gold, as well on that part which was over the whitened half of the photograph as over that which was unchanged.

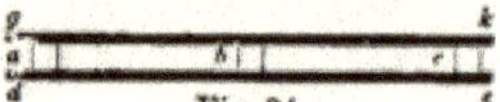

But as it might happen that the mercury diffused itself laterally past the imperfect obstacle b, I made the following decisive trials:

I iodized three silver plates, A, B, C, each three inches by four in surface, conducting the processes for each in the same way; and having exposed each for two minutes to a faint daylight, I laid them aside in the dark, to be presently used as test-plates in lieu of the gilded plate ($g\ k$).

Then I took three other plates, D, E, F, of the same size, and conducting the preparatory processes for each as before, I iodized D in the dark, and mercurialized it forthwith at 170° Fahr., taking the utmost care that not a ray of light should be suffered to impinge upon it.

E was iodized and exposed for two minutes to diffused daylight, and then mercurialized at 170° Fahr.

F was iodized and exposed to the sun until it began to turn brown, an effect occurring almost at once. It was then mercurialized at 170° Fahr.

All these plates then had their sensitive coating removed by hyposulphite, and were thoroughly washed in distilled water and dried.

I had, therefore, three plates, representing accurately the conditions proposed to be investigated. D was in the condition of the most perfect shadows; E in that of the highest lights, and F solarized. In appearance, D was black, E was white, and F bluish-gray.

Upon D, E, F, I placed A, B, C, respectively, separating each pair of plates one sixteenth of an inch, or thereabouts, by slips of glass. Then I laid them on the level surface of the sand-bath, the test-plates being kept cool by sponging occasionally with water. Temperature of the sand, 200° Fahr.; duration of the experiment, fifteen minutes.

On examination, A, B, C, were all found powerfully mercurialized, nor did there seem to be any difference between them.

I consider, therefore, that the shadows, the demi-tints, the lights, and the solarized portions of a daguerreotype, are covered with mercury; for at a temperature of 200° Fahr. they all evolve it alike, a sufficiency of vapor rising from the parts that have not been exposed to the light to bring a plate that has been so exposed to its maximum of whiteness.

In Memoir XIV. I described a remarkable effect which
I had noticed in these investigations, that if an object
such as a wafer be laid upon a piece of cold glass or
metal, and you breathe once on it, and as soon as the
moisture has disappeared remove the object and breathe
again on the glass, a spectral image of the wafer will
make its appearance. The impression thus communicated to the surface, under certain conditions, remains
there a long time. During the cold weather last winter
I produced such an image on the mirror of my heliostat;
it could be revived by breathing on the metal many
weeks afterwards, nor did it finally disappear until the
end of several months.

I do not at present know what is the explanation of
this result, but the analogy between it and the arrangement of mercurial globules covering the surface of a daguerreotype is too striking to be overlooked. It proves
that surfaces may assume such a condition as to affect
the deposition of vapors upon them, so as to produce
the reproduction of appearances of external forms. I
gave, therefore, particular attention to this point, but
eventually found that silver exists in an ordinary daguerreotype, in connection with the mercury all over the
plate, in a less proportion in the shadows, and in a
greater proportion in the lights. This result was, however, only obtained after the following fact was discovered — that the mucilage of gum-arabic, when slowly
dried in a thin layer on the surface of a daguerreotype,
splits up in shivers, bringing along with it the white
portions of the picture, and leaving the plate clean.

Having therefore prepared three plates, D, E, F, exactly as before, I poured on them a solution of gum,
drained them so as to leave only a small quantity, and
let them dry slowly over the sand-bath. The gum separated readily, and lay in chips on the surface of each

P

plate; it was easily removed to three sheets of paper, by tapping with the finger on the back of the plate. Each was then treated alike as follows:

The gummy matter was incinerated on a platinum leaf, and the remaining ashes transferred to a test-tube half an inch in diameter. One drop of nitric acid and one drop of water were added; it was boiled over a small flame, and diluted with a little water. Dilute hydrochloric acid was now added, and the chloride of silver immediately fell. In repeating this, it is necessary to attend to the state of dilution of the acid, for if too strong it wholly dissolves the minute quantity of chloride of silver generated.

As, from the minuteness of that quantity, it was impossible to obtain a direct quantitative analysis, I adopted the foregoing method, and added the dilute acid to all three tubes at the same time. In D there was a faint opalescence, in E and F a cloud; but I could not always determine whether the deposit of E or F was most copious, sometimes the one and sometimes the other appearing to have a slight advantage.

I conclude, therefore, that while the whole surface of the plate is coated with mercury, it exists as silver amalgam chiefly in the lights, and as uncombined mercury chiefly in the shadows, and in a mixed proportion in the demi-tints; and that when a plate is solarized, both free mercury and amalgam are present.

Such is the state of surface in a daguerreotype, *recently formed*. In the course of time, however, a great portion of the mercury that is in the shadows, and also free in the lights, evaporates away. When the picture has thus changed, the shadows are metallic silver, and the lights silver amalgam.

2d. *That in an iodized daguerreotype, as taken from the mercury-bath, there is no order of superposition of the parts, that is to say, the iodide is neither upon nor beneath the mercury, but both are, as it were, in the same plane.*

Soon after I had ascertained the action of gum-arabic, some of it was applied to the surface of a plate on which an impression had just been formed in the mercury-bath. This was without removing the coat of iodine. On drying it, the gum chipped up, as was expected, bringing away with it all the lights of the picture, and leaving a uniform coat of yellow iodide of silver beneath. It seems, therefore, that the film of iodide coheres more strongly to the metal plate than the amalgam; and, further, from this result we should judge that the amalgam is *on the surface* of the iodide.

But this is not true; for on three different occasions I found that when Russian isinglass was employed instead of gum for the purposes presently to be related, the isinglass, from its stronger cohesive power, chipped off in the act of drying, tearing up the yellow film from end to end of the plate, and leaving the amalgam constituting the lights undisturbed. It is here to be understood that this action takes place without the *smallest disturbance* of the lights and demi-tints, the plate remaining in all the beauty and brilliancy and perfection that it would have had if it had been carefully washed in hyposulphite of soda.

This is a result, however, which cannot be produced with uniformity. Most commonly the lights are torn up with the iodide. Had it occurred but once, I should still have cited it with decision, for from the very character of it, it is impossible to be mistaken or to commit an error of judgment. It proves that the film of iodide may be mechanically *torn off* from the metallic surface as perfectly as it can be *dissolved off* by chemical agents —a singular fact.

This result, therefore, proving that we can tear off

the film of iodide and leave the amalgam, can only be co-ordinated with that by gum-water, in which the amalgam is removed and the iodide left, by supposing that there is not anything like a direct superposition in the case, and that the particles of amalgam and iodide lie, as it were, side by side.

3d. *That when a ray of light falls upon the surface of this*, etc.

There is no difficulty in proving this directly, and the indirect evidence is copious. If we lay a piece of paper imbued with starch on an iodized plate, and expose it to the sun, although the plate presently assumes a dark olive-green color, the starch remains uncolored.

This dark substance is probably a subiodide of silver; the iodine therefore which has been disengaged from it, not having been set free, must have necessarily united with the adjacent metallic silver—this, for very obvious reasons, there is no difficulty in admitting.

Now, therefore, when a photogenic impression existing on the surface of a plate in an invisible state is brought out by the action of mercury vapor, we easily understand how this is effected. No iodine is ever evolved. But each atom of iodide of silver that has been acted on by the light yields to the attraction of the mercury its atom of silver, and the iodine thus set free unites with the me-tallic silver particles around it, reproducing the same yellow iodide by *a direct corrosion* of the plate: the proofs that we have of this are two in number:

1st. Dry some mucilage of gum-arabic on a daguerre-otype just brought from the mercury-bath; when it has split up, we perceive that the white amalgam of silver is removed, and a uniform coat of yellow iodide of sil-ver, of the very same thickness as at first, as is proved by its color, is left.

2d. Dry upon the same plate a solution of Russian

isinglass, and when it has split up, it will be seen that it uniformly rends off with it the yellow iodide, leaving the metallic plate with an exquisite polish; and wherever the light has touched, *there it is corroded.*

These two facts, taken together, prove that in mercurializing a plate no iodine is evolved, but that a new film of iodide of the same thickness is formed, at the expense of the metallic surface.

From these facts we readily gather that on the presence of the metallic silver the sensitiveness of this preparation mainly depends, for to the tendency which the light has impressed on the elements of the iodide to separate is added the strong attraction of metallic silver for nascent iodine.

This corrosion or biting in of the silver plates, by the conjoint action of the mercury and iodine, gives rise to etchings that have an inexpressible charm. Could any plan be hit upon of forcing the iodine to continue its action, the problem of producing *engraved* daguerreotypes would be solved. By another process, which will be described hereafter, I have succeeded in producing deep etchings from daguerreotypes.

UNIVERSITY OF NEW YORK, *September,* 1841.

•

MEMOIR XVII.

ON SOME ANALOGIES BETWEEN THE PHENOMENA OF THE CHEMICAL RAYS AND THOSE OF RADIANT HEAT.

From the Philosophical Magazine, September, 1841.

CONTENTS:—*The chemical rays are absorbed.—Photographic effects are transient.—The chemical rays are not conducted ; they become latent. —Optical qualities control chemical action.—The active rays are absorbed and the complementary reflected.—Relation of optical conditions and chemical affinities.*

It is the object of this Memoir to establish some striking analogies existing between the phenomena of the chemical rays and those of radiant heat.

Without saying anything respecting the laws of reflection, refraction, polarization, and interference, to which the chemical rays are subject, the study of which I commenced more than five years ago on paper rendered sensitive by bromide of silver, further than that a general similitude holds in all these cases between the rays of heat and the chemical rays, I shall at present confine my observations to establishing the following propositions:

1st. That the chemical action produced by the rays of light depends upon the *absorption* of those rays by sensitive bodies; just as an increase of temperature is produced by the absorption of those of heat.

2d. That as a body warmed by the rays of the sun gradually loses its heat by radiation, or conduction, or contact with other bodies, so likewise, by some unknown process, photographic effects produced on sensitive surfaces are only transient, and gradually disappear.

3d. That, as when rays of heat fall on a mass of cold ice its temperature rises degree by degree until it reaches 32° Fahr. and there stops until a certain molecular change (liquefaction) is accomplished, and after that rises again, so also the chemical rays impress certain changes proportional to their quantity, up to a certain point, and there a pause ensues—a very large amount of light being now rendered latent or absorbed, without any indication thereof being given by the sensitive preparation (as the heat of fluidity is latent to the thermometer); a molecular change then setting in, the increments of the quantity of light are again indicated by changes in the sensitive preparation.

4th. That it depends on the CHEMICAL nature of the ponderable material what rays shall be absorbed.

5th. That while the *specific rays* thus absorbed depend upon the chemical nature of the body, the *absolute amount* is regulated by its OPTICAL qualities, such as depend on the condition of its surfaces and interior arrangement.

6th. It will be proved from this that the SENSITIVE-NESS of any given substance depends on its chemical nature and optical qualities conjointly, and that it is possible to exalt or diminish the sensitiveness of any chemical compound by changing the character of its optical relations.

7th. That, as when radiant heat falls on the surface of an opaque body, the number of rays reflected is the complement of those that are absorbed, so in the case of a sensitive preparation, the number of rays reflected from the surface is the complement of those that are absorbed.

I now commence with the proofs of the propositions of this Memoir.

1st. *That the chemical action produced by the rays*

*of light depends upon the absorption of those rays by
sensitive bodies,* etc.

I iodized a plate to a golden yellow color, and exposed
it to the diffused light of day, setting it in such a position that it reflected specularly the light falling upon
it from the window to the objective of the camera-obscura, which formed an image upon a second sensitive plate.
The rays falling upon the sensitive plate of course exerted their usual influence upon the iodide, which, after
the lapse of a short time, began to turn brown. As
soon as this effect was observed, I closed the aperture of
the camera, and, taking out its plate, mercurialized it;
but it was found that the rays reflected from the sensitive plate, although they had been converged by a lens
four inches in diameter, and formed a very bright image,
had lost the quality of changing the iodide of silver.

We see, therefore, that a ray of light which has impinged on the surface of yellow iodide of silver has lost
the power of causing any further change on a second
similar plate on which it may fall.

In the practice of photography this observation is of
much importance, especially when lenses having large
apertures are used; the rays converging upon the sensitive plate are reflected by it in all directions, and the
camera is full of light; its sides reflect back again in all
directions on the surface of the plate these rays, which,
if they were effective, must stain the plate in the shadows. But if the plate has been iodized to the proper
tint, this light is wholly without action, and hence the
proof comes out neat and clean.

Upon an iodized plate I received a solar spectrum
formed by a flint-glass prism, the ray being kept motionless by reflection from a heliostat, and the plate so arranged as to receive the refracted rays perpendicularly.
After five minutes it was mercurialized, and the resulting

proof exhibited the place of the more refrangible colors in the most brilliant hues. The less refrangible colors had also left their impress of a whitish aspect, but the region of the yellow was unaltered. All the different rays, therefore, except the yellow, have the power of changing this particular preparation. Now when several pieces of cloth of different colors are placed in the sunbeam, they absorb heat in proportion as their color is deeper. A black cloth, which does not reflect any of those calorific rays, becomes presently hot; and in the same way Daguerre's sensitive preparation absorbs all the rays having any chemical action on it, and reflects the yellow only, which does not affect it. In this particular lies the secret of its sensitiveness, compared with the common preparations of the chloride and bromide of silver.

2d. *That as a body warmed by the rays of the sun*, etc.

After a beam of light has made its impression on the iodide, if the plate be laid aside in the dark before mercurializing, that impression decays away with more or less rapidity; first the faint lights disappear, then those that are stronger.

Having brought three plates to the same condition of iodization, and received the image of a gas-flame in the camera on each for three minutes, I

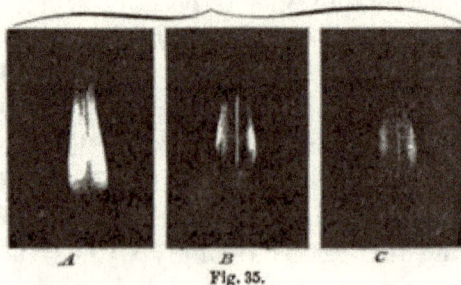
Fig. 35.

mercurialized one, *A*, forthwith; the second, *B*, I kept an hour, the third, *C*, forty-eight hours. The relative appearance of these three images is represented in Fig. 35.

Those who are in the habit of taking daguerreotypes know how much they suffer when the process of mercurialization is deferred. To show this effect in the extreme, I took four plates, and having prepared all alike, I exposed half of the surface of each to a bright sky for eight seconds.

No. 1 mercurialized immediately,		came out	black solarized.
" 2	" in five hours,	"	white.
" 3	" " twenty-two hours,	"	same effect.
" 4	" " one hundred and forty-four hours, no effect.		

This last plate, on being submitted twice more to the vapor of mercury, gave an indistinct mark. On exposing a corner of it to the sun it blackened instantly, these results showing that the peculiar condition brought on by the action of the light gradually disappears, the compound all the time retaining its sensitiveness.

Similar results are mentioned by Daguerre in the case of the changes produced on surfaces of resinous bodies, and I have noticed them in a variety of other cases. Now to whatever cause these phenomena are due, whether to anything analogous to radiation, conduction, etc., it is most active during the first moment after the light has exerted its agency, but it must also take effect even at the very time of exposure; and it is for these reasons that it comes to pass that when light of a double intensity is thrown upon a metallic plate the time required to produce a given effect is less than one half.

I could conceive the intensity of a ray so adjusted that in falling upon a given sensitive preparation the loss from this cause, this casting off of the active agent, should exactly balance the primitive effect, and hence no observable change result. Hereafter we shall find that one cause of the non-sensitiveness of a number of bodies is to be traced directly to the circumstance that they yield up these rays as fast as they receive them.

It needs no other observation than a critical examination of the sharp lines of a daguerreotype proof with a magnifying glass to show that the influence of the chemical rays is not propagated laterally on the yellow iodide of silver. Of the manifestations which these rays may exhibit, after they have lost their radiant form and become absorbed, we know but little. If they conform to the analogous laws for heat, and if the absorbing action of bodies for this agent is inversely as their conducting power, we perceive at once why a photographic effect produced on yellow iodide of silver retains the utmost sharpness without any lateral spreading; the absorbing power is almost perfect, the conducting should therefore be zero.

3d. *That, as when rays of heat fall on a mass of cold ice,* etc.

Although in the sun the iodide of silver blackens at once, this is only the result of a series of preliminary operations.

When we look at a daguerreotype, we are struck with the remarkable gradation of tint, and we naturally infer that the amount of whitening induced by mercurialization is in direct proportion to the amount of incident light; otherwise it would hardly seem that the gradation of tones could be so perfect.

But in truth it is not so. When the rays begin to act on it, the iodide commences changing, and is capable of being whitened by mercury. Step by step this process goes on, an increased whiteness resulting from the prolonged action or increased brilliancy of the light, until a certain point is gained, and now the iodide of silver apparently undergoes no further visible change; but another point being gained, it begins to assume, when mercurialized, a pale blue tint, becoming deeper and deeper, until it at last assumes the brilliant blue of a watch-spring. This incipient blueness goes under the technical name of solarization.

The successful practice of the art of daguerreotyping, therefore, depends on limiting the action of the sun-ray to the first moments of change in the iodide; for if the exposure be continued too long, the high lights become stationary, while the shadows increase unduly in whiteness, and all this happens long before solarization sets in.

Let us examine this important phenomenon more minutely. Having carefully cleaned and iodized a silver plate, three inches by four in size, it is to be kept in the dark an hour or two.

By a suitable set of tin-foil screens, rectangular portions of its surface, half an inch by one eighth, are to be exposed at a constant distance to the rays of an Argand gas-burner (the one I have used is a common twelve-holed burner), the first portion being exposed fifteen seconds, the second thirty seconds, the third forty-five seconds, the fourth sixty seconds, etc.

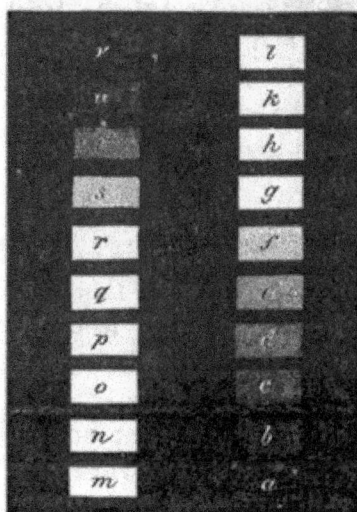

Fig. 36.

We have thus a series of spaces upon the plate, a, b, c, d, Fig. 36, each of which has been affected by known quantities of light; b being affected twice as much as a, having received a double quantity of light; c thrice as much as a, having received a triple quantity, etc.

The plate is now exposed to the vapor of mercury at 170° Fahr. for ten minutes; the spaces all come out in their proper order, and nothing remains but to remove the iodide.

An examination of one of these plates thus prepared

shows* that, commencing with the first space, a, we discover a gradual increase of whitening effect until we
reach the seventh; that a perfect whiteness is there attained; that, passing on to the sixteenth, no increase of
whitening is to be perceived, although the quantities of
light that have been incident and absorbed have been continually increasing; but as soon as the light thus latent
has reached a certain quantity, visible decomposition sets
in, indicated by a blueness, and the sensitive surface once
more renders evident the increments of incident light.

Or, by presenting a plate covered with a screen to a
sky that is clear or uniformly obscured, and with a regu-
lar motion withdrawing the screen deliberately from one
end to the other, and then
suddenly screening the
whole, it is plain that those
parts first uncovered will
have received the greatest
quantity of light, and the
others less and less. On
mercurializing, it will be
seen that a stain will be
evolved on the plate, as is
represented in Fig. 37;
from a to b the changes
have been successive; from
b to c no variation in the
amount of whitening is perceptible; at d solarization

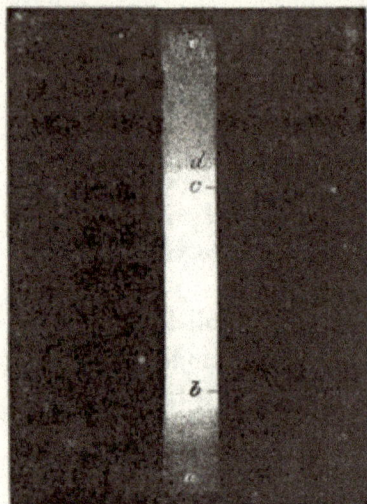

Fig. 37.

is commencing, which becomes deeper and deeper to the
end, e, of the stain.

---

* It is impossible to represent these changes in a drawing which is simply black
and white; it will be understood that the characteristic distinction of the spaces
from the sixteenth to the twentieth, for example, depends on their assuming a *blue*
tint, which continually deepens in intensity.

The plate from which the drawing of Fig. 37 is taken gives from a to b ten parts, from b to c seventeen parts, from d to e twelve parts; we perceive, therefore, how large an amount of light is absorbed, and its effects rendered latent, between the maximum of whiteness being gained and solarization setting in.

4th. *That it depends on the* CHEMICAL *nature of the ponderable material what rays shall be absorbed.*

I had prepared a number of observations in proof of this, very much of the same kind as those which have some time ago been published in the *Phil. Trans.* by Herschel. These refer chiefly to the variable lengths of the stains impressed by the prismatic solar spectrum on different chemical bodies, and the points of maximum action noticed in them. For the present I content myself with referring to that Memoir for proofs substantiating this proposition.

5th. *That while the specific rays thus absorbed depend upon the chemical nature of the body, the absolute amount is regulated by its* OPTICAL QUALITIES, *such as depend on the condition of its surfaces and interior arrangement.*

I took a polished silver plate, and having exposed it to the vapor of iodine, found that it passed through the following changes of color: 1st, lemon yellow; 2d, golden yellow; 3d, reddish yellow; 4th, blue; 5th, lavender; 6th, metallic; 7th, yellow; 8th, reddish; 9th, green, etc., the differences of color being produced by the differences of thickness in the film of iodide, and not by any difference of chemical composition.

It is a common remark, originally made by Daguerre, that of these different tints that marked 2 is the most sensitive, and photogenic draughtsmen generally suppose that the others are less efficient from the circumstance of the film of iodide being too thick. Some suppose, indeed, that the first yellow alone is sensitive to light. We shall see in a few moments that this is very far from being the case.

Having brought nine different plates to the different colors just indicated, I received in the camera on each the image of a uniform gas-flame, treating all as nearly alike as the case permitted. I readily found that in No. 1 there was a well-marked action, No. 2 still stronger, but that the rays had less and less influence down to No. 6, in which they appeared to be almost without action; but in No. 7 they had recovered their original power, being as energetic as in No. 2, and from that declining again. This is shown in Fig. 38.

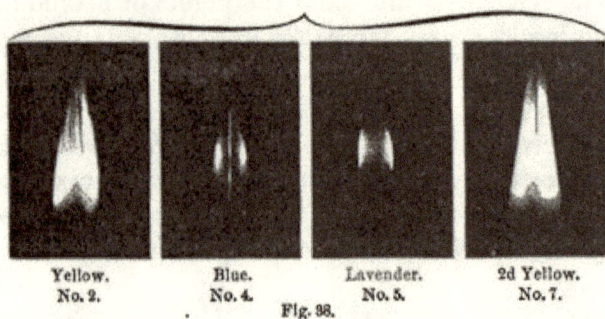

Yellow.   Blue.   Lavender.   2d Yellow.
No. 2.   No. 4.   No. 5.   No. 7.

Fig. 38.

Hence we see that the sensitiveness of the iodide of silver is by no means constant; that it observes period-ical changes depending on the optical qualities of the film and not on its chemical composition; and that by bringing the iodide into those circumstances that it re-flects the blue rays we greatly reduce its sensitiveness, and still more so when we adjust its thickness so as to give it a gray metallic aspect. But the moment we go beyond this, and restore by an increased thickness its original color, we restore also its sensitiveness. Here, then, in this remarkable result we again perceive a cor-roboration of our first proposition.

I may, however, observe in passing, that although I am describing these actions as if there were an actual absorption of the rays, and that films on metallic plates exhibit colors, not through any mechanism like interference, but simply because they have the power of absorbing this or that ray, there is no difficulty in translating these observations into the language of that hypothesis. When the diffracted fringes given by a hair or wire in a cone of diverging light are received on these plates, corresponding marks are obtained, a dark stripe occupying the place of a yellow fringe, and a white that of a blue. I found, more than four years ago, that this held in the case of bromide of silver paper, and have since verified it in a more exact way with this French preparation. Similar phenomena of interference may be exhibited with the chloride of silver.

We have it, therefore, in our power to exalt or depress the sensitiveness of any compound by changing its optical character. Until now, it has been supposed that the amount of change taking place in different bodies, by the action of the rays of light, depends wholly on their chemical constitution, and hence comparisons have been instituted as to the relative sensitiveness of the chlorides, bromides, oxides, and iodides of silver, etc. But it seems the liability to change depends also on other principles which, being liable to variation, the sensitiveness of a given body varies with them. Thus this very iodide of silver, when in a thin yellow film, is decomposed by the feeblest rays of a taper, and even moonlight acts with energy; yet simply by altering the thickness of its film it becomes sluggish, blackening even in the sunlight tardily, and recovering its sensitiveness again on recovering its yellow hue.

We have now no difficulty in understanding how, in the preparation of ordinary sensitive paper, great variations ensue, by modifying the process slightly, and how even on a sheet which is apparently washed uniformly over, large blotches appear, which are either inordinately sensitive or not sensitive at all. If, without altering the chemical composition of a film on metallic silver, or even its mode of aggregation, such striking changes result by difference of thickness, how much more may we expect that the great changes in molecular condition, which apparently trivial causes must bring about on sensitive paper, should elevate or depress its capability of being acted on by light. If I mistake not, it is upon these principles that an explanation is to be given of the successful modes of preparation which Talbot and Hunt have described, and the action of the mordants of Herschel.

I therefore infer—

6th. *That the* SENSITIVENESS *of any given preparation depends on its chemical nature and its optical qualities conjointly, and that it is possible to exalt or diminish the sensitiveness of a given compound by changing its optical relations.*

7th. *That, as when radiant heat falls on the surface of an opaque body, the number of rays reflected is the complement of those that are absorbed, so in the case of a sensitive preparation, the number of chemical rays reflected from the surface is the complement of those that are absorbed.*

This important proposition I prove in the following way: I take a plate, A G, Fig. 39, three inches by four, and by partially screening its surface while in the act of iodizing with a piece of flat glass, I produce upon it five transverse bands, *b, c, d, e, f;* the fifth, *f*, which has been longest exposed, is of a pale lavender color, the fourth a bright blue, the third a red, the sec-

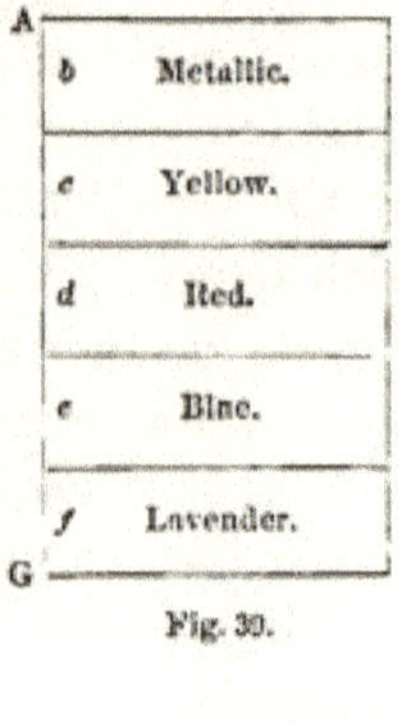

A	
b	Metallic.
c	Yellow.
d	Red.
e	Blue.
f	Lavender.
G	

Fig. 39.

Q

ond a golden yellow, and the first uniodized metal; the
object of this arrangement being to expose at the same
time and on the same plate a series of films of different
colors and of different thicknesses, and to examine the
action of the rays impinging on them and the rays reflected by them.

Having prepared a second plate, B, and iodized it uniformly to a yellow, I deposit it in the camera, and now,
placing the first plate, A G, so that the rays coming on it
from the sky through the window shall be specularly
reflected to the object-glass of the camera, and the image
of A G form upon B, I allow the exposure to continue
until the yellow of A G is beginning to turn brown;
then I shut the camera and mercurialize both plates.

In accordance with what has been said, it will be
readily understood that of the bands on A G, the first
one, which is the bare metal, does not whiten in the mercury vapor; the second, which is yellow, mercurializes
powerfully; the third, which is red, is less affected; the
fourth, which is blue, still less; and the fifth, which is
lavender, hardly perceptibly.

But the changes on B, which have been brought about
by the rays reflected from A G, are precisely the converse; the band which is the image of b is mercurialized
powerfully; that of c is untouched and absolutely black,
d faintly stained, e whitened, and f mercurialized but little less than b.

It follows from this that a white stripe on B corresponds to a black one on A G, and the converse; and for
the depth of tint of the intermediate stripes those of the
one are perfectly complementary to the corresponding
ones of the other.

By the aid of these results we are now able to give
an account of the variability of sensitiveness in photogenic preparations; the yellow iodide of silver is exces-

sively sensitive, because it absorbs all the chemical rays
that can disturb it, while the lavender is insensitive, because it reflects them. Under this point of view, sensitiveness therefore is directly as absorption and inversely
as reflection.

The superiority of Daguerre's preparation over common sensitive paper may now be readily understood. It
absorbs all the rays that can affect it, but the chloride
of silver, spread upon paper, reflects many of the active
rays. The former, when placed in the camera, gives rise
to no reflections that can be injurious; the latter fills it
with active light, and stains the proof all over. Hence
the daguerreotype has a sharpness and mathematical accuracy about its lines, and a depth in its shadows, which
is unapproachable by the other. Moreover, the translu-
cency of the white chloride of silver, as well as its high
reflecting power, permits of particles lying out of the
lines of light being affected, the light becoming diffused
in the paper.

The fact, therefore, that a given compound remains unchanged even in the direct rays of the sun is no proof
that light cannot decompose it; it may reflect or transmit the active rays as fast as it receives them. It results
from this that optical conditions can control and even
check the play of chemical affinities. While thus it appears that there are points of analogy between this chemical agent and radiant heat, we must not too hastily infer
that the laws which regulate the one obtain exclusively
also with the other. As is well known, there are striking analogies between radiant heat and light, but there
are also points of difference, the convertibility of heat of
one degree of refrangibility to another does not occur
with light; there are also dissimilitudes in the phenomena of radiation and its consequences.

From the phenomena of the interference of these rays,

of the sensitiveness or non-sensitiveness of the same chemical compound being determined merely by the fact of its thickness or thinness, these, and many other similar results obviously depending upon mechanical principles, it seems to me that very powerful evidence may be drawn against the materiality of light, and its entering into chemical union with ponderable atoms. Those philosophers who have adopted the undulatory theory will probably find in studying these subjects evidence in favor of their doctrines.

MEMOIR XVIII.

DESCRIPTION OF THE CHLOR-HYDROGEN PHOTOMETER.

From the American Journal of Science and Art, Vol. XLVI. ; Philosophical Magazine, December, 1843.

CONTENTS :—*Properties of a mixture of chlorine and hydrogen.—It is acted upon by lamplight, an electric spark at a distance, etc.— The gases unite in proportion to the amount of light.—Mode of measuring out known quantities of radiations.—The maximum action is in the indigo space.—Construction of the instrument.—The gases are evolved by electricity and combined by light.— Theoretical conditions of equilibrium.— Preliminary adjustment.—Method of continuous observation.—Method of interrupted observation.—Remarkable contraction and expansion.*

I HAVE invented an instrument for measuring the force of the chemical rays found at a maximum in the indigo space, and which from that point gradually fade away to each end of the spectrum. The sensitiveness, speed of action, and exactitude of this instrument will bring it to rank as a means of physical research with the thermo-multiplier of Melloni.

The methods hitherto available in optics for measuring intensities of light by a relative illumination of spaces or contrast of shadows are admitted to be inexact. The great desideratum in that science is a photometer which can mark down effects by movements over a graduated scale. With those optical contrivances may be classed the methods hitherto adopted for determining the force of the chemical rays by stains on daguerreotype plates or the darkening of sensitive papers. As deductions drawn in this way depend on the opinion of the observer, they can never be perfectly satisfactory, nor bear any comparison with thermometric results.

Impressed with the importance of possessing for the study of the properties of the chemical rays some means of accurate measurement, I have resorted in vain to many contrivances; and, after much labor, have obtained at last the instrument which it is the object of this Memoir to describe.

This photometer consists essentially of a mixture of equal measures of chlorine and hydrogen gases, evolved from and confined by a fluid which absorbs neither. This mixture is kept in a graduated tube, so arranged that the gaseous surface exposed to the rays never varies in extent, notwithstanding the contraction that may be going on in its volume, and the hydrochloric acid resulting from its union is removed by rapid absorption.

The theoretical conditions of the instrument are, therefore, sufficiently simple; but, when we come to put them into practice, obstacles appearing at first sight insurmountable are met with. The means of obtaining chlorine are all troublesome; no liquid is known which will perfectly confine it; it is a matter of great difficulty to mix it in the true proportion with hydrogen, and have no excess of either. Nor is it at all an easy affair to obtain pure hydrogen speedily, and both these gases diffuse with rapidity through water into air.

Without dwelling further on the long catalogue of difficulties thus to be encountered, I shall first give an account of the capabilities of the instrument in the form now described, which will show to what an extent all those difficulties are already overcome. In a course of experiments on the union of chlorine and hydrogen, some of which were read at the last meeting of the British Association, I found that the sensitiveness of their mixture had been greatly underrated. The statement made in the books of chemistry, that artificial light will not affect it, is wholly erroneous. The feeblest gleams of a taper produce a change. No further proof of this is required than the tables given in this communication, in which the radiant source was an oil-lamp. For speed of action no compound can approach it: a light which perhaps does not endure the millionth part of a second affects it energetically, as will be hereafter shown.

Proofs of the sensitiveness of the instrument.—The following illustrations will show that this instrument is promptly affected by rays of the feeblest intensity and of the briefest duration.

When, on the sentient tube, the image of a flame formed by a convex lens is caused to fall, the liquid instantly begins to move over the scale, and continues its motions as long as the exposure is continued. It does not answer to expose the tube to the direct emanations of the lamp without first absorbing the radiant heat, or the calorific effect will mask the true result. By the interposition of a lens this heat is absorbed, and the chemical rays alone act.

If the photometer be exposed to daylight coming through a window, and the hand or a shade of any kind be passed in front of it, its movement is in an instant arrested; nor can the shade be passed so rapidly that the instrument will fail to give the proper indication.

The experimenter may further assure himself of the extreme sensitiveness of this mixture by placing the instrument before a window and endeavoring to remove and replace its screen so quickly that it shall fail to give any indication: he will find that it cannot be done.

Charge a Leyden-jar, and place the photometer at a little distance from it, keeping the eye steadily fixed on the scale; discharge the jar, and the rays from the spark

will be seen to exert a very powerful effect, the movement taking place and ceasing in an instant.

This remarkable experiment not only serves to prove
sensitiveness, but also brings before us new views of the
powers of that extraordinary agent electricity. That
energetic chemical effects can thus be produced at a
distance by an electric spark in its momentary passage,
effects which are of a totally different kind from the
common manifestations of electricity, is thus proved;
these phenomena being distinct from those of induction
or molecular movements taking place in the line of discharge, they are of a radiant character; and we are led at
once to infer that the well-known changes brought about
by passing an electric spark through gaseous mixtures,
as when oxygen and hydrogen are combined into water,
or chlorine and hydrogen into hydrochloric acid, arise
from a very different cause than those condensations and
percussions by which they are often explained—a cause
far more purely chemical in its kind. If chlorine and
hydrogen can be made to unite silently by an electric
spark passing outside the vessel which contains them, at
a distance of several inches, there is no difficulty in understanding why a similar effect should take place with
a violent explosion when the discharge is made through
their midst, nor how a great many mixtures may be
made to unite under the same treatment. A flash of
lightning cannot take place, nor an electric spark be discharged, without chemical changes being brought about
by the radiations emitted.

*Proofs of the exactness of the indications of this pho-
tometer.*—The foregoing examples may serve to illustrate
the extreme sensitiveness of this instrument. I shall
next furnish proofs that its indications are exactly proportional to the quantities of light incident on it.

As it is necessary, owing to the variable force of day-light, to resort to artificial means of illumination, it will be found advantageous to employ the following method of obtaining a flame of suitable intensity.

Let A B, Fig. 40, be an Argand oil-lamp of which the

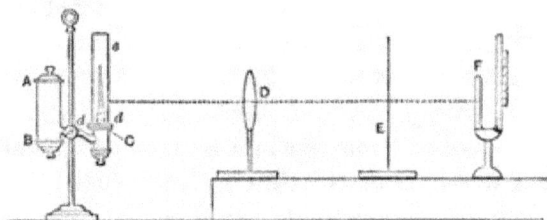

Fig. 40.

wick is C.   Over the wick, at a distance of half an inch or thereabouts, place a plate of thin sheet-copper, three inches in diameter, perforated in its centre with a circu-lar hole of the same diameter as the wick, and concentric therewith.   This piece of copper is represented at $d\,d$; it should have some contrivance for raising or depressing it through a small space, the proper height being deter-mined by trial.   On this plate the glass cylinder, e, an inch and three quarters in diameter and eight or ten inches long, rests.

When the lamp is lighted, provided the distance be-tween the plate $d\,d$ and the top of the wick be properly adjusted, on putting on the glass cylinder the flame in-stantly assumes an intense whiteness; by raising the wick it may be elongated to six inches or more, and becomes exceedingly brilliant.   Lamps constructed on these principles may be purchased in the shops.   I have, however, contented myself with using a common Argand study-lamp, supporting the perforated plate $d\,d$ at a proper height by a retort stand.   It will be easily un-derstood that the great increase of light arises from the circumstance that the flame is drawn violently through the aperture in the plate by the current established in the cylinder.

As much radiant heat is emitted by this flame, in order to diminish its action and also to increase the chemical effect I adopt the following arrangement: Let A B (Fig. 40) be the lamp; the rays emitted by it are received on a convex lens, D, four inches and three quarters in diameter, that which I use being the large lens of a lucernal microscope. This, placed at a distance of twenty-one inches from the lamp, gives an image of the flame at a distance of thirteen inches, which is received on the sentient tube F; between it and the lens there is a screen, E.

Things being thus arranged, and the lamp lighted so as to give a flame about three inches and a half long, the experiments may be proceeded with. It is convenient always to work with the flame at a constant height, which may be determined by a mark on the glass cylinder. At a given instant, by a seconds watch, the screen E is removed, and immediately the liquid begins to descend. When the first minute is elapsed the position on the scale is read off and registered; at the close of the second minute the same is done, and so on with the third, etc. And now, if these numbers be compared, casting aside the first, they will be found equal to one another, as the following table of experiments, made at different times and with different instruments, shows.

From this it will be perceived that, taking the first experiment as an example, if at the end of 30ˢ the photometer has moved 7.00, at the end of 60ˢ it has moved 8.00 more; at the end of 90ˢ, 7.50 more; at the end of 120ˢ, 7.75 more; the numbers set down in the vertical column representing the amount of motion for each 30ˢ. And, when it is recollected that the readings are all made with the instrument in motion, the differences between the numbers do not greatly exceed the possible errors of observation.   It may be remarked that the third and fourth experiments were made with a different lamp.

TABLE I.

Showing that when the radiant source is constant, the amount of move-ment in the photometer is directly proportional to the times of exposure.

Time.	Experiments.				
	1.	2.	3.	4.	5.
s.					
30	7.00	7.00	10.25	...	5.25
60	8.00	7.75	11.50	11.75	6.50
90	7.50	8.00	11.50	...	6.25
120	7.75	7.75	11.50	13.00	6.00
150	7.75	7.25	...	...	6.00
180	...	...	...	12.00	6.00
210	...	...	...	...	6.00
Mean...	7.60	7.55	11.19	12.25	6.00

Though a certain amount of radiant heat from a source so highly incandescent as that here used will pass the lens, its effects can never be mistaken for those of the chemical rays.   This is easily understood when we re-member that the effect of such transmitted heat would be to expand the gaseous mixture, but the chemical ef-fect is to contract it.

Next, the indications of the photometer are strictly proportional to the quantity of rays that have impinged upon it; a double quantity producing a double effect, a triple quantity a threefold effect, etc.

A slight modification in the arrangement (Fig. 40) enables us to prove this in a satisfactory way.   The lens, D, being mounted in a square wooden frame, can easily be converted into an instrument for delivering at its focal point, where the sentient tube is placed, measured quantities of the chemical rays, and thus becomes an invaluable auxiliary in those researches which require known and predetermined quantities of radiations to be

measured out. The method of doing this will be described in a subsequent part of this Memoir.

In order, therefore, to prove that the indications of
the photometer are proportional to the quantity of impinging rays, place this *measuring lens* in the position
D, setting its screens at an angle of 90°. Remove the
screen E, and determine the effect on the photometer for
one minute. At the close of the minute, and without
loss of time, turn one of the screens so as to give an
angle of 180°, and now the effect will be found double
what it was before, as in the following table:

TABLE II.

*Showing that the indications of the photometer are proportional to the
quantity of incident rays.*

Quantities.	Experiment 1.		Experiment 2.	
	Observed.	Calculated.	Observed.	Calculated.
90	2.18	2.22	2.69	2.75
180	4.27	4.45	5.75	5.50
270	6.70	6.67	8.25	8.25
360	8.90	8.90	11.00	11.00

I have stated in the commencement of this paper that
the action upon the photometer is limited to a ray which
corresponds in refrangibility to the indigo, or, rather, that
in the indigo space its maximum action is found. The
table on the following page serves at once to prove
this fact, and also to illustrate the chemical force of the
different regions of the spectrum.

In this table the spaces are equal; the centre of the
red, as insulated by cobalt blue glass, is marked as
unity; the centre of the yellow, insulated by the same,
being marked 3; the intervening region being divided
into two equal spaces, and divisions of the same value
carried on to each end of the spectrum.

As instruments will no doubt be hereafter invented
for measuring the phenomena of different classes of rays,

TABLE III.

Showing that the maximum for the photometer is in the indigo space of the spectrum.

Space.	Ray.	Force.	Space.	Ray.	Force.
0	Extreme red....	.33	8	Blue-indigo.....	204.00
1	Red............	.50	9	Indigo...........	240.00
2	Orange.........	.75	10	Violet..........	121.00
3	Yellow.........	2.75	11	Violet..........	72.00
4	Green..........	10.00	12	Violet..........	48.00
5	Green-blue.....	54.00	13	Violet..........	24.00
6	Blue...........	108.00	14	Extra-spectral..	12.00
7	Blue...........	144.00			

it may prove convenient to designate the precise ray to which they apply. Perhaps the most simple mode is to affix the name of the ray itself. Under that nomenclature the instrument described in this paper would take the name of indigo-photometer.

There is no difficulty in adapting this instrument to the determination of questions relating to absorption, reflection, and transmission. Thus I found that a piece of colorless French plate-glass transmitted 866 rays out of 1000.

Description of the Instrument. First, of the Glass Part. — The chlor-hydrogen photometer consists of a glass tube bent into the form of a siphon, in which chlorine and hydrogen can be evolved from hydrochloric acid containing chlorine in solution by the agency of a voltaic current. It is represented by Fig. 41, where *a b c* is a clear and thin tube four tenths of an inch in external diameter,

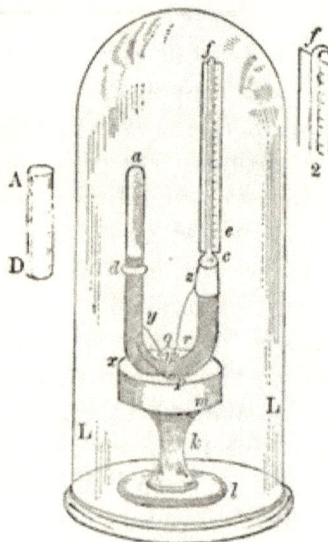

Fig. 41.

closed at the end a. At d, a circular piece of metal an inch in diameter, which may be called the stage, is fastened on the tube, the distance from d to a being 2.9 inches. At the point x, which is two inches and a quarter from d, two platinum wires, x and y, are fused into the glass, and entering into the interior of the tube, are destined to furnish the supply of chlorine and hydrogen; from the stage d to the point b, the inner bend of the tube, is 2.6 inches, and from that point to the top of the siphon c the distance is three inches and a half. Through the glass at z, three quarters of an inch from c, a third platinum wire is passed; this wire terminates in the little mercury-cup r, and x and y in the cups p and q respectively.

A stout tube, six inches long and one tenth of an inch interior diameter, e f, is fused on at c. Its lower end opens into the main siphon tube; its upper end is turned over at f, and is narrowed to a fine termination so as barely to admit a pin, but is not closed. This serves to keep out dust, and in case of a little acid passing out, it does not flow over the scale and deface the divisions. At the back of this tube a scale is placed, divided into tenths of an inch, being numbered from above downwards. Fifty of these divisions are as many as will be required. Fig. 2 shows the termination of the narrow tube bent over the scale.

From a point one fourth of an inch above the stage d, downwards beyond the bend, and to within half an inch of the wire z, the whole tube is carefully painted with India-ink, so as to allow no light to pass; but all the space from a fourth of an inch above the stage d to the top of the tube a is kept as clear and transparent as possible. This portion constitutes the sentient part of the instrument. A light metallic or pasteboard cap, A D, closed at the top and open at the bottom, three inches long and six tenths of an inch in diameter, blackened on its interior, may be dropped over this sentient tube; it being the office of the stage d to receive the lower end of the cap when it is dropped on the tube so as to shut out the light.

The foot of the instrument, $k\ l$, is of brass; it screws into the block m, which may be made of hard wood or ivory; in this three holes, p, q, r, are made to serve as mercury-cups; they should be deep and of small diameter, that the metal may not flow out when it inclines for the purpose of transferring. A brass cylindrical cover, L M, L M, may be put over the whole when it is desirable to preserve it in total darkness.

Things being thus arranged, the instrument is filled with its fluid, prepared as will presently be described; and as the tubes $a\ b, b\ c$ are not parallel to each other, but include an angle of a few degrees, in the same way that Ure's endiometer is arranged, there is no difficulty in transferring the liquid to the sealed side. Enough is admitted to fill the sealed tube and the open one partially, leaving an empty space to the top of the tube at c of two and three quarter inches.

Secondly, of the Fluid Part.—The fluid from which the mixture of chlorine and hydrogen is evolved, and by which it is confined, is yellow commercial hydrochloric acid, holding such a quantity of chlorine in solution that it exerts no action on the mixed gases as they are produced. From the mode of its preparation it always contains a certain quantity of chloride of platinum, which gives it a deep golden color, a condition of considerable incidental importance.

When hydrochloric acid is decomposed by voltaic electricity its chlorine is not evolved, but is taken up in very large quantity and held in solution; perhaps a bichloride of hydrogen results. If through such a solution hydrogen gas is passed in minute bubbles, it removes with it a certain proportion of the chlorine. From this, therefore, it is plain that hydrochloric acid thus decomposed will not yield equal measures of chlorine and hydrogen unless it has been previously impregnated with a certain volume of the former gas. Nor is it possible to obtain that degree of saturation by voltaic action, no matter how long the electrolysis is continued, if the hydrogen be allowed to pass through the liquid.

Practically, therefore, to obtain the photometric liquid we are obliged to decompose commercial hydrochloric acid in a glass vessel, the positive electrode being at the bottom of the vessel and the negative at the surface of the liquid. Under these circumstances, the chlorine as it is disengaged is rapidly taken up, and the hydrogen being set free without its bubbles passing through the mass, the impregnation is carried to the point required.

Although this chlorinated hydrochloric acid cannot of course be kept in contact with the platinum wires without acting on them, the action is much slower than might have been anticipated. I have examined the wires of photometers that had been in active use for four months, and could not perceive the platinum sensibly destroyed. It is well, however, to put a piece of platinum foil in the bottle in which the supply of chlorinated hydrochloric acid is kept; it communicates to it slowly the proper golden tint.

The liquid being impregnated with chlorine in this manner until it exhales the odor of that gas is to be transferred to the siphon *a b c* of the photometer, and its constitution finally adjusted as hereafter shown.

Thirdly, of the Voltaic Battery.—The battery which will be found most applicable for these purposes consists of two Grove's cells, the zinc surrounding the platinum.

The following are the dimensions of the pairs which I use. The platinum plate is half an inch wide and two inches long; it dips into a cylinder of porous biscuit-ware of the same dimensions, which contains nitric acid. Outside this porous vessel is the zinc, which is a cylinder one inch in diameter, two inches long, and two tenths thick; it is amalgamated. The whole is contained in a cup, two inches in diameter and two deep, which also receives the dilute sulphuric acid.

The force of this battery is abundantly sufficient both for preparing the fluid originally and for carrying on the photometric operations. It can decompose hydrochloric acid with rapidity, and will last with ordinary care a long time.

Before passing to the mode of using this photometer, it is absolutely necessary to understand certain theoretical conditions of its equilibrium. These in the next place I shall describe.

Theoretical Conditions of Equilibrium.—This photometer depends for its sensitiveness on the exact proportion of the mixed gases. If either one or the other is in excess a great diminution of delicacy is the result. The comparison of its indications at different times depends on the certainty of evolving the gases in exact, or, at all events, known proportions.

Whatever, therefore, affects the constitution of the sentient gases alters at the same time their indications. Between those gases and the fluid that confines them certain relations subsist the nature of which can be easily traced. Thus, if we had equal measures of chlorine and hydrogen, and the liquid not saturated with the former, it would be impossible to keep them without change, for by degrees a portion of chlorine would be dissolved and an excess of hydrogen remain; or if the liquid was overcharged with chlorine, an excess of that gas would accumulate in the sentient tube.

It is absolutely necessary, therefore, that there should be an equilibrium between the gaseous mixture and the confining fluid.

As has been said, when hydrochloric acid is decomposed by a voltaic current, all the chlorine is absorbed by the liquid and accumulates therein; the hydrogen bubbles, however, as they rise withdraw a certain proportion, and hence pure hydrogen passed up through the photometric fluid becomes exceedingly sensitive to the light.

There are certain circumstances connected with the constitution and use of the photometer which continually tend to change the nature of its liquid. The platinum wires immersed in it by slow degrees give rise to a chloride of platinum. It is true that this takes place very gradually, and by far the most formidable difficulty arises from a direct exhalation of chlorine from the narrow tube ef, for each time that the liquid descends, a volume of air is introduced, which receives a certain amount of chlorine which with it is expelled the next time the battery raises the column to zero; and this going on time after time finally impresses a marked change on the liquid. I have tried to correct it in various ways, as by terminating the end f with a bulb; but this entails great inconvenience, as may be discovered by any one who will reflect on its operation.

When by the battery we have raised the index to its zero point, if the gas and liquid are not in equilibrium that zero is liable to a slight change. If there be hydrogen in excess, the zero will rise; if chlorine, the zero will fall.

In making what will be termed "interrupted experiments," we must not too hastily determine the position of the index on the scale at the end of a trial. It is to be remembered that the cause of movement over the scale arises from a condensation of hydrochloric acid, but that condensation, though very rapid, is not instantaneous. Where time is valuable and the instrument in perfect equilibrium, this condensation may be instantaneously effected by simply inclining the instrument so that its liquid may pass down to the closed end a, but not so much as to allow gas to escape into the other side—the inclination of the two sides to each other makes this a very easy manipulation—and the gas thus brought into contact with an extensive liquid surface yields up its hydrochloric acid in an instant.

Directions for using the Photometer. Preliminary adjustment.—Having transferred the liquid to the sealed end of the siphon, and placed the cap on the sentient extremity, the voltaic battery being prepared, the operator dips its polar wires into the cups p, q, which are in connection with the wires x, y. Decomposition immediately takes place, chlorine and hydrogen rising through the liquid and gradually depressing it, while, of course, a corresponding elevation takes place in the other limb. This operation is continued until the liquid has risen to the zero. It takes but a few seconds for this to be accomplished.

The polar wires having been disengaged, the photometer is removed opposite a window, care being taken that the light is not too strong. The cap is now lifted off the sentient extremity $a\,d$, and immediately the liquid ascends. This exposure is allowed to continue, and the liquid suffered to rise as much as it will to the end a. And now, if the gases have been properly adjusted, an entire condensation will take place, the sentient tube $a\,d$ filling completely. In practice this precision is not however obtained, and if a bubble as large as a pepper-corn be left, the operator will be abundantly satisfied with the sensitiveness of his instrument. Commonly, at first, a large residue of hydrogen gas, occupying perhaps an inch or more, will be left. It is to be understood that even this large surplus will disappear in a few hours by absorbing chlorine. But this is not to be waited for; as soon as no further rise takes place, in a minute or two, the siphon is to be inclined to one side, and the residue turned into the open side.

Now, recurring to what has been said on the equilibrium, it is plain that this excess of hydrogen arises from a want of chlorine in the photometric liquid. A proper quantity must therefore be furnished by proceeding as follows:

The sentient tube being filled with the liquid by inclination, connect the polar wires with p, q, as before. These may be called *generating wires*. Allow the liquid to rise in $b\ c$ until the third platinum wire z, which may be called the *adjusting wire*, is covered an eighth of an inch deep. Then remove the negative wire from the cup p into the cup r, and now the conditions for saturating the liquid are complete; hydrogen escaping from the surface of the liquid at z, and chlorine continually accumulating and dissolving between x and d. This having been carried on for a short time, the gas in $a\ d$ is to be turned out by inclination and the instrument recharged. That a proper quantity is evolved is easily ascertained by allowing total condensation to take place, and observing that only a small bubble is left at a.

It will occasionally happen in this preliminary adjustment that an excess of chlorine may arise from continuing the process too long. This is easily discovered by its greenish-yellow tint, and is to be removed by inclining the instrument and turning it out.

Thus adjusted, everything is ready to obtain measures of any effect, there being two different methods by which this can be done: 1st, by continuous observation; 2d, by interrupted observation.

Of the method of continuous observation.—This is best described by resorting to an example. Suppose, therefore, it is required to verify Table I., or, in other words, to prove that the effect on the photometer is proportional to its time of exposure.

Put on the cap of the sentient tube *a d*, connect the polar wires with *p, q*, and raise the liquid to zero.

Place the photometer so that its sentient tube will receive the rays properly.

At a given instant, marked by a seconds watch, remove the cap A D, and the index-liquid at once begins to descend. At the end of the first minute read off the division over which it is passing. Suppose it is 7. At the end of the second do the same: it should be 14; at the end of the third, 21, etc. This may be done until the fiftieth division is reached, which is the terminus of the scale.

Recharge the tube by a momentary application of the polar wires; but it is convenient first to remove any excess of hydrochloric-acid gas in the sentient tube by allowing it time for condensation; or, if that be inadmissible, by inclining a little to one side, so as to give an extensive liquid contact.

Of the method of interrupted observation.—It frequently happens that observations cannot be had during a continuous descent, as when changes have to be made in parts of apparatus or arrangements. We have then to resort to interrupted observations.

This method requires that the gas and liquid should be well adjusted, so that no change can arise in volume when extensive contact is made by inclination.

The photometer being charged, place it in a proper position. At a given instant remove its cap, and the liquid descends. When the time marked by a seconds watch has elapsed, drop the cap on the sentient tube. The liquid simultaneously pauses in its descent, but does not entirely stop, for a little uncondensed hydrochloric acid still exists, which is slowly disappearing in the sentient tube. Now, incline the instrument for a moment to one side, so that the liquid may run up to the end a, but not so much as to let any gas escape. Restore it to its position and read off on the scale. It is then ready for a second trial.

The difference between continuous and interrupted observation is this, that in the latter we pause to wash out the hydrochloric acid, and though this is effected by the simplest of all possible methods, continuous observations are always to be preferred when they can be obtained.

I have extended this Memoir to so great a length that many points on which remarks might have been made must be passed over. It is scarcely necessary to say that the sentient tube must be uniformly and perfectly clean. As a general rule also, the first observation may be cast aside, for reasons which I will presently give. Further, it is to be remarked, as it is an essential principle, that during different changes of volume of the gas its exposed surface must never vary in extent, the liquid is not to be suffered to rise above the blackened portion at d. If the measures of the different parts be such as have been here given, this cannot take place, for the liquid will fall below the fiftieth division before its other surface rises above d.

The same original volume of gas in a d will last for a long time, as we keep replenishing it as often as the fiftieth division is reached.

The experimenter cannot help remarking that, on suddenly exposing the sentient tube to a bright light, the liquid for an instant rises on the scale, and on dropping the cap in an instant falls. This important phenomenon, which is strikingly seen under the action of an electric spark, I shall consider hereafter.

In conclusion, as to comparing the photometric indication at different times, if the gases have the same constitution the observations will compare; and if they have not, the value can from time to time be ascertained by exposure to a lamp of constant intensity. To this method I commonly resort.

From the space occupied in this description the reader might be disposed to infer that this photometer is a very complicated instrument and difficult to use. He would form, however, an erroneous opinion. The preliminary adjustment can be made in five minutes, and with it an extensive series of measures obtained. These long details have been entered into that the theory of the instrument may be known, and optical artists construct it without difficulty. Though surprisingly sensitive to the action of the indigo ray, it is as manageable by a careful experimenter as a common differential thermometer.

UNIVERSITY OF NEW YORK, *Sept.* 26, 1843.

NOTE. (From *Harper's Magazine*, No. 328.) — Professors Bunsen and Roscoe, in their photo-chemical researches, made at the University of Heidelberg, and communicated to the Royal Society of London, 1856, say: "The first and only attempt which has been made to refer the chemical action of light to a standard measure is to be found in the researches of Draper.

The description of his instrument and mode of observation employed by him was published in 1843.   Even
with this instrument, which, as we shall show, is in many
respects defective, Draper has succeeded in establishing
experimentally some of the most important relations of
the chemical action of light.   In these experiments Draper collected hydrogen, evolved by electrolysis over hydrochloric acid saturated with chlorine, and to this hydrogen he admitted so much chlorine, either by diffusion
from the saturated acid or by electrolysis, that the mixture consisted of nearly equal volumes of the two gases,
and entirely or almost entirely disappeared on exposure
to light.   The alteration in the volume of the gaseous
mixture arising from the absorption of the hydrochloric
acid formed by the action of the light was read off on a
scale, and being within certain limits proportional to the
time of exposure, served as a measure of the chemical
rays."

Professors Bunsen and Roscoe, having modified this
instrument to suit the objects they had in view, accordingly used it in their very exhaustive and important
series of researches.

The measuring lens referred to in previous paragraphs
is constructed upon this principle: If half the surface of
a convex lens be screened by an opaque body, as a piece
of blackened card-board, of course only half the quantity
of rays will pass which would have passed had the
screen not been interposed; if one fourth of the lens be
left uncovered, only one fourth of the quantity will pass.
But in all these instances the focal image remains of the
same size as at first.   Therefore by adjusting upon the
frame of the lens two screens, the edges of which pass
through its centre and are capable of rotation thereupon,
we shall cut off all light when the screens are applied

edge to edge opposite each other. We shall have 90°
when they are rotated so as to be at right angles, and
180° when they are superposed with their edges coinciding, or one of them be taken away. Thus, by setting
them in different angular positions, we can have all
quantities, from 0° up to 180°, and by removing them
entirely, reach 360°. The lens will thus give an image
of a visible object always of the same size, its brilliancy
or intensity varying at pleasure in a known proportion.

In Fig. 42, A, B, B, D is a double convex lens set in a
wooden frame, F. Its face can be cov-
ered by two semicircles of blackened
card-board, one of which revolves on the
centre at *c*. In the figure they are represented as set at right angles, and the
quarter of the lens at A is uncovered.

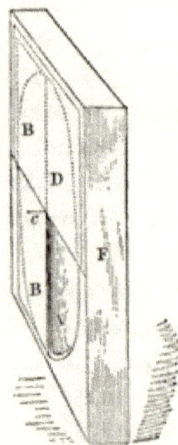

To the foregoing description of the
chlor-hydrogen photometer I may add a
reference to another which I have very
advantageously used when extreme sensitiveness was not required. It depends
on the employment of an aqueous solution of ferric oxalate. This substance,

Fig. 42.

which is of a golden-yellow color, may be kept for many
years without undergoing any change, if in total darkness; but on exposure to a lamp or the daylight it
decomposes, carbonic acid escaping, and lemon-yellow
ferrous oxalate precipitating. If set in the sunshine, it
actually hisses through the escape of the gas. The ray
which chiefly affects it is the indigo, the same which
affects the chlor-hydrogen photometer and the silver
compounds used in photography. This ray, to produce
its effect, undergoes absorption, as might be anticipated
from what has been previously said in this Memoir, and

as is easily proved by causing a sunbeam to pass through two parallel strata of the oxalate, when it will be found that the light which has gone through the first portion is inoperative on the second.

Other properties which this solution of ferric oxalate possesses strongly recommend it as a photometric agent. Unlike solution of chlorine, it may be very conveniently confined in glass tubes by mercury. In its use there are two points which must be attended to: (1) the lemon-yellow ferrous oxalate must not be permitted to incrust the side of the glass exposed to the light, and thereby injure its transparency; (2) the ferric solution must be kept nearly at a constant temperature, for its color changes with the heat. At the freezing of water it is of an emerald-green tint; at the boiling, of a brownish-yellow. With these variations of tint its absorptive action varies, and therefore its liability to be changed. In an extensive series of experiments made with it, but which I have not yet published, I found that it is greatly improved by the addition of an aqueous solution of ferric chloride.

It may be remarked that the oxalate is an excellent photographic substance. A piece of tissue-paper, made yellow by being dipped into a neutral solution of it, when dried in the dark is very sensitive. Its invisible impression may be developed by a weak solution of nitrate of silver, two grains dissolved in an ounce of water answering very well. A weak solution of chloride of gold is a still more sensitive developer. I have in my possession photographs made by both these methods more than thirty years ago, which have undergone apparently no deterioration.

In the application of ferric oxalate to photometry several methods may be followed. The course I have most commonly taken has been to determine the quantity of carbonic acid produced, sometimes by volume, sometimes by weight. It is to be understood that before any carbonic acid can be disengaged the solution must become saturated therewith, and that before we can correctly measure the quantity of light by the quantity of acid produced this dissolved portion must be ascertained. In one of my photometers the expulsion of the dissolved gas is accomplished by exposure to a bath of boiling water, in another by a stream of hydrogen. Both yield satisfactory results.

But this method by the determination of the produced carbonic acid is only one of numerous plans; for instance, we might use the weight of certain metals which the solution after exposure will precipitate. Thus a portion which has been made and kept in the dark may be mixed with chloride of gold without any action ensuing, but if it has been illuminated, the weight of metallic gold precipitated is in proportion to the incident light. On this principle I commenced an attempt to determine the hourly and diurnal illumination of a certain locality. At the bottom of a metal tube, arranged as a polar axis, was placed a bulb containing a standard solution of the iron salt, and at the close of the proposed periods the weight of gold it could reduce was ascertained. There is something fascinating in determining the quantity of light which the sun yields by the quantity of gold it can produce. Upon the whole, however, I would recommend those who are disposed to renew these attempts to select a method depending on the volume of carbonic acid, for it is always easier to make an observation than an experiment.

Among the important results which may be expected from these new modes of photometry are the hourly, diurnal, and annual quantities of sunlight. These are important not only in a meteorological point of view, but also as respects physical geography and the great interests of agriculture. The sum of vegetable organization is in all climates and localities a function of the light distributed thereto. And so far as heat is concerned, it is not the intensity only, but the absolute quantity, which is to be measured. To each plant, from the moment of its germination to the moment of its maximum development and the completion of its life, a definite quantity of heat and of light must be given. As respects the heat in such inquiries, it is not only the thermometer but the calorimeter which must be considered; and as to the light, the photometers herein described determine its quantity but not its brilliancy, and therefore answer the indications required. And since it is not merely the temperature of a locality, but also the light of the sun, which is the effective condition of vegetable growth, we see how important even in agriculture itself these proposed determinations really are.

To those who would devote themselves to such inquiries I recommend as a photometric means a mixture of chlorine and hydrogen where great sensitiveness is required, and in other cases ferric oxalate.

The chlor-hydrogen and the ferric oxalate photometers act by selective absorption, on the principle of the calorimeter; that is to say, they measure the *quantity* of the radiations they select.

I may, perhaps not inappropriately, close this Memoir with a brief allusion to an instrument I formerly used very much. It measures the *intensity* of the radiations it is made to select, and these radiations may be varied at pleasure. I will describe it first as adjusted for the radiations of which chlor-hydrogen and ferric oxalate take charge. The description is extracted from a paper I published in the *Philosophical Magazine*, August, 1844.

Fig. 43.

Let a wooden box, A B (Fig. 43), six inches long, two wide, and two deep, with perforations at A and B in its ends, be provided; in the centre of its top an aperture three quarters of an inch in diameter is to be made. The box must be blackened interiorly, and a rectangular prism of wood, C, be placed in it, with its right angle in such a position that its edge bisects as a diameter the circular aperture; over this wooden prism a piece of white paper is pasted, care being taken that where it bends over the right angle of the prism it is folded sharp. So far the reader will recognize in this Ritchie's photometer. Upon the aperture in the top of the box a glass trough, *g h*, is placed; it is made by cutting a circular hole an inch in diameter in a piece of plate-glass one third of an inch thick, and laying on each side of it a thin piece of plate-glass. This forms a circular trough, in which a strong solution of sulphate of copper and ammonia may be enclosed. Over the trough a tube, *d*, eight or ten inches long, is placed so that the eye may see distinctly through the aperture in the top of the box the disk of paper, and more especially its dividing diameter.

Lights, E, F, set at the opposite ends of the box, A B, may therefore be compared as regards their photographic intensity, the calculations being made by the common photometric law. And by changing the liquid in the absorbing trough, *g h*, any radiations may be selected for examination. The instrument may therefore be designated as " the selective absorption photometer."

From the facts presented in this and the preceding paper, as conveying the modern conception of the relation of luminous and calorific radiations, it may be concluded that the thermometer with a blackened bulb is an *absolute* photometer, and that, in accordance with the principles set forth, many other *selective* photometers and thermometers may be devised.

MEMOIR XIX.

ON MODIFIED CHLORINE.

From the Philosophical Magazine, July, 1844.

[This Memoir was read at the meeting of the British Association held at Cork, 1843. The concluding paragraphs were subsequently added.]

CONTENTS :—*Description of the experiment.—The change in the chlorine is not transient.—There are two stages in the phenomenon.—Rays are absorbed in producing this change.—The indigo ray is absorbed.—The action is positive from end to end of the spectrum.—The indigo ray forms hydrochloric acid and also produces the preliminary modification.—Change in other elementary bodies.—Verification of these results with the chlor-hydrogen photometer.*

CHLORINE gas which has been exposed to the daylight or to sunshine possesses qualities not possessed by chlorine made in the dark.

This is shown by the circumstance that chlorine which has been exposed to the sunshine has obtained from that exposure the property of speedily uniting with hydrogen gas, a property not possessed by chlorine made and kept in the dark.

This quality gained by the chlorine arises from its having absorbed chemical rays corresponding in refrangibility to the indigo. It is not a transient, but apparently a permanent property, the rays so absorbed becoming latent, and the effect lasting for an unknown period of time. The facts here presented will be interesting to chemists, because they plainly lead us to suspect that the descriptions we have of the properties of all elementary and compound bodies are either inaccurate or confused. These properties are such as bodies exhibit after they have been exposed to light; we still require to know what are the properties they possess before exposure to such influences.

Natural philosophers will also find an interest in these phenomena, for they finally establish for the chemical rays two important facts: 1st, that those rays are absorbed by ponderable bodies; and, 2d, that they become latent after the manner of heat. Some years ago I endeavored to prove that these things held for a compound substance, the iodide of silver (*Phil. Mag.*, September, 1841).

For reasons which will be obvious as the description proceeds, I shall speak of chlorine which has been exposed to the beams of the sun as *modified chlorine.*

I. *Description of the Experiment.*

In two similar glass tubes place equal volumes of chlorine, made from peroxide of manganese and hydrochloric acid by lamplight, and carefully screened from access of daylight. Expose one of the tubes to the full sunbeams for some minutes, or, if the light be feeble, for a quarter of an hour: the chlorine in it becomes modified. Keep the other tube during this time carefully in a dark place; and now, by lamplight, add to both equal volumes of hydrogen gas. These processes are best carried on in a small porcelain or earthenware trough, filled with a saturated solution of common salt, which dissolves chlorine slowly; and to avoid explosions operate on limited quantities of the gases. Tubes that are eight inches long and half an inch in diameter will answer very well. The two tubes now contain the same gaseous mixture, and only differ in the circumstance that one is modified and the other not. Place them, therefore, side by side before a window, through which the entrance of daylight can be regulated by opening the shutter; and now, if this part of the process be conducted properly, it will be seen that the modified chlorine commences to unite with the hydrogen and the salt water rises in that tube. But the unmodified chlorine shows no disposition to unite with its hydrogen, and the liquid in its tube remains motionless for a long time. Finally, as it becomes slowly modified by the action of the daylight impinging on it, union takes place. From this, therefore, we perceive that chlorine which has been exposed to the sun will unite promptly and energetically with hydrogen; but chlorine that has been made and kept in the dark shows no such property.

As I doubt not this remarkable experiment will be repeated by chemists, I will add that the only point to which attention in particular is to be given is in the final exposure to the light. This must not be too feeble, or the action will be tedious; but the direct sunbeam must be sedulously excluded or an explosion will result. A room illuminated by one small window looking to the north answers very well. It need scarcely be added that care must be taken that both tubes are illuminated alike.

II. *The Change in the Chlorine is not Transient.*

Now it might be supposed that this apparent exaltation of the electro-negative properties of the chlorine is only a transient thing, which would speedily pass away, the gas reverting to its original condition.

To show that this is not so, modify some chlorine in a tube as before. Place it for an hour or two in the dark along with the tube of unmodified chlorine with which it is to be compared, then to both add hydrogen. Expose them as in the former experiment to the daylight, and the result will turn out as before; the modified chlorine forming hydrochloric acid at once, and the unmodified refusing to do so.

S

This, therefore, shows that the change which the sun-beams impress upon chlorine is to a certain extent a permanent change, and, unlike a calorific effect, it does not spontaneously and rapidly pass away.

III. *There are Two Stages in the Phenomenon.*

Let us now make inquiry into the nature of the change thus impressed on the chlorine. This, I shall show, rests in the circumstance of the absorption of rays which correspond in refrangibility to the indigo, and appear to become latent.

In a tube, over salt water, mix together equal volumes of unmodified chlorine and hydrogen gas. Expose it to the daylight, marking the time at which the exposure commences. Watch the level of the liquid in the tube narrowly, and, though stationary for a considerable time, after a certain period has elapsed it will be seen on a sudden to start and commence rising. Observe now how far it will rise during a period equal to the time that elapsed between the first exposure and the beginning of the rise, and it will be seen that one fourth or half the gases will disappear.

It is obvious that from the first moment of exposure the rays must have been exerting their influences on the mixture. As will presently be proved, absorption has been all along taking place. There are, therefore, two distinct phenomena exhibited by this experiment. There is a period during which, though large quantities of the dark rays are disappearing, no visible change is produced; there is a second period, during which absorption is accompanied by a remarkable chemical effect— the production of hydrochloric acid. From these things we gather that a definite amount of the chemical rays must disappear and become latent before hydrochloric acid can form. The phenomenon is not unlike that of the disappearance of a definite quantity of heat in the passage of ice into the condition of water.

A mixture of chlorine and hydrogen does not, therefore, instantly give rise to the production of hydrochloric acid on exposure to the light, but as a preliminary condition a certain definite amount of absorption must take place.

Now if this were a mere molecular disturbance, such as might be brought about by the action of heat, we should expect to find it transient and speedily passing away. Such, however, is far from being the case. As with simple chlorine, so with this mixture: after it has been modified it loses its quality very slowly. I have observed that after a week or more has elapsed since it was first exposed to the light, it commences to contract when placed in a feeble gleam.

IV. *Rays are Absorbed in Producing this Change.*

I have thus far assumed that the rays which bring about these changes are absorbed; the following is the proof which I have to offer:

Over a tube half an inch in diameter and six inches long, closed at its upper extremity and open at its lower, invert a jar of the same length and one inch and a half in diameter. Fill the tube and the jar at the salt-water trough, about two thirds full, with the same mixture of chlorine and hydrogen. Expose them to diffuse daylight. Now it is clear that no rays can gain access to the tube except after having passed through the gaseous mixture in the jar. After a certain space of time the level of the liquid in the jar commences to rise, but that in the tube will remain much longer wholly stationary.

It therefore appears that a beam which has passed through a mixture of chlorine and hydrogen has lost, to a great extent, the quality of bringing about the union of a second portion of the mixed gases through which it may be caused to traverse. The active rays have been absorbed; they disappear from the beam and are lost in producing their first effect.

A beam of light loses its energy in producing a chemical effect; the beam, as well as the medium on which it acts, becomes changed. I have a series of results which proves that this takes place for a great variety of compound bodies.

V. *It is the Indigo Ray which is Absorbed.*

As has been said, it is a ray corresponding in refrangibility to the indigo which produces these results.

In a small porcelain trough I inverted, side by side, ten tubes, each of which was three inches long and one third of an inch in diameter, the trough being filled with salt water. I passed into each tube a certain quantity of unmodified chlorine and hydrogen. A beam of the sun, being directed by a heliostat into a dark room, was dispersed horizontally by a flint-glass prism, and the trough with its tubes so placed as to offer an exposure to the different colored rays. The aperture admitting the beam was about half an inch in diameter. For a while no movement was observed in any of the tubes; but as soon as the preliminary absorption previously described was over, the level of the liquid began to rise. In the red and in the orange no movement could be perceived, in the violet only after a time; but first of all the tube that was immersed in the indigo light was in action, and exhibited finally a very rapid rise; this was soon followed by the tube that was in the space where the indigo and violet joined, then by that in the violet and that in the blue; the tube in the green was next in order. The following table gives the numerical results obtained by observing the time which elapsed before movement took place in each tube:

TABLE I.

Name of ray.	Time.	Name of ray.	Time.
Extreme red.......	*	Indigo..	1.50
Red and orange....	+100.00	Indigo and violet..	2.00
Yellow and green..	52.00	Violet..............	2.25
Green and blue....	4.00	Violet..............	5.00
Blue...............	2.33	Extreme violet....	5.50

Many years ago M. Berard made experiments on the explosion of chlorine and hydrogen, and concluded from his results that it was brought about by the violet ray. This was at a time when the methods of making these experiments were less exactly known. It is a very easy matter to prove that in reality the indigo is the active ray, and that from a maximum point which is in the indigo, but towards the blue, the effect gradually diminishes to each end of the spectrum.

The following table gives the calculated approximate intensity of the chemical force for each ray, deduced from the foregoing experiment:

TABLE II.

Name of ray.	Force.	Name of ray.	Force.
Extreme red.......	....	Indigo............	66.60
Red and orange....	1.00	Indigo and violet..	50.00
Yellow and green..	1.90	Violet.............	44.40
Green and blue....	25.00	Violet.............	20.00
Blue...............	42.90	Extreme violet....	18.10

There is a great advantage which experiments conducted in this way possess over those depending for their indication on the stains impressed on daguerreotype plates or sensitive papers. In those cases we obtain merely a comparative contrast for different regions of the spectrum; in this we have absolute measures determined by a definite chemical effect and the rise of a

---

* Even after the longest exposure I had the means of giving it, no movement took place in the tube in the extreme red, and I am doubtful about that in the red and orange.

liquid in a graduated tube; and from this we gain just-
er views of the true constitution of the spectrum.   On
studying the numbers in the foregoing table, or, better
still, if we project them, it will appear what an enormous
difference there is in the chemical force of the different
rays.   In the experiment from which I have deduced
this table it appears that the force of the indigo ray
exceeds that of the orange in a greater ratio than 66 to
1; and from the circumstances under which the experiment is made this difference must be greatly underrated.
There is always diffused light in the room coming from
the intromitted beam, and this accelerates the rise in the
less refrangible tubes; then, again, it is impossible that
the tube which gives the greatest elevation shall coincide mathematically with the maximum point and express the maximum effect.

From some estimates I have made I am led to believe
that in point of chemical force, for this mixture of chlorine and hydrogen, the indigo ray exceeds the red in a
higher ratio than 500 to 1.

VI. *The Action is Positive from End to End of the Spectrum.*

M. Becquerel found that for an iodized silver plate the
red, the orange, and the yellow rays possess the quality
of continuing the action begun by the more refrangible
colors; he therefore names these "*rayons continuateurs.*"
For the same compound I found that those rays, acting
conjointly with the diffused daylight, exerted a negative
agency.   It is therefore desirable to understand whether,
with respect to the gases now under consideration, the
less refrangible rays exert anything in the way of an action of depression or hindrance to union.   By direct experiment I found that this was not the case, the action
being positive from end to end of the spectrum.   This

can be shown by removing the tubes, after they have
been in the spectrum for an hour or two, into the gleams
of daylight. One by one they exhibit after a time a
rise, the order being the green first, then the yellow and
the orange, and at last the red. And if at the same time
a tube which has been kept in the dark be exposed
along with them, they will all rise before it, showing
that modification had set in and been going on in them
all; that it had been more active in the green than in
the yellow, in the yellow than in the orange, in the orange than in the red; and, had the exposure to the spectrum been long enough, the liquid in every one of the
tubes would have risen.

VII. *The Indigo Ray forms the Hydrochloric Acid as well as produces the Preliminary Modification.*

It only remains now to inquire whether the rays causing the production of the hydrochloric acid are those
which effect the modification of the chlorine; in other
words, whether the first stage of the process is brought
about by the same agent which carries on the second.
The experiment I have just described shows that modification is most actively produced by the indigo ray, and
it is easy to show that it is the same ray which carries
on the second part of the process; for, if before placing
the tubes in the prismatic spectrum we expose them to
the daylight, so that the liquid has just commenced to
rise in each, and then to the spectrum, it will be found
that the liquid of the tube in the indigo rises most rapidly, and the others in the order stated before. Therefore we perceive that the same ray commences, carries
on, and completes the process.

Few substances can exceed in sensitiveness to light a
mixture of chlorine and hydrogen previously modified.
Brought into the obscure daylight of a gloomy chamber,

it is remarkable how promptly the level of the liquid in the tube rises; how, when the shutters are successively thrown open, the action becomes more and more energetic; and how, in an instant, it stops when the instrument is shaded by a screen.

I have not recorded in this communication a multitude of experiments of detail supporting the conclusions here drawn. It has been my object on this occasion to call attention to the fact that chlorine, an elementary body, undergoes a change after exposure to the light; a change which appears to produce an exaltation of its electro-negative properties, as is shown by its power of uniting more energetically with hydrogen. This change must not be confounded with those transient elevations of activity due to increased temperature, inasmuch as this is more permanent in its character. It arises from the absorption of rays existing most abundantly in the indigo space of the spectrum. That the phenomenon is due to a true absorption is fully shown by the circumstance that a beam which has produced this effect has lost the quality of ever after producing a similar result. This is borne out by what we observe to take place when a feeble light falls on a mixture of chlorine and hydrogen prepared in the dark. A certain space of time elapses before any formation of hydrochloric acid occurs, during which the absorption in question is going on; and when that is completed, and the mixture is modified, union of the gases begins and hydrochloric acid forms. From end to end of the spectrum the action is positive, and differs only in intensity; but this difference in intensity opens before us new views of the constitution and character of the solar beam.

UNIVERSITY OF NEW YORK, *June* 20, 1843.

Paragraphs subsequently added.

Chlorine is not the only elementary substance in which the radiations produce a change. In his chapter on phosphorus, Berzelius remarks: "Light produces in it (phosphorus) a peculiar change, of which the intimate nature is unknown; and which, so far as we can judge at present, does not alter its weight. It makes it take a red tint. This phenomenon occurs not only in a vacuum, even in that of a barometer, but also in nitrogen gas, in carburetted hydrogen, under water, alcohol, oil, and other liquids. When we expose to the sunlight phosphorus dissolved in ether, oil, or hydrogen gas, it instantly separates under the form of red phosphorus; it undergoes very rapidly this modification in *violet light*, or in glass vessels of a violet color. The light of the sun makes it easily enter into fusion in nitrogen gas, but it does not melt in hydrogen, and in the torricellian vacuum it sublimes in the form of brilliant red scales" (Berzelius, *Traité*, tom. i., p. 258).

Again, when speaking of phosphuretted hydrogen, he says: "Exposed to the influences of the direct solar light this gas is decomposed, a part of the phosphorus separates under the form of red phosphorus, and is deposited on the interior surface of the glass. If we cover the vessel which contains the gas imperfectly, no phosphorus is deposited on the covered spaces" (*Ib.*, tom. i., p. 265).

As Berzelius does not give these experiments as his own, and I do not know to whom we are indebted for them, I repeated some of them. Among other corroborative results it appeared that a piece of phosphorus of a pale or whitish color, in a vessel filled with pure and dry carbonic-acid gas, placed in the sunshine, rapidly exhibited the phenomenon in question. Eventually the phosphorus became of a deep blood-red color, and on the sides of the glass towards the light feathery crystals formed, the tint of which bore a close resemblance to that of the red prussiate of potash.

Since the invention of the chlor-hydrogen photometer I have been able to observe more closely the habitudes of chlorine. In the description given of that instrument it is recommended to cast aside the first observation, because it never gives an accurate estimate of the true effect. When a mixture of chlorine and hydrogen is exposed, hydrochloric acid does not immediately form; but a preliminary absorption is necessary, and then at the end of a certain period contraction begins to take place.

Such a photometer exposed to the daylight is much too powerfully affected to allow the successive stages of change to be distinctly made out; the preliminary modification is accomplished so rapidly that the indications of it are merged and lost in the contraction which instantly follows. It is necessary therefore that we should operate with a small lamp-flame.

To such a flame I exposed a mixture of chlorine and hydrogen, and marked the number of seconds which elapsed before contraction, arising from the production of hydrochloric acid, took place. The first indications of movement occurred at the close of 600 seconds.

The index then moved through the first degree in 480 seconds;

"	"	"	second	"	165	"
"	"	"	third	"	130	"
"	"	"	fourth	"	95	"
"	"	"	fifth	"	93	"
"	"	"	sixth	"	93	"

and continued to move with regularity at the same rate.

These observations, therefore, prove that a very large amount of radiant matter is absorbed before chemical combination takes place, and that in the case of chlorine and hydrogen the total action is divisible into two periods: the first during which a simple absorption is taking place without a chemical effect, the second during which absorption is attended with the production of hydrochloric acid.

The facts which I am endeavoring to set forth prominently in this Memoir are—1st, the preliminary modifications just discussed; and, 2d, the persistent character of the change impressed upon chlorine when it has been exposed to the sun, an effect wholly unlike a calorific effect, which would soon disappear.

By resorting to the chlor-hydrogen photometer we obtain information equally distinct upon the second point, that the preliminary modification is not a transient effect which at once passes away, but is, on the contrary, a persistent change.

I modified the chlorine and hydrogen contained in the instrument, and kept it in the dark for ten hours. On exposure to the lamp-rays it moved after a few seconds, showing, therefore, that the change impressed on the chlorine was not lost. In the former case 600 seconds had elapsed before any movement was visible.

When, however, we remember that the invisible images on daguerreotype plates, and even photographic impressions on surfaces of resin, and probably all other similar changes, are slowly effaced, it would be premature to conclude that modified chlorine does not revert to its original condition. I have sometimes thought that there were in several of my experiments indications that this was taking place, but would not be understood as asserting it positively. Whether it be so or not, one thing is certain, that the taking on of this condition and the loss of it is a very different affair from any transient exaltation of action due to a temporary elevation of temperature, or the contrary effect produced by cooling.

April 26, 1844.

.

MEMOIR XX.

ON THE ALLOTROPISM OF CHLORINE AS CONNECTED WITH THE THEORY OF SUBSTITUTIONS.

From the American Journal of Science and Art, Vol. XLIX.; Philosophical Magazine, Nov., 1845.

Contents :—*Chlorine exists in two states, active and passive.—Decomposition of water by it in the sunlight.—Facts connected with this decomposition.—The relations of chlorine and hydrogen.—The allotropism of chlorine.—Connection of these facts with the theory of substitutions.*

The researches of M. Dumas on chemical types have shown that between chlorine and hydrogen remarkable relations exist, indicating that the electrical characters of elementary atoms are not essential, but rather incidental properties. The extension of these researches has given much weight to the opinion that the electrochemical theory may be regarded as failing to account for the replacement of such bodies as hydrogen by chlorine, bromine, oxygen, etc. I do not know that as yet any direct evidence has been offered that the electrical character of an atom is not an essential quality, but one that changes with circumstances. It appears to be rather a matter of inference than of absolute demonstration.

It is the object of this Memoir to furnish such direct evidence, and to show that chlorine, the substance which has given rise to the discussions connected with the theory of substitutions, under the very circumstances contemplated, has its electro-chemical relations changed.

More than two years ago I brought before the British

Association some of the facts. The connection of these experiments with the discussion between the theory of substitutions and the electro-chemical theory is obvious.

Very recently M. Berzelius has published an important paper on the allotropism of simple bodies, the object of which is to point out that many of those bodies can assume different qualities by being subjected to certain modes of treatment. Thus carbon furnishes three forms—charcoal, plumbago, and diamond.

To a certain extent these views coincide with those which have offered themselves to me from the study of the properties of chlorine. They are not, however, altogether the same. M. Berzelius infers that elementary bodies can, as has been said, assume under varying circumstances different qualities. The idea which it is attempted to communicate in this Memoir is simply this, that a given substance, such as chlorine, can pass from a state of high activity, in which it possesses all its well-known properties, to a state of complete inactivity, in which even its most energetic affinities disappear. And that between these extremes there are innumerable intermediate points. Between the two views there is, therefore, this essential difference: from the former it does not appear what the nature of the newly assumed properties may be; from the latter they must obviously be of the same character, and differ only in intensity or degree—diminishing from stage to stage until complete inactivity results.

In the case of chlorine the same activity which is communicated by the indigo rays can also be communicated by a high temperature, or by the action of platinum.

The points which this Memoir is intended to establish are—

I. That chlorine gas can exist under two forms. In the same way that metallic iron can exist as active or passive iron, chlorine can assume the active or passive state.

II. Having established the fact of the allotropism of chlorine, I shall then show its connection with the theory of substitutions of M. Dumas, and how the most remarkable points in that theory may be easily accounted for.

The time, perhaps, has not yet arrived for offering a complete mechanical explanation of the assumption of an active or passive state. It may be remarked that a very trivial modification of our admitted views of the relation between atoms and their properties is all that is required to give a consistent explanation of every one of these facts. Instead of regarding the specific qualities of an atom as appertaining equally to the whole of it in the aggregate, we have merely to assume that there is a relation between its properties and its sides, and that any force which can make it change its position upon its own axis will throw it into the active or passive state. But this is nothing more than the well-known idea of the polarity of atoms.

Phenomena of the Decomposition of Water by Chlorine in the Rays of the Sun.

From the various facts which might be employed as offering the means of establishing the allotropism of chlorine, I shall select those which arise from an examination of the phenomena of the decomposition of an aqueous solution of chlorine by the rays of the sun.

For many years it has been known that an aqueous solution of chlorine undergoes decomposition by the action of the solar rays. Several of the most remarkable phenomena connected with this decomposition appear to have been overlooked. Among such may be mentioned the singular fact that chlorine which has been thus influenced by the sun has obtained the quality of effecting this decomposition subsequently, to a measured extent, even in the dark. Not to anticipate what I have to offer on this point, I shall now proceed in the first place to establish the various facts connected with the decomposition in question.

Having provided a number of small glass vessels, consisting of a bulb and neck of the capacity of from 1.5 to 2 cubic inches, I filled them with a solution of chlorine in recently boiled water, and inverted them in small glass bottles containing the same solution, as shown in Fig. 44. With these bulbs the following experiments were made:

I. An aqueous solution of chlorine does not decompose in the dark.

One of the bulbs was shut up in a dark closet, and kept there for a week, being examined from time to time. No decomposition was perceptible, for no gas collected in the upper part of the bulb.

Fig. 44.

II. An aqueous solution of chlorine decomposes in the light.

One of the bulbs was placed in a beam of the sun reflected into the room by a heliostat. For sixteen minutes no change was perceptible, then small bubbles of gas made their appearance; they increased in quantity for a time, but finally the speed of decomposition became uniform. On analysis by explosion with hydrogen, after washing out any chlorine contained in it, this gas was found to contain 97 per cent. of oxygen.

III. The rapidity of this decomposition depends on the quantity of the rays and on the temperature.

In various repetitions of these experiments on different days I soon convinced myself that the rate of evolution of the oxygen depended on the quantity of the rays. Among other proofs I may mention this: After

ascertaining the rate of decomposition in the reflected
beam, if the bulb be set in the direct sunshine the bubbles increase in number; the total quantity of oxygen
evolved becoming greater in the same space of time, an
effect obviously due to the difference of intensity of the
reflected and incident beams. When a certain point is
gained, apparently no further increase of effect takes
place on increasing the brilliancy of the light, as I found
by employing a convex lens.

With respect to the influence of temperature. If,
while one of the bulbs is actively evolving gas in the
sun-rays, it be warmed by the application of a spirit-
lamp, the amount of gas thrown off becomes very much
greater. A difference of a few degrees produces a striking effect. As an illustration of this I placed in the sunshine two bulbs which were nearly alike, except that
one of them was painted black with India-ink on that
portion which was farthest from the sun. The rays
coming through the transparent part had access to the
solution, and then, impinging on the dark side, raised its
temperature. On measuring the quantity of gas collected, it was found—

In the transparent bulb........................ 3.46
In the half-blackened bulb.................... 6.19

IV. The decomposition of water, once begun in the
sunbeams, goes on afterwards in the dark.

1st. This very important fact may be established in a
variety of ways. Thus, if a bulb be removed from the
sunshine while it is actively evolving gas, and be placed
in the dark after all the gas has been turned out of it, a
slow evolution continuously goes on, the gas collecting
in the upper part of the bulb.

2d. A bulb, A, Fig. 45, having a neck, b, the end of
which was bent at c upwards at an angle of about 45
degrees, was employed. After exposure to the sun, by

inverting the bulb, and with one finger closing the extremity c, the gas disengaged could be transferred to a graduated vessel and measured. I satisfied myself by several variations of this arrangement that the small quantity of water introduced from time to time when the gas bubble escaped from the end of the tube c exerted no essential influence on the phenomenon. The following table shows the amount of gas evolved in the dark during the periods indicated.

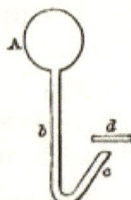

Fig. 45.

The bulb having been exposed to the sunshine, in ten minutes the evolution of gas commenced, and in an hour, 0.107 cubic inch having collected, this was thrown away and the arrangement placed in the dark. To prevent the undue escape of the chlorine, the flat piece of glass d was laid on the open end of the tube c. In each successive hour the quantity of gas given in the following table was then evolved:

First hour	0.0162
Second "	0.0159
Third "	0.0086
Fourth "	0.0060
Fifth "	0.0038
Sixth "	0.0031

And for four days afterwards gas was collecting in the bulb in diminished quantities.

V. This evolution of gas in the dark is not merely a gradual escape of oxygen, originally formed while the solution was exposed to the sun, but is traceable to an influence continuously exerted by the chlorine arising in properties it has acquired during its exposure to the rays.

If a bulb which has been exposed to the sun be raised by a spirit-lamp to such a temperature that its gaseous constituents are rapidly evolved, its extremity dipping beneath some of the solution in the bottle, after allowing

T

a sufficient space of time for the disengaged chlorine to be redissolved, and the oxygen be turned out of the bulb, it will be found on keeping the arrangement in the dark that oxygen will slowly disengage as before.

Now there is every reason to believe that any small amount of oxygen dissolved in the liquid would be expelled with the chlorine at a high temperature. We therefore have to infer that the chlorine after this treatment still retains the quality of causing the decomposition to go steadily forward.

The oxygen which thus accumulates in the course of time in the dark, after an exposure to the sun, does not arise from any portion of that gas held in a state of temporary solution, nor from peroxide of hydrogen, nor from chlorous acid in the liquid undergoing partial decomposition. From any of these states a high temperature would disengage it.

VI. The evolution of gas is not of the nature of a fermentation; for when it once sets in, the molecular motion is not propagated from particle to particle, but affects only those originally exposed to the rays.

Let a bulb be filled with chlorine-water which has been exposed to the sun, and in a second bulb place a quantity of the same liquid equal to about one third of its capacity. Fill up the remaining two thirds with chlorine-water which has been made and kept in the dark; and after keeping both bulbs in obscurity for some days, measure the volumes of gas they contain. If the qualities of chlorine which has been changed by exposure were communicable by contact or close proximity from atom to atom, we might expect that both the bulbs would yield the same quantity of gas; but this is far from being the case, and in such an experiment I found that the bulb containing the mixture gave only one fourteenth of the gas found in the other.

VII. The quantity of gas thus collecting in the dark depends on the intensity of the original disturbance, which in its turn depends on the time of exposure to the rays, to their intensity, and other such conditions. In other words, the rays are perfectly definite in their action—a long exposure giving a larger amount of subsequent decomposition, and short exposure a less amount.

On exposing a bulb filled with chlorine-water to the rays until bubbles of gas began to appear, and a second one until the decomposition had been actively going on for a quarter of an hour, and then transferring both to the dark and measuring the oxygen collected at the end of a day, I found in the former one twelfth of what was collected in the latter.

VIII. In a given quantity of chlorine-water, the decomposition in the dark corresponding to a given exposure to the light having been performed, and the proper quantity of oxygen evolved and the phenomenon ended, it can be re-established from time to time, as long as any chlorine is found in the liquid, by a renewed exposure to the sun.

In a glass vessel, like Fig. 46, which, indeed, was nothing more than one of Liebig's drying apparatus, I placed a sufficient quantity of chlorine-water to fill the larger vessel, and the vertical tubes half full. After exposing this to the light for a certain time,

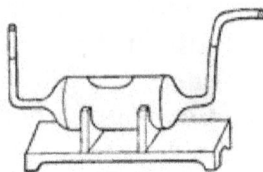

Fig. 46.

until decomposition had fairly set in, I placed it in the dark and found that for several days it gave off gas—the quantity continually diminishing. Finally, no more gas was evolved. But the liquid still contained free chlorine, as was shown by its color. I therefore again exposed it to the sun, and, repeating the former observation, found that it evolved gas for several days in the dark. A third exposure was followed by the same result.

The form of this vessel renders it very convenient for these experiments; because when sufficient gas has collected for the purpose of observation, it is easily removed by inclining the instrument, without the necessity of introducing fresh quantities of liquid.

Having found, as has been said, that the rapidity of the decomposition depended to a certain extent on the temperature, it seemed desirable to determine whether heat alone could bring about the change.

IX. The decomposition of water by chlorine is not brought about by mere elevation of temperature when the liquid is set in the sunbeam: although heat accelerates, it does not give rise to the phenomenon.

1st. I raised by a spirit-lamp the temperature of one of the bulbs nearly to its boiling-point, until so much gas was given off that all the liquid was expelled from the tube to the bottle beneath. If at this temperature, which probably was higher than 200° Fahr., chlorine had been able to decompose water, an equivalent quantity of oxygen would have been produced; but on allowing the apparatus to cool, all the gas was reabsorbed with the exception of a small bubble, amounting in volume to $\frac{1}{1087}$ of the water. This bubble, which was left after the chlorine was recondensed, I found in three different experiments contained 32, 33, and 36 per cent. of oxygen, the remainder being nitrogen; but this being nearly the constitution of the gas dissolved in ordinary water, the source from which the small bubble came was inferred to be the water used in these experiments.

2d. One of the bulbs was painted black all over with India-ink. Its temperature now rose much higher than in former experiments when it was set in the sun, but not a bubble of oxygen appeared.

X. When chlorine-water has been exposed to the sun, the oxygen accumulated in it is readily expelled by raising the temperature.

Having exposed one of the bulbs used in the last experiment until it was actively evolving gas, I raised its temperature with the spirit-lamp until the bulb was full of gas. But on cooling this gas did not all condense as in the last instance, a large quantity remained behind. This was oxygen.

These ninth and tenth facts are of further interest, as bearing upon a question much discussed by chemists—the nature of the bleaching compounds of chlorine. The chloride of lime, and other such substances, probably have the same theoretical constitution as chlorine-water. Berzelius and Balard suppose that in this solution chlorous or hypochlorous acid exists. It might be inquired, if this be the condition of things, why does not an exposure to heat alone evolve oxygen, for chlorous acid is exceedingly liable to decomposition by slight elevation of temperature, and we should be justified in inferring that if any of this acid is to be found in chlorine-water it would be decomposed at the boiling-point. M. Millon adopts the view that the bleaching compounds are metallic chlorides analogous to the corresponding peroxides. But the ninth fact seems incompatible with this view. If chlorine-water be analogous to peroxides of hydrogen, and this last be what its name imports, and not merely oxygenated water, it is difficult to understand why, when chlorine-water is thus boiled, oxygen is not given off. If the atom of chlorine and the atom of oxygen in this body are placed under the same relations to the atom of hydrogen, it seems necessary that the chlorine atom at 212° Fahr. should expel the oxygen atom and hydrochloric acid form. It is probable, indeed, that the two oxygen atoms in peroxide of hydrogen are re-

lated to their hydrogen atom with different degrees of
affinity, and that one of them is retained far more loosely
than the other.   But this would correspond to our ideas
of oxygenized water and not of peroxide of hydrogen,
and lead us to the conclusion that the solution employed
in this Memoir is strictly a solution of chlorine in water.

XI. The decomposition of chlorine-water, when placed
in the sunbeam, does not begin at once, but a certain
space of time intervenes, during which the chlorine is
undergoing its specific change.

I need quote no further instance of the truth of this
than the experiment given in support of the second fact.
This is the same phenomenon which takes place when
chlorine and hydrogen are exposed together; they do
not begin to unite at once, but a certain space of time
elapses, during which the preliminary absorption is taking place, and when that is over union begins.

On the Relations of Chlorine and Hydrogen.

We have thus traced the cause of the decomposition
of water, in the case before us, to a change impressed on
the chlorine by exposure to the rays of the sun.   In this
decomposition three elementary bodies are involved—
chlorine, oxygen, and hydrogen.

We can therefore reduce the problem under discussion
to simple conditions, and study the relations of each of
these substances to each other and to the solar rays successively.

When a mixture of oxygen and hydrogen gases, in the
proportion to form water, is exposed to the most brilliant
radiation converged upon it by convex lenses, union does
not ensue; the reason being, as I have shown, that those
gases are perfectly transparent to the rays, and do not
possess either real or ideal coloration.   For the same
cause, water exposed alone for any length of time to the

sun, or to the influence of a large convex lens, does not decompose. It is transparent, and cannot absorb any of the rays.

But, as is well known, a mixture of chlorine and hydrogen unites, under the same circumstances, with an explosion. I have formerly proved that this depends on the absorption of the indigo rays. For in the indigo space of the spectrum the action goes on with the greatest activity.

If, therefore, this phenomenon be due to absorption taking place by the mixture, it is easy to determine the function discharged by each of its ingredients.

I transmitted a ray of light through hydrogen gas contained in a tube seven inches long, the ends of which were terminated by pieces of flat glass; and then, dispersing the ray by a flint-glass prism, received the resulting spectrum on a daguerreotype plate. Simultaneously, by the side of it, I received the spectrum of a ray which had not gone through hydrogen, but through a similar tube filled with atmospheric air. On comparing the impressions together, I could find no difference between them.

I therefore infer that hydrogen gas does not exert any absorptive action on the solar rays.

In one of the foregoing tubes I placed dry chlorine gas, the other containing atmospheric air as before, and receiving the two spectra side by side on the same daguerreotype plate, I found that a powerful absorption had been exercised by the chlorine. All the chemical rays between the fixed line H and the violet termination of the spectrum were removed, and no impression corresponding to their place was left upon the plate. On repeating this experiment so as to determine with precision the rays which had been absorbed, I found that chlorine absorbs all the rays of the spectrum included between the fixed line i and the violet termination, and is probably affected by all those waves the lengths of which are between 0.00001587 and 0.00001287 of a Paris inch; and inasmuch as it absorbs luminous rays included between the same limits, it is to this absorption that its yellow color is due.

In these Memoirs the same result is established by me in another way. I found that a ray which had passed through a given thickness of a mixture of equal volumes of chlorine and hydrogen lost by absorption just half as much of its original intensity as when it passed through the same thickness of pure chlorine gas; a result which obviously leads to the conclusion that when chlorine and hydrogen unite under the influence of the sun, they discharge different functions—the chlorine an active, and the hydrogen a passive function. The primary action or disturbance takes place upon the chlorine, and a disposition is communicated to it enabling it to unite readily with the hydrogen.

By arranging in the spectrum a series of tubes containing a mixture of these gases, it was found that the gases placed in the indigo space went into union first.

These various experiments enabling me thus to trace to the chlorine the source of disturbance, I have next to remark that chlorine which has been exposed to the rays of the sun has gained thereby a tendency to unite with hydrogen not possessed by chlorine made and kept in the dark. In proof of this fact I may cite an experiment from Memoir XIX.:

"In two similar glass tubes place equal volumes of chlorine made from peroxide of manganese and hydrochloric acid by lamplight, and carefully screened from access of daylight Expose one of the tubes to the full sunbeams for some minutes, or, if the light be feeble, for a quarter of an hour: the chlorine in it becomes modified. Keep the other tube during this time carefully in a dark place; and now, by lamplight, add to both equal volumes of hydrogen gas. These processes are best carried on in a small porcelain or earthenware trough, filled with a saturated solution of common salt, which dissolves chlorine slowly, and to avoid explosions operate on limited quantities of the gases. Tubes that are eight inches long and half an inch in diameter will answer very well. The two tubes now contain the same gaseous mixture, and only differ in the circumstance that one is modified and the other not. Place them, therefore, side by side before a window, through which the entrance of daylight can be regulated by opening the shutter; and now, if this part of the process is conducted properly, it will be seen that the modified chlorine commences to unite with the hydrogen, and the salt water rises in that tube. But the unmodified chlorine shows no disposition to unite with its hydrogen, and the liquid in its tube remains motionless for a long time. Finally, as it becomes slowly modified by the action of daylight impinging on it, union takes place. From this, therefore, we perceive that chlorine which has been exposed to the sun will unite promptly and energetically with hydrogen; but chlorine which has been made and kept in the dark shows no such property."

This form of experiment may be supposed imperfect, since the chlorine is in a moist condition and confined by water. I have therefore made the following variation:

I took a tube, A, Fig. 47, six inches long and half an inch in diameter, closed at one end and open at the other, and cemented its open end on a piece of flat plate-glass, M N, one inch wide and two long, ground on both sides, and having a hole, p, one sixth of an inch in diam-

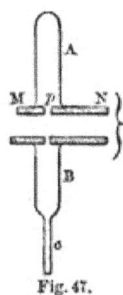

Fig. 47.

eter perforated through it. This hole was not in the centre of the glass, but towards one side, as shown in the figure. The interior of the tube was perfectly clean and dry.

A second tube, B, consisting, as shown in Fig. 47, of two portions; a wide portion, B, and a narrower tube, c, was cemented on another piece of ground plate-glass, similar to the foregoing in all respects. The tube c was open at its lower extremity, and the entire capacity of B and c conjointly was adjusted so as to be equal to the capacity of A.

Next I filled A with dry chlorine, and B c with dry hydrogen, and kept them from mixing until the proper time by operating in the following way: I placed the ground glasses face to face, as shown in Fig. 48, with a small quantity of soft tallow between them, arranging them in such a way that the aperture which led to the interior of A was open.

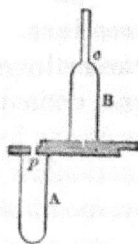

Fig. 48.

Through this aperture dry chlorine was conveyed. It was generated by a mixture of peroxide of manganese and hydrochloric acid in the flask D, Fig. 49, and passed along a tube, E, filled with chloride of calcium. A slender glass tube, f,

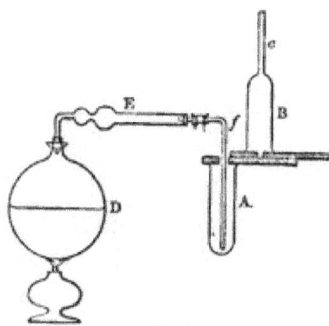

Fig. 49.

conveyed it to the bottom of A, which was then filled by displacing the atmospheric air. When A was supposed to be full of chlorine, it was slowly lowered so as to bring the tube out of the aperture, and as soon as it was disengaged the glass plates were moved in such a manner by sliding them on one another that the aperture leading into A was shut, but that leading into B was open. The vessel A was thus filled with dry chlorine and securely closed.

In the next place I filled B with dry hydrogen, which was done as follows: To a bottle, G, Fig. 50, containing dilute sulphuric acid and zinc, a drying-tube, K, of chloride of calcium was adjusted, and at its upper end a cork, h, arranged so as to receive tightly the tube c. In a short time, therefore, B became full of dry hydrogen, the surplus escaping through the open aperture p. The two ground-glass plates were now moved on one another in such a manner that they mutually closed one another. The vessel A was therefore filled with dry chlorine, and the vessel B c with an equal volume of dry hydrogen, without communicating for the present with one another.

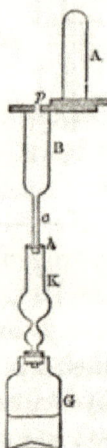

Fig. 50.

I had provided two sets of these tubes as nearly alike as they could be made, and operated with them in the following manner:

In a dark room I filled the tube A of each of them with dry chlorine in the manner just described, and confined it by sliding the plates. One of the tubes was retained in the dark room and kept carefully screened from the light, but the other was set for half an hour in the sunbeams. The chlorine which was in it underwent the specific change—the object of this Memoir to describe.

After restoring this tube to the dark room and waiting a few minutes for it to gain the same temperature as the other, the tubes B c of each set were filled with dry hydrogen in the manner described. In each instance, as soon as the plates were moved on each other

so as to confine the hydrogen, and they were released
from the cork h of the drying-tube K, Fig. 50, the lower
extremity of each was dipped beneath the surface of
some water contained in a saucer, P, Fig. 51; the two

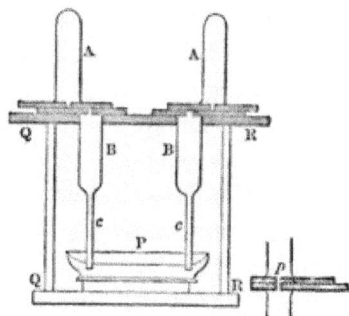

Fig. 51.

sets of tubes being held
steadily in a proper position by the aid of a wooden
frame, Q R. The tubes now
differed from one another
in nothing but the circumstance that the chlorine of
one had been exposed to
the sun, and that of the
other had not.

The gases were now brought in contact. This was
easily done by sliding each pair of ground glasses until
their apertures coincided, as shown at p in Fig. 51. The
hydrogen now rose through the hole into the upper vessel, the chlorine descending through it, mutual and perfect diffusion of the two gases rapidly taking place.
This was done by lamplight in the dark room. And
now it could be ascertained that the gases were at the
same temperature in the different tubes, and that the
experiment had thus far been carried on successfully, by
the water retaining its level at the same point in the
tubes c of both sets. If that which had been in the
sunshine was warmer than the other, as soon as the
apertures coincided a bubble of gas would have escaped
through the water, or at all events the level would have
changed.

It remained now to open the shutter of the dark room,
the tubes having been previously set in such a position
that the light would fall equally on both. As soon as
this was done, the chlorine which had been exposed to
the sun united at once with its hydrogen, and the water

rose in the tube c. But in the other, which had not been exposed to the sun, no movement took place until the gases had had time to be modified by the light coming through the open shutter.

When care has been taken to have the gases made quite dry, and, owing to the narrowness of the tube c, no aqueous vapor has had time to contaminate the gas in B, so that no water is present to condense the hydrochloric acid as it forms, a little delay may be occasioned in the liquid rising in the tube, the chlorine of which was exposed to the sun. But after a time a mist arises in the neighborhood of the water in the narrow tube, due to the hydrochloric acid condensing, and then the process goes forward with regularity.

It appears, therefore, that chlorine by exposure to the sun contracts a tendency to unite with hydrogen which is not possessed by chlorine which has been kept in the dark.

On the Allotropism of Chlorine, or its Passive and Active States.

In what, then, does this remarkable change impressed by indigo rays upon chlorine consist? This is the question immediately arising from the phenomena we have had under consideration.

To this I answer that when chlorine has been thus influenced its electro-negative properties are exalted, and it has passed from an inactive to an active state.

It is now fully established that a great number of the elementary bodies undergo similar modifications. Many of them can exist in no less than three different states, and these peculiarities are impressed on the compounds to which they give rise. To these peculiarities Berzelius directed the attention of chemical philosophers in his Memoir "On the Allotropism of Simple Bodies, and its Relation with Certain Cases of Isomerism in their Combinations." He shows that of the elementary bodies now known, many undoubtedly exist in several allotropic states, and infers that all are liable to analogous modifications. He indicates that the isomerism of compound bodies is due sometimes to the different modes in which the atoms of which their constituent molecules consist are grouped, and sometimes to the different allotropic states in which one or the other of those elements is found. Thus, as M. Millon has remarked, the intrinsic difference between carburetted hydrogen gas (CH) and ottar of roses (CH), which are isomeric bodies, may perhaps consist in this, that in the former the carbon is under the form of common charcoal, and in the latter under the form of diamond.

The following instances from Berzelius may serve as examples of these allotropic states:

Carbon is known under three forms—charcoal, plumbago, and diamond. They differ in specific gravity, in specific heat, and in their conducting power as respects caloric and electricity. In their relations to light, the first perfectly absorbs it, the second reflects it like a metal, the third transmits it like glass. In their relations with oxygen they also differ surprisingly; there are varieties of charcoal that spontaneously take fire in the air, but the diamond can only be burned with difficulty at a high temperature in pure oxygen gas. The second and third varieties do not belong to the same crystalline form.

Silicon exists also under two forms. In its first it burns with facility in the air under a slight elevation of temperature. But if it be previously exposed to a strong red heat it changes into the second variety and becomes incombustible, so that it will not oxidize when placed with nitrate of potash in the hottest part of a blowpipe flame. As is well known, there are two forms of silicic acid: one soluble in water and hydrochloric acid, but passing into the insoluble state by being previously made red-hot. The silicon therefore carries in its combination the same properties that it exhibits in the free state.

In the same manner it might be shown that sulphur, selenium, phosphorus, titanium, chromium, uranium, tin, iridium, osmium, copper, nickel, cobalt, and a variety of other bodies, exist under several different forms, with distinctive properties that are often well marked. In several of them the influence of this allotropic condition is plainly carried into the compounds, as is well shown in the two varieties of arsenic which give rise to the two arsenious acids.

The passage from one allotropic state to another takes place commonly through the agency of apparently very trivial causes, such as a slight elevation of temperature and the contact of certain bodies. Thus iron, which is so easily oxidized under ordinary circumstances, appears to lose its affinity for oxygen after it has been touched under the surface of nitric acid by a piece of platinum. It then puts on the attributes of a noble metal, and simulates the properties of platinum and gold.

This remarkable instance of the passage from an active to a passive state, as Berzelius remarks, may lead to a conjecture respecting the true condition of certain gases. No one can reflect on the inactivity of nitrogen gas under ordinary circumstances, contrasted with its equally extraordinary activity as a constituent of organic bodies, without being struck with the apparent connection of that phenomenon with these of allotropism. And though Berzelius with his customary caution merely insinuates that nitrogen can exist under two forms, the facts here developed in relation to chlorine appear to

show that that opinion rests on something more solid
than conjecture.   The habitudes of many of the gaseous
bodies strengthen this conclusion.   Oxygen refuses to
unite when mixed with hydrogen precisely in the manner of chlorine, and it requires a certain modification to
be made in the electro-negative element before water or
hydrochloric acid can result.

Just, therefore, in the same manner that so many elementary bodies can put on under the influence of external causes an active or passive condition, I infer, as the
result of the experiments brought forward in this Memoir, that chlorine is one of these allotropic bodies, having
a double form of existence.   That, as commonly prepared,
it is in its passive state; but that on exposure to the
indigo rays or other causes it changes and assumes an
active form.   That, in this latter state, its affinity for
hydrogen becomes so great that it decomposes water
without difficulty, as in the experiment which this Memoir is designed to illustrate.

On the Relation of the Preceding Conclusions with the Theory of Substitutions.

Having thus explained the facts which appear to indicate the allotropism of chlorine, I shall now offer some
considerations on its connection with the theory of substitutions of M. Dumas.

Admitting the fact that the electro-negative qualities
of chlorine are exalted upon its exposure to the indigo
rays, and that the resulting effect is not a temporary
change, but one lasting for a considerable period of time,
we can give a very plain and simple account of the decomposition of water by this gaseous substance under
the influence of sunshine.

On the same principle that a mixture of chlorine and
hydrogen may be kept in the dark without union for a

long time, so may a solution of chlorine in water be preserved. The chlorine is in an inactive state.

But if anything be done to make the chlorine take on
its other form and pass to the active condition—if it be,
for example, set in the sunshine—its affinity for hydrogen is exhibited and decomposition is the result.

The qualities thus communicated to the chlorine not
being of a transient kind, but remaining for a length of
time, we see how it is that after an exposure to the sun
decomposition is subsequently carried forward in the
dark.

The indisposition of chlorine to unite with carbon,
which has been regarded as a singular quality, is not
more remarkable than its indisposition to unite with hydrogen in the dark.

If the power assumed by chlorine of uniting with hydrogen and carbon depends on a change in its electrical
relations—a passage from the passive to the active state
—we might expect that those various causes which in
the case of other elementary bodies bring about analogous changes, and throw them from one allotropic condition to another, would here also exercise a perceptible
action. Among such causes we may enumerate the action of a high temperature and the contact or presence
of other bodies.

It may be remarked, in the instances to which Berzelius has referred, that exposure to a high temperature is
one of the most frequent causes of allotropic change. In
the case of chlorine the remark holds good, for, as is
well known, when a mixture of chlorine and hydrogen is
passed through a red-hot tube, hydrochloric acid forms
with rapidity. The high temperature, therefore, impresses on chlorine the same tendency to unite with hydrogen
which is communicated by the solar rays.

But the contact of other bodies frequently determines

U

in a given substance an allotropic change. Thus, when a piece of iron is placed in nitric acid in contact with platinum, the iron becomes less electro-positive, or, what is the same thing, more electro-negative, than it was before, and the acid can no longer oxidize it. The contact of the very same substance, platinum, determines an analogous change in chlorine—giving it at once the capacity of uniting with hydrogen. The porous condition of spongy platinum is not essential to the result, for clean platinum foil exhibits the same phenomenon.

In the case of iron, the action of a high temperature or the contact of platinum throws the metal from the active to the passive state; in the case of chlorine the same causes apparently produce the opposite result, throwing the gas from the passive to the active state. But the difference is rather in appearance than in reality. In both cases it amounts to the same thing, and is an exaltation of the electro-negative qualities of either substance respectively.

The same causes, therefore, which produce allotropic changes in other bodies produce analogous changes in chlorine.

Now, among the physical facts connected with the theory of types and substitutions, two are prominent: 1st. The union of chlorine with hydrogen, giving rise to the removal of that hydrogen as hydrochloric acid. 2d. The subsequent function discharged by the chlorine, which has entered as an integrant portion of the molecules, and occupies the place of the hydrogen removed. This function is in many instances that of the hydrogen itself, and it is this fact which is the remarkable point in the phenomena of substitution — that an intensely electro-negative body can act the part of a positive body. It is this fact which is leading chemists to the conclusion that the properties of compound bodies arise as much from the mode of grouping of their constituent atoms as from the qualities of those atoms themselves.

But, if it be admitted that the experiments related in this Memoir establish the allotropism of chlorine, then it is plain that a very different and perhaps satisfactory account of the phenomena of substitution may be given.

As has been already said, no difficulty can arise in accounting for the removal of hydrogen from organic bodies, or for the first fact just alluded to. This removal will ensue whenever processes are resorted to which bring the chlorine into an active state. When we expose acetic acid and chlorine to the sun, the latter becomes active, gains the quality of uniting with hydrogen, and chlor-acetic acid forms. Probably the same change could be brought about by the aid of spongy platinum and heat.

Upon the second fact—the similarity of function discharged by the chlorine which has replaced the hydrogen atoms with the function of those atoms themselves— a flood of light is thrown by other phenomena of allotropism. If a piece of iron be dipped in hydrated nitric acid, though it may be acted on for a few moments, it rapidly becomes passive. And so with the chlorine atoms which have substituted the hydrogen. In the circumstances in which they are placed they rapidly revert from the active to the passive state. They are no longer endued with an intense electro-negative quality —they have assumed the condition of inactivity. The fact that chlorine in chlor-acetic acid simulates the functions of hydrogen in acetic acid is not more remarkable than that iron touched by platinum under nitric acid simulates the properties of that noble metal.

Do not, therefore, these circumstances seem to point out that if we admit the fact that simple substances can exist in different states, in a passive and an active form, the phenomena of substitution are deprived of much of their singularity.

Thus, to recall once more the example to which I have before referred, and which has been so well illustrated by the researches of M. Dumas, the transmutation of acetic into chlor-acetic acid exhibits a double phenomenon. 1st. The existence of active chlorine, expressed by the removal of hydrogen, activity having been communicated by the rays of the sun, or by some other appropriate method. 2d. The existence of passive chlorine in the particles of chlor-acetic acid.

I consider that, were no other instances known, the two cases cited by Berzelius of the double forms of silicic acid and arsenious acid establish the fact that a given allotropic condition may be continued by an elementary atom when it goes into union with other bodies. And I regard the various cases in which hydrogen is replaced by iodine, bromine, etc., in which, in the resulting compound, those energetic electro-negative elements fail to give any expression of their presence and activity, as analogous to other common and too much overlooked facts. Chlorine which is in the dark may be kept in contact with hydrogen without exhibiting any of its latent energies. Touched by an indigo ray, it instantly assumes the active state, and a violent explosion is the result.

To use, therefore, the same nomenclature to which Berzelius has resorted in the case of other allotropisms, we may designate the ordinary form of chlorine made by the action of hydrochloric acid on peroxide of manganese as $Cl\beta$, and admit that this passes into the condition $Cl\alpha$ by the action of the solar rays, contact of platinum, or a high temperature; and that in any case of substitution the hydrogen is removed under the condition $Cl\alpha$, and the resulting compound contains $Cl\beta$, the assumption of the passive state disguising the presence of the electro-negative atom.

The explanation here given of the phenomena of substitutions involves the position that chlorine when brought in relation with carbon under certain circumstances is thrown into the passive state, the state Clβ. We naturally look for direct evidence that this is the case. It seems to me that there are many well-known chemical facts tending to establish the passive condition. In the first case to which we turn, the chlorides of carbon, the inactive state is established in a striking manner. The affinity existing between chlorine and carbon is apparently feeble; yet when these bodies have once united the chlorine is brought into such a condition that it has lost the quality of being detected by the ordinary tests which determine its presence. How strongly does this contrast with the case of hydrochloric acid! A feeble affinity unites carbon and chlorine, an intense affinity unites hydrogen and chlorine; yet in the former case the chlorine is undiscoverable by the commonest tests, in the latter it yields to them all. And the causes are obvious; in the one case it is in the passive, in the other in the active condition.

I have hitherto spoken of the active and passive states as though they were fixed points in elementary bodies, and as though the transition from one to the other were abrupt and sudden. I have done this that the views here offered might be unembarrassed and distinct. But there are many facts which serve to show that the passage from a state of complete activity to a state of complete inactivity takes place through gradual steps. Thus, in carbon itself, there are undoubtedly many intermediate stages between the almost spontaneously inflammable varieties and diamond, which, under common circumstances, is incombustible. Berzelius ad-

mits three allotropic conditions of this body, Ca, $C\beta$, $C\gamma$.
Between the first and last terms of this series it is probable that several intermediate bodies besides plumbago
might be found, their existence establishing the gradual
passage from one to the other state.

For similar reasons, in this Memoir the illustrations
and arguments given have for the most part been restricted to one subject, chlorine. It need scarcely be
pointed out, in conclusion, that if the views here offered
be true, very much of this reasoning may be transferred
to other bodies, as oxygen, nitrogen, hydrogen, sulphur,
etc. When oxygen and hydrogen are mixed, there is no
disposition exhibited by them to unite; and this does
not arise from their happening to have the gaseous form.
As in the instance we have been considering, if they are
exposed to a high temperature, or to the influence of
platinum, the active condition is assumed with promptitude, and union takes place.

The power possessed by carbon of throwing bodies
into a completely passive state is far from being limited
to chlorine. It reappears in the case of sulphur. The
sulphide of carbon yields to none of the tests to which
we commonly resort for determining the presence of sulphur, for the simple reason that its sulphur is in an inactive state. This substance, moreover, serves to illustrate
what has been said of the gradual passage of bodies
from a state of complete activity to one of complete inactivity. Berzelius recognizes for it three different allotropic states—an alpha, beta, and gamma condition. In
none of these is it in that condition of absolute inactivity which it assumes in the sulphide of carbon.*

---

* For these examples, the chloride and sulphide of carbon, I am indebted to M.
Millon's paper, "Remarks on the Elements which Compose Organic Substances, and
on their Mode of Combination," in the *Comptes Rendus*, t. xix., p. 799. That chemist, however, gives a very different explanation of the phenomena involved.

In offering these experiments and arguments to the consideration of chemists, I am fully aware of the magnitude of the change which would be impressed on the science generally, and especially on several modern theories, by their reception. The long-established idea of the immutability of the properties of elementary bodies would, to a certain extent, be sacrificed; and it is probable that before these results are conceded more cogent evidence of the main principle will be required. In the meantime, however, it is plain that the admission of these doctrines throws much light on theories now extensively attracting attention, and for that reason they commend themselves to our consideration. I have offered no opinion here on the atomic mechanism involved in these changes from an active to a passive state, though it is impossible to deal with these things without the reflection arising in our minds that here we are on the brink of an extensive system of evidence connected with the polarity of atoms—an idea which, under a variety of forms, is now occurring in every department of natural philosophy.

UNIVERSITY OF NEW YORK, *July* 20, 1845.

MEMOIR XXI.

ON THE INFLUENCE OF LIGHT UPON CHLORINE, AND SOME
REMARKS ON ALCHEMY.

From the Philosophical Magazine, November, 1857.

CONTENTS :—*Modification of chlorine by the sun-rays.—The modification is not transient.—Alchemical attempts to modify metals.—Exposure of silver chloride to a burning-lens.—The resulting silver is not acted upon by nitric acid.*

SEVERAL years ago I observed that when a mixture of chlorine and hydrogen is exposed to light, union does not occur at once, but that a certain time must elapse during which absorption takes place, the combination then proceeding in a uniform manner.

It is by the chlorine that this absorptive agency is exercised, the indigo ray being mainly concerned. And not only is it that ray which is thus absorbed : to it must be attributed also the subsequent combination.

Among several other facts connected with this subject, which may be found in the *Philosophical Magazine* (July, 1844), *The American Journal of Science*, Vol. XLIX., and other publications of that time, there is one to which I would particularly direct attention. Chlorine which has been exposed to the sun has obtained properties not possessed by chlorine which has been made and kept in the dark, and the change is by no means transient. It lasts for many hours, and even days.

In their recent examination of this fact, Professor Bunsen and Dr. Roscoe do not appear to regard the modification in question as being of so permanent a nature.

Perhaps it may have been that the insolation to which they submitted the chlorine was not continued sufficiently long, or perhaps the light was not sufficiently intense. My opinion was founded on three different conditions of the experiment—1st. On the behavior of chlorine itself confined over salt water; 2d. On the effect of a mixture of chlorine and hydrogen in equal volumes as disengaged from hydrochloric acid by a voltaic current; 3d. On the action of a solution of pure chlorine in distilled water. In each of these instances, the active properties imparted to the chlorine by exposure to light were plainly perceptible for a long time after. Indeed, I infer from the experiments of those chemists that they found the effects to continue for a certain brief period. If they do so continue, though only in a momentary manner, after the light has been shut off, I do not see in what other manner we are to explain the result than on the principle of a change in the relations of the chlorine. In this interpretation it is very well known that Berzelius coincided in his account of my experiments in the *Annual Report* for 1847.

At first I thought that there was a general analogy between the case of chlorine thus thrown into an active state and that of iron in its passive condition. An iron wire which has been made passive will quickly revert to the condition of activity if submitted to any jarring, vibration, or other trivial disturbance; its passive state being in one sense permanent, though very easily lost. But subsequently I found many reasons for supposing that the impression is of a much more lasting nature, and resembles that on phosphorus after a similar exposure to the indigo rays. As an illustration of what is here meant, I may relate that having obtained a thin stratum of perfectly white phosphorus between two pieces of glass, I exposed it to a motionless solar spec-

trum, and found that it turned to a dark-brown color in
those spaces on which the more refrangible rays fell, the
effect reaching a maximum under the indigo ray.   The
fixed lines of Fraunhofer were very prettily depicted, as
white streaks, particularly the large ones at H.   I kept
the sample of phosphorus for several years without its
showing any disposition to resume the active state.

Professor Bunsen and Dr. Roscoe dwell very appro-
priately on the disturbing effects of minute quantities of
extraneous gases mingled with chlorine on photo-chemical induction.   No one who has used a chlor-hydrogen
photometer can have failed to make a similar remark.
My attention has been directed to that subject in its
more general aspect, and I will ingenuously confess that
I have made several attempts at the transmutation of
metals, on the principle of compelling them, by the aid
of solar light, to be disengaged from states of combination in the midst of resisting or disturbing media.

The following is a description of one of these alchemical attempts: In the focus of a burning-lens, twelve
inches in diameter, was placed a glass flask, two inches
in diameter, containing nitric acid diluted with its own
volume of water.   Into the nitric acid were poured alternately small quantities of a solution of nitrate of silver
and of hydrochloric acid, the object being to cause the
chloride of silver to form in a minutely divided state, so
as to produce a milky liquid, into the interior of which
the brilliant converging cone of light might pass, and the
currents generated in the flask by the heat might drift
all the chloride successively through the light.   The
chloride, if otherwise exposed to the sun, blackens merely upon the surface, the interior parts undergoing no
change; this difficulty I therefore hoped to avoid.   The
burning-glass promptly brings on a decomposition of the
salt, evolving on the one hand chlorine, and disengaging

a metal on the other. In one experiment the exposure lasted from 11 A.M. to 1 P.M.; this was, therefore, equal to a continuous mid-day sun of seventy-two hours. The metal was disengaged readily. But what is it? It cannot be silver, since nitric acid has no action upon it. It burnished in an agate mortar, but its reflection is not like the reflection of silver: it is yellow. The light must, therefore, have so transmuted the original silver as to enable it to exist in the presence of nitric acid. In 1837 I published some experiments on the nature of this decomposition in the *Journal of the Franklin Institute.*

Though this experiment, and several modifications of it which I might relate, fail to establish any permanent change in the metal under trial, in the sense of an actual transmutation, it does not follow that we should despair of final success. It is not likely that Nature has made fifty elementary substances of a metallic form, many of them so closely resembling each other as to be with difficulty distinguished. Moreover, chlorine and other elementary substances can be changed by the sunlight in some respects permanently; and if silver has not thus far been transmuted into a more noble metal, as platinum or gold, it has at all events been made transiently into something which is not silver. Those who will reflect a little on the matter cannot fail to observe that the sun-rays really possess many of the powers once fabulously imputed to the powder of projection and the philosopher's stone.

•

MEMOIR XXII.

ON THE ACTION OF GLASS AND QUARTZ ON THE RADIATIONS THAT PRODUCE PHOSPHORESCENCE.

From the Philosophical Magazine, Aug., 1844.

CONTENTS: — *Phosphorescence by rays from a Leyden spark through quartz.—From the voltaic arc.—It is occasioned by the more refrangible rays.—Imperviousness of glass.—Professor Henry's experiments.—Comparison of the chemical and phosphorogenic rays.*

AT the distance of six inches from the terminations of two blunt wires, between which the spark from a Leyden-jar was caused to pass, I placed a lens of quartz, the focus of which for parallel rays was six inches, and then intercepted the resulting beam by a diaphragm, with a circular aperture in it one third of an inch in diameter. I had caused an equiangular prism of quartz to be cut and polished from a large and faultless crystal; it was cut transverse to the axis. This prism I placed in such a way that, in dispersing the beam coming through the circular aperture, I got rid of double refraction and obtained only one spectrum; this was received on a metal plate, which, having been washed over with gum-water and sulphide of lime dusted on it, offered a uniform phosphorescent surface which might be set in a vertical plane. When the spark passed, I saw that the plate was phosphorescing on those portions where the more refrangible rays had fallen.

But the transient light of a Leyden spark did not last long enough, nor was the phosphorescence it produced powerful enough to enable me to conduct the experiment in a way entirely satisfactory. I resorted, therefore, to the brilliant light obtained when a piece of metal, or, what is better, the hard variety of carbon obtained from gas-retorts, is lowered upon mercury entirely filling a very small open porcelain cup, and the continuous discharge of a voltaic battery passed. The battery used contained fifty pairs of Grove's cells, but a smaller number would have been amply sufficient.

As soon as the light was emitted, I marked on the lime sulphide the beginning of the red, the centre of the yellow, and the termination of the visible violet ray. Then, stopping the current, I examined on what parts the plate was phosphorescing. The commencement of the glow was between the indigo and the blue; towards the blue it extended far beyond the visible boundaries of the spectrum. I could not see any divisions or points of maximum in it. The surface of the plate shone all over except in the region of the less refrangible rays, and there were traces of the negative action which M. Becquerel has illustrated in the cases of the solar emanations—rays which, however, were first observed in the last century.

It is necessary to remark that the rays from the voltaic discharge resemble those from an electric spark in their inability to pass through glass. On this observation all the value of the foregoing experiment depends.

But it can nevertheless be easily proved that although glass is impervious to the phosphorogenic emanation coming from the voltaic deflagration of any metallic bodies, the observation applies to transient discharges only. A voltaic light which lasts but a moment fails to cause phosphorescence through glass in the same way that an electric spark does; but if the discharge be continued the surface presently begins to glow, and if maintained for several minutes it shines as brightly as though a piece of quartz had been used.

The inability of an electric spark to cause phosphorescence is connected with its transient duration. The
voltaic light enables us at pleasure to imitate the effects
of an electric spark or those of the sun. The phosphorogenic rays, whether they originate in an electric spark
or from the sun, occupying thus the same place in the
spectrum, and even exhibiting the same peculiarities as
the chemical rays on iodide of silver, we have next to determine whether this is an apparent or a positive identity.

Professor Henry, of Princeton, read a paper before the
American Philosophical Society, in May, 1843, in which
he discussed all the leading mechanical properties of the
phosphorogenic rays, and, among other important experiments, made some with a view of determining this particular question. A daguerreotype plate and some lime
sulphide were simultaneously exposed to the sky: the
plate was stained, but no effect was produced on the
lime. A daguerreotype plate and some lime sulphide
were exposed to the light of an electric spark: the lime
was observed to glow, but no impression was produced
on the plate. When the plate was exposed to a succession of sparks for ten minutes with a sheet of mica interposed, an impression was made. Lime exposed to the
moon did not phosphoresce, but a sensitive plate under
the circumstances is said to be stained. In view of
these different facts, Professor Henry observes: "These
experiments, although not sufficiently extensive, appear
to indicate that the phosphorogenic emanation is distinct
from the chemical, and that it exists in a much greater
quantity in the electric spark than either the luminous
or chemical radiation."

From Wilson's experiments it appears that he was
aware that when the phosphorescent surface is warmed
so as to hasten the disengagement of light, the moonbeams may be found to have left traces of action upon

it, feeble, it is true, but nevertheless very apparent. We have seen, also, that the peculiarity of an electric spark is due to its transient duration. Before, therefore, a final decision can be obtained on this point, we are required to examine the effects of the chemical rays and phosphorogenic emanations under circumstances which are precisely similar as to intensity and time.

For the transient rays of an electric spark, quartz is transparent and glass is nearly opaque. Having prepared a bromo-iodized silver plate so as to be exceedingly sensitive, I set in front of it at the distance of about one third of an inch a disk of quartz and one of crown-glass of equal thickness, and between a pair of copper wires, the interval of which was three eighths of an inch, I passed the spark of a Leyden-jar fifteen times; the distance between this spark and the sensitive plate was about two inches. On mercurializing this plate, it was deeply whitened all over, equally so through the glass, through the quartz, and on the uncovered spaces; but a spot of sealing-wax which I had put on the glass left its shadow on the plate beautifully depicted, so also were the edges of the glass and the quartz. The two disks overlapped one another to a certain extent, but the corresponding portion of the silver plate was as deeply stained there as anywhere else.

Next I put a surface of lime sulphide in the place of the daguerreotype plate, everything else remaining as before. On passing fifteen sparks the lime phosphoresced powerfully under the quartz, but not under the glass, so that the difference between its shadow and that of the spot of wax could not be distinctly seen.

For these reasons, therefore, I adopt the view expressed by Professor Henry, that the phosphorogenic emanation and the chemical rays are distinct. Under the same circumstances glass is transparent to the one, to the other it is opaque.

.

.

MEMOIR XXIII.

ON A REMARKABLE DIFFERENCE BETWEEN THE RAYS OF INCANDESCENT LIME AND THOSE EMITTED BY AN ELECTRIC SPARK.

From the Philosophical Magazine, December, 1845.

CONTENTS :—*Non-permeability of glass to spark radiations.—Permeability to calcium-light radiations.—Different refrangibility of the spark and the calcium-light radiations.—Shadows imbedded in Canton's phosphorus.—Evolution of old shadows in an order of succession.*

SOME years ago M. Becquerel discovered that the rays of an electric spark, if transmitted through a screen of glass, could not excite the phosphorescence of lime sulphide.

To make this experiment, wash a metallic plate over with gum-water. Dust upon it from a fine sieve a quantity of Canton's phosphorus (oyster-shells calcined with sulphur), and allow the plate to dry. A uniform surface is thus obtained suitable for these purposes. Place before that surface a piece of glass and a piece of polished quartz, and discharge a Leyden-jar a few inches off, so that the rays of its spark may fall on the plate. It will be found that under the quartz the phosphorus will shine as much as on the spaces that have not been covered, but under the glass it will remain almost entirely dark.

Last winter I observed the curious fact that when this experiment is made with a piece of lime incandescing in a stream of oxygen directed through the flame of a spirit-lamp, the glass, so far from being unable to transmit the rays, appears to be as transparent to them as quartz or atmospheric air.

A screen of glass is opaque to the phosphorogenic rays of an electric spark, but it is quite transparent to those of incandescent lime.

It might be supposed that the very brief duration of an electric spark has something to do with this phenomenon, but the voltaic arc passing between charcoal points gives the same results. I caused its rays to impinge on a plate for thirty seconds, and observed the obstructing effect of glass in a very satisfactory manner. That a certain portion of the rays passes through may be shown by continuing the light for a few minutes, when the phosphorus will begin to shine under the glass.

I came to the conclusion, also, that the transient duration of the light had nothing to do with the phenomenon, because the lime light occasions phosphorescence through glass in the space of a single second, but in that time the rays from a voltaic arc could not traverse a piece of glass so as to produce a sensible effect; the phosphorus beneath it appearing quite dark, and yet this light is incomparably brighter than the lime light.

The blue light emitted when a platinum wire in connection with one pole of the voltaic battery is brought down upon some mercury in connection with the other, and the green light obtained when the copper wires are the medium of discharge, appear to produce the same effect as charcoal points.

It is therefore neither the color nor the duration of the light that determines this result. It seems to depend on a peculiarity of the electric discharge.

Some time ago I determined the refrangibility of the rays of an electric spark which excite phosphorescence in lime sulphide; they are found at the violet extremity of the spectrum. I have made attempts to ascertain the

X

position of the active rays of incandescent lime. They cannot pass through the blue solution of ammonio-sulphate of copper, but through the red solution of sulphocyanide of iron, and also through a strong solution of bichromate of potash, they pass; in the latter case almost as copiously as through atmospheric air.

The phosphorogenic rays of an electric spark are in the violet space, but those of incandescent lime are at the other extremity of the spectrum.

An Argand lamp, when made to burn very brightly, emits phosphorogenic rays which traverse glass. As has been proved long ago, the sun-rays possess the same property.

Thus, therefore, the rays of incandescent lime, of an oil-lamp, and of the sun can excite phosphorescence through glass, and differ from those of an electric spark or voltaic discharge, in which that peculiarity is deficient.

I have also remarked some curious cases of spectral appearances. They are analogous to those instances to which I first drew public attention in 1840, and which at a later date M. Moser brought before the British Association. They are interesting as affording an ocular proof of secondary radiation. The following experiments may serve as illustrations:

Place a key or any other opaque object before a sensitive phosphorescent surface, and having made that surface glow intensely by a voltaic discharge between charcoal points continued for two or three minutes, on removing the key an image of it will of course be seen. This image in a short time will disappear. Then shut the plate in a dark place, where no light can have access to it in the daytime. If in a day or two the surface be carefully inspected in the dark, no trace of anything will be visible upon it; but if it be laid on a piece of warm iron, a spectral image of the key is suddenly evolved.

It is still more curious that a number of these latent images may co-exist on the same surface. Provide a phosphorescent surface on which the latent image of a key impressed a day or two before by a voltaic discharge is known to exist. Take some other object, as a metal ring, and setting it before the surface, discharge at a short distance a Leyden-jar. The phosphorus shines all over, save on those portions shaded by the ring; it exhibits, therefore, an image of that body. This image soon fades away and totally disappears. Set the plate now upon a piece of warm iron; it soon begins to glow, and the image of the ring is first reproduced, and as it declines away the spectral form of the key gradually unfolds itself, and after a time it totally vanishes.

A series of spectral images may thus exist together on a phosphorescent surface, and after remaining there latent for a length of time, they will come forth in their proper order on raising the temperature of the surface.

The idea that phosphorescence is merely the light of an electric discharge from particle to particle seems to me wholly incompatible with such results.

MEMOIR XXIV.

ON THE ELECTRO-MOTIVE POWER OF HEAT.

From the Philosophical Magazine, June, 1840; Harper's Magazine, No. 328.

CONTENTS:—*Experimental arrangement to determine the electro-motive power.—Temperatures calculated from quantities of electricity.—Increase of tension with increase of temperature.—Depends on increased resistance to conduction.—Quantity of electricity independent of heated surface.—In thermo-electric piles the quantity of electricity is proportional to the number of pairs.—Best forms of construction of thermo-electric pairs.*

FROM the Memoir of M. Melloni "On the Polarization of Heat," inserted in the second part of the first volume of Taylor's "Scientific Memoirs," we learn that M. Becquerel, as well as himself, had made experiments to determine the quantities of electricity set in motion by known increments of heat. From these experiments they conclude that through the whole range of the thermometric scale those quantities are directly proportional to each other.

But as thermo-electric currents are now employed in a variety of delicate physical investigations, and as there appears to be much misconception as to their character, I propose in this Memoir to show—

1st. That equal increments of heat do not set in motion equal quantities of electricity.

2d. That the tension undergoes a slight increase with increase of temperature, a phenomenon due to the increased resistance to conduction of metals when their temperature rises.

3d. That the quantity of electricity evolved at any

given temperature is independent of the amount of heated surface; a mere point being just as efficacious as an indefinitely extended surface.

4th. That the quantity of electricity evolved in a pile of pairs is directly proportional to the number of the elements.

First, then, as to the comparative march of electric development with the rise of temperature in the case of pairs of different metals.

The experimental arrangement which I have employed is represented in Fig. 52. A A is a glass vessel, about three inches in diameter, with a wide neck, through which can be inserted a mercurial thermometer, *b*, and one extremity of a pair

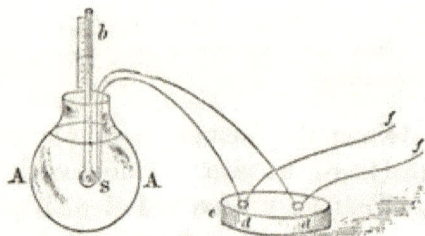

Fig. 52.

of electro-motoric wires. The wires I have employed have generally been a foot long and one sixteenth of an inch in diameter. The extremities, S, of the wires thus introduced into the vessel ought to be soldered with hard solder; their free extremities dip into the glass cups *d*, *d*, filled with mercury, and immersed in a trough, *e*, containing water and pounded ice. By means of the copper wires *f f*, one sixth of an inch thick, communication is established with the mercury-cups of the galvanometer. The coil of this galvanometer is of copper wire one eighth of an inch thick, and making twelve turns only round the needles, which are astatic. The deviations were determined by the torsion of a glass thread.

It is surprising to those who have never before seen the experiment with what promptitude and accuracy a

copper and iron wire soldered thus together will indicate temperatures.

In the arrangement now described, when an experiment has to be made, the vessel A A is to be filled
two thirds full of water, the bulb of the thermometer
being so adjusted as to be in its middle, the soldered
extremity S of the two wires being placed in contact
with it, and a small cover with suitable apertures adjusted on the top of the vessel, so that the steam as it
is generated may rush up alongside of the tube of the
thermometer and bring the mercurial column in it to
a uniform temperature. If the extremity of the thermo-
electric pair be allowed to rest on the bottom of the
glass vessel, no accurate results can be obtained; the
pair does not then indicate the temperature of the
water. The communicating wires ff are then placed in
the cups, the trough e filled with water and pounded ice,
and carefully surrounded with a flannel cloth. The water in the vessel A A is then gradually raised to the
boiling-point by means of a spirit-lamp, and kept at that
temperature until the galvanometer needles and the thermometer are quite steady. The same plan must be followed when any other temperature than 212° is under
trial, for the thermo-electric wires changing their temperature more rapidly than the mercury in the thermometer, it is absolutely necessary to continue the experiment for some minutes to bring both to the same state
of equilibrium.

When a temperature higher than 212° Fahr., but under a red heat, is required, I substitute in place of the
vessel A A a tabulated retort, the tubulure of which is
large enough to allow the passage of the bulb of the
thermometer and the wires. A quantity of mercury,
sufficient to fill the retort half-full, is then introduced,
and the tubulure being closed by appropriate pieces of

soapstone, the neck of the retort is inclined upwards, so that the vapor as it rises may condense and drop back again without incommoding the operator. As in the former case, it is here also necessary to continue each experiment for a few minutes, to bring the thermometer and thermal pair to the same condition. There is not much difficulty in obtaining any required temperature, by raising or lowering the wick of the lamp.

The metals I have tried were in the form of wires. They were in the state found in commerce, and therefore not pure; they were obtained in the shops of Philadelphia.

TABLE I.

Names of the Pairs of Metals.	Temperatures (Fahr.).				
	32° F.	122° F.	212° F.	662° F.	
Copper and iron............	0	93	176	233	Quantities of Electricity.
Silver and palladium......	0	65	147	613	
Iron and palladium........	0	112	223	631	
Platinum and copper......	0	11	26	122	
Iron and silver.............	0	89	137	244	
Iron and platinum.........	0	28	56	248	

In this table I have estimated the temperature of boiling mercury at 662° Fahr. The quantities of electricity evolved, as estimated by the torsion of a glass thread, are ranged in columns under their corresponding temperatures. Each series of numbers is the mean of four trials, the differences of which were often imperceptible, and hardly ever amounted to more than one degree.

Now if this table be constructed, the temperatures being ranged along the axis of abscissas, and the quantities of electricity being represented by correspond-

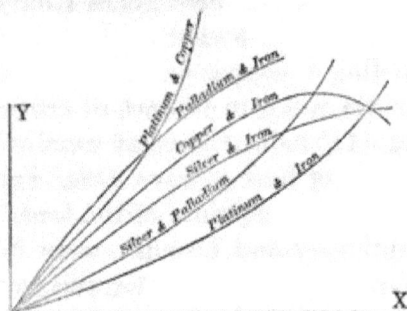

Fig. 53.

ing ordinates, we shall have results similar to those given
in Fig. 53, in which it is to be observed that the curves
given by the systems of silver and iron, copper and iron,
and palladium and iron are concave to the axis of abscissas; but those given by platinum and copper, silver
and palladium, and platinum and iron are convex.

Let us now apply the numbers obtained by these several pairs for the calculation of temperatures, which will
set their action in a more striking point of view. The
following table contains such a calculation, on the supposition that for the 90 degrees from 32° Fahr. to 122°
Fahr. the increments of electricity are proportional to the
temperatures.

TABLE II.

		Temperatures by the Mercurial Thermometer.			
		32° F.	122° F.	Water boils.	Mercury boils.
Temperature by a pair of	Copper and iron......	32	122	202	257
	Silver and palladium..	32	122	235	880
	Iron and palladium...	32	122	211	539
	Platinum and copper..	32	122	244	1030
	Iron and silver........	32	122	170	279
	Iron and platinum....	32	122	212	829

We are therefore led to the general conclusion that in
*these six different systems of metals the developments of
electricity do not increase proportionally with the temperatures, but in some with greater rapidity and in others
with less.*

The results here given I have corroborated in a variety of ways and with a variety of wires. A pair consisting of copper and platinum gave for the temperature
of tin when in the act of congealing 452° Fahr. instead
of 442° Fahr., the point usually taken. For the melting-
point of lead it gave 942½° Fahr. instead of 612° Fahr.
The melting-points of tin, lead, zinc, and occasionally of
antimony and bismuth, were in this manner employed,
for they allow time for the working of the torsion balance, and with the exception of bismuth, their tempera-

ture appears to be steady all the while they are in a granular condition before they finally solidify. The action of these metals on the thermo-electric pair is easily prevented by dipping it into a cream of pipe-clay.

A pair of copper and platinum gave for a dull red heat 1416° Fahr., and for a bright red 2103° Fahr.

A pair of palladium and platinum gave for a dull red 1850° Fahr., and for a bright red 2923° Fahr.

Some of the combinations into which iron enters as an element give rise to remarkable results. Thus, if we project the curve given by a system of copper and iron we shall find it resembling Fig. 54, where the maximum ordinate b occurs at a temperature of about 650° Fahr.; the point c ap-

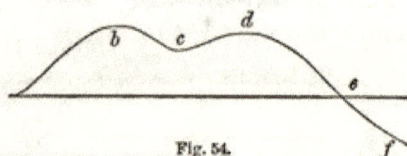

Fig. 54.

pears to be given between 700 and 800 degrees; d by a dull red heat; e is very nearly the point at which an alloy of equal parts of brass and silver melts, for if the pair be soldered with this substance, it fuses when the needles have returned almost exactly to the zero point. With harder solders or with wires simply twisted, the curve may be traced on the opposite side of the axis towards f, its ordinates increasing with regularity. At 60° Fahr., taking the length of the ordinate corresponding to a temperature of 212° Fahr. as unity, the length of the maximum ordinate at b is 1.85, very nearly.

A system of silver and iron gives also a similar curve, the point b occurring at a temperature rather higher than the analogous one for the preceding system, but still below the boiling-point of mercury.

Now, all these things serve to show that we cannot determine with accuracy unknown temperatures by the aid of thermo-electric currents, on the supposition that the increments of the quantities of electricity are proportional to the increments of temperature throughout the range of the mercurial thermometer.

Let us now pass to the second proposition : "That the tension undergoes a slight increase with increase of temperature—a phenomenon due to the increased resistance to conduction of metals when their temperature rises."

It will be seen, on consulting the following table, that pairs of *different metals* at the same temperature have tensions which are apparently very different.

The currents the tensions ot which are here indicated were generated by keeping one end of the thermal pair in boiling water, the other ends being maintained at a temperature of 32° Fahr.

TABLE III.

A pair of	Tension.	A pair of	Tension.
Antimony and bismuth..	.137	Platinum and iron........	.470
Copper and iron..........	.183	Copper and platinum.....	.473
Silver and lead	.307	Platinum and palladium..	.500
Lead and palladium.....	.313	Tin and iron.............	.518
Silver and platinum......	.380	Platinum and tin.........	.567

We perceive, therefore, that there apparently exist specific differences in the qualities of electric currents derived from different sources. If, for example, we take a pair of platinum and palladium and expose it to a temperature which shall generate a current capable of deflecting the torsion balance through 1000°, and then obstruct it by a wire of such dimensions as to stop one half, or only allow 500° to pass, and repeat the experiment with a current generated by bismuth and antimony, the temperature being still so adjusted as to give a deflection of 1000°, on making this pass through the same intercepting wire, perhaps not much more than one eighth of it will go through the galvanometer.

This peculiarity of thermo-electric currents depends

on the conducting resistance of the system that generates them. It is possible to give a current a higher or a lower tension by simply making use of thin or thick wires to generate or to carry it. In the foregoing table the current from platinum and palladium had a high tension, because slender wires of those metals happened to be used to generate it; and the current from antimony and bismuth had a low tension because thick bars of those substances were employed. In the former case, the conducting resistance was greater than in the latter, and hence the tension of the current was higher.

That this is strictly true will appear on examining the current evolved by any number of systems under the same condition of resistance to conduction.

In a great number of trials I failed in getting any trustworthy results as respects tension of currents at high temperatures, on account of the difficulty of maintaining the thermo-electric pair at the same degree without variation. By employing, however, a small black-lead furnace, to which was adapted a covered sand-bath into which the wires could be plunged, I succeeded at last; for with this arrangement a regulated temperature could be kept up for a length of time.

The experiment was made with care in the case of two systems of metals: 1st, copper and platinum; 2d, copper and iron.

1st. At the boiling-point of water, a pair of copper and platinum, the unexcited extremity of which was carefully maintained at 67° Fahr., evolved as a mean of four trials, three of which were absolutely identical, 123° of electricity, of which 23° could pass a secondary wire.

Then, by the aid of the furnace and sand-bath, the temperature was raised until the pair evolved 783° as a mean of four trials; of these 163° could pass the secondary wire. Now,

As 783 : 163 :: 123 : 25¼ instead of 23,

showing, therefore, a slight rise of tension.

2d. The pair of *copper* and *iron* gave at the boiling-point of water 300°, of which 57° passed the secondary wire. The temperature was now raised, with the following results:

490 degrees passing the primary, 95 the secondary wire.
553   "   "   "   113   "   "
545   "   "   "   112   "   "
493   "   "   "   110   "   "

It will be understood that although the quantities of electricity indicated in the first column do not regularly increase, the temperatures were, notwithstanding, going regularly upwards: to this peculiarity of the systems into which iron enters I have already alluded. Let us now compare these measures with those obtained for the boiling-point of water:

As 490 : 95 :: 300 : 58 instead of 57.
    553 : 113 :: 300 : 61   "
    545 : 112 :: 300 : 61   "
    493 : 110 :: 300 : 67   "

We find, therefore, that in the case of both these systems of metals the tension slowly rises with increase of temperature, being much better marked in the latter than in the former instance.

The increase of tension here detected depends unquestionably on increased resistance to conduction, which the wires exhibit as their temperature rises, as the following experiments show.

A pair of copper and iron evolved a current at the boiling-point of water, which, passing through a wire of copper eight feet long, was determined at the galvanometer to be 176°. Having twisted a part of this wire into a spiral, so as to go over the flame of a spirit-lamp, eight inches of it were thereby brought to a red heat; the de-

viation of the needle fell now to 165°, being a deficit of
11°. In this experiment care was taken that no heat
should be transmitted along the wire to the connecting
cups.

The same was repeated with an iron wire of the same
length and under the same circumstances. The current
at first being 90°, as soon as the spiral was made red-hot
it fell to 61°, being a deficit, therefore, of nearly one third
the whole amount.

To the increased resistance to conduction, occasioned
by an increased temperature, we are to impute the slight
rise of tension observed in thermo-electric currents. The
quantities are of the same order.

We have next to show "that the quantity of electricity evolved at any given temperature is independent of
the amount of heated surface, a mere point being just
as efficacious as an indefinitely extended surface."

The quantities of electricity evolved by hydro-electric
pairs increase with the surfaces, but it is not so in thermo-electric arrangements. A pair of disks of copper and
iron, two inches in diameter, were soldered together;
they had continuous straps projecting from them, which
served to connect them with the galvanometer cups. At
the boiling-point of water they gave 62°; on being cut
down to half an inch in diameter, they still gave 62°.
On the disk being entirely removed, and the copper
made to touch the iron by a mere point, its extremity
being roughly sharpened, the deflection was still 62°.

By means of a common deflecting multiplier, I obtained the following results: 1st. A copper wire being
placed in a bath of mercury, the temperature of which
was 240° Fahr., I dipped into it a second copper wire,
the temperature of which was about 60° Fahr.; the galvanometer needles moved through 15°.

2d. The cold wire being sharpened to a point and
plunged deliberately into the mercury to the bottom of
the bath, the deflection was 19°.•

3d. But when I touched the surface of the mercury
with the *very point* of the cold wire, there was a deflection of 60°.

Having laid a plate of tinned iron upon the surface
of some hot mercury, it was touched with the point of
the cold wire. There was a strong deflection of the
needles in the opposite direction to what would have
been the case had the mercury been touched and not the
iron. The under surface of the iron was, therefore, acting as a hot face, and the parts round the point as a cold
face, being temporarily chilled by the touch of the wire.

These results explain the anomalies observed by some
of those who investigated the course of thermo-electric
currents by means of small metallic fragments.

It would therefore seem that when wires of the same
metal are used as electromotors, the quantity of electricity evolved depends on the quantity of caloric that can
be communicated in a given time. Time, therefore, under
these circumstances, must enter as an element of thermo-
electric action. In the case of a single metal, the maximum effect would be produced when the hot element is
a mass and the cold one a point.

And, lastly, "that the quantities of electricity evolved
in a pile of pairs are directly proportional to the number
of the elements."

In the first trials I made to determine the effect of increasing the number of pairs in a pile, the results obtained were contradictory; by operating, however, in the
following way, the proposition was at last satisfactorily
determined.

1st. The resistance to conduction was made nearly

constant by uniting all the pairs intended to be worked with at once. The current, therefore, whether generated by one, two, three, four, or more pairs, had always to run through the same length of wire, and experienced in all cases a uniform resistance.

2d. By making each individual element of considerable length, the liability of transmission from the hot to the cold extremity was diminished.

Having, therefore, taken six pairs of copper and iron wires, $\frac{1}{18}$ of an inch thick and each element 38 inches long, I formed them, by soldering their alternate ends, into a continuous battery. Then I successively immersed in boiling water one, two, three, etc., of the extremities, their length allowing freedom of motion, and the other extremities not differing perceptibly from the temperature of the room.

The following table exhibits one of this series of experiments:

TABLE IV.

No. of Pairs.	Calculated Deviations.	Observed Deviations.
1	55	55
2	110	111
3	165	165
4	220	220
5	275	272
6	330	332

Hence there cannot be any doubt that the quantities of electricity evolved by compound batteries at the same temperature are directly proportional to the number of the pairs.

With some general remarks, arising from the foregoing subjects, I shall conclude this Memoir.

1. It is of importance to remember that thermo-electric currents traverse metallic masses only on account of differences of temperature existing at different points.

2. When a current of electricity, flowing from the poles of a battery, is made to traverse a metallic sheet, the whole of it does not pass in a straight line from one pole to the other, but diffuses itself through the metal, diverging from one point and converging to the other. The greater part of the current is found, however, to take the shortest route.

3. Combining, therefore, the foregoing observations (1, 2), we perceive that there are certain forms of construction which will give to thermo-electric arrangements peculiar advantages. For example, the surfaces united by soldering must not be too massive. Let A, Fig. 55, be a semi-ring of antimony, and B a semi-ring of bismuth; let them be soldered together along the line C H, and at the point H let the temperature be raised: a current is immediately excited; but this does not pass around the ring A B, inasmuch as it finds a shorter and readier channel through the metals, between H and d, circulating therefore as indicated by the arrows. Nor will the whole current pass round the ring until the temperature of the soldered surface has become uniform.

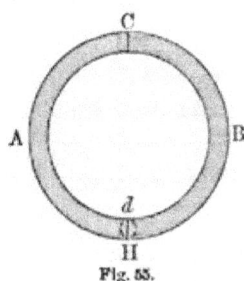

An obvious improvement in such a combination is shown at H in Fig. 56, which consists of the former arrangement cut out along the dotted lines: here the whole current so soon as it exists is forced to pass along the ring. And because the mass of metal has been diminished at the line of junction, such a pair will change its temperature very quickly.

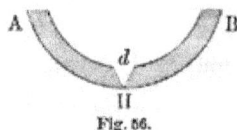

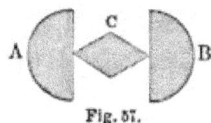

One of the best forms for a thermo-electric couple is given in Fig. 57, where

A is the section of a semicylindrical bar of antimony, B of one of bismuth, united together by the opposite corners of a lozenge-shaped piece of copper, C. From its exposing so much surface, the copper becomes hot and cold with the greatest promptitude, and from its good conducting power it may be made very thin without injury to the current. With a pair of bars three fourths of an inch thick, and a circular copper plate, as at D, Fig. 58, having both surfaces blackened, I have repeated the greater part of those experiments which M. Melloni made with his mul-

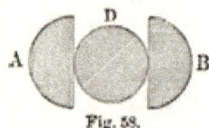

Fig. 58.

tiplier. A platinum wire, as at E, Fig. 59, may sometimes be very advantageously used.

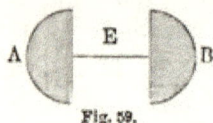

Fig. 59.

4. The currents circulating in a steel magnet are to all appearance perpetual. I thought for some time it might be possible to procure similar perpetual currents by compound thermo-electric arrangements. Let a, b, c be wires of three different metals soldered together so as to form a triangle. Now if these metals were selected, so that a and b could form a more powerful thermo-electric pair than a and c, or b and c, it might be expected that at all temperatures an incessant current would run round the system. Such, however, will not be found to be the case. In effect, any one of these three serves simply as a connecting solder to the other two, and hence no current is excited; for the ends that have the third metal between them, although that metal intervenes, are under exactly the same condition as the other ends which are in contact.

Y

.

MEMOIR XXV.

ON MICROSCOPIC PHOTOGRAPHY.

CONTENTS:—*Method of making microscopic photographs by condensed sunlight.—Specimens of the art.*

WHEN giving courses of lectures on Physiology in the University of New York, I found it very desirable to have photographic representations of various microscopic objects. There are many such objects, costing much time and trouble in their preparation, which it is very difficult, if not impossible, to preserve. A photograph is their best substitute.

I used as the sensitive surface daguerreotype plates, for this was previous to the invention of the collodion process. The daguerreotype plate is, however, much less sensitive than the collodion film. My first attempts were to copy the objects by lamplight, but this was found to be not sufficiently intense. Even after very long exposures with low powers the impressions obtained are unsatisfactory.

On employing a beam of sunlight impressions could be quickly obtained, and by concentrating the beam with a condensing lens the object could be sufficiently illuminated to suit any magnifying power. A fine microscope was used to give the image.

Two difficulties were, however, to be provided for: 1st. If the condensing lens be large, the heat at its focal point may be so high as to injure or even destroy the object. 2d. It is not easy to find the chemical focus, for it does not coincide with the visual one.

But these difficulties may be completely overcome by causing the illuminating beam to pass through a solution of sulphate of copper and ammonia. This transmits all the radiations that affect a photographic surface, but absorbs all others. With the absorbed rays the heat so nearly disappears that the most delicate preparation may be left at the focal point for any length of time without risk.

The solution of sulphate of copper and ammonia also enables us to ascertain the focal point with precision. On receiving the image on the ground glass of a camera, it is, of course, of a blue color, and when brought to a sharp focus its photograph will be equally sharp.

The best results are obtained when the blue image of the sun, formed by the condenser, coincides with the object.

In Fig. 60, *a a* is a heliostat mirror reflecting the sun's

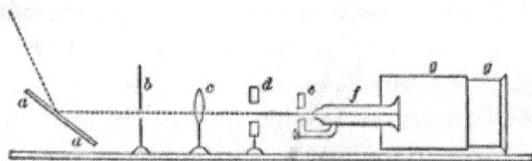

Fig. 60.

rays horizontally; *c* a condensing lens three inches in aperture; there are adapted to it different diaphragms, *b*, to regulate the quantity of light used, since it is not well to have too bright an image on the photographic plate; *d* is a glass cell, formed by cutting a circular hole three inches in diameter in a glass plate a quarter of an inch thick; on each face of this plate a flat piece of glass is laid, thus forming a receptacle in which the sulphate of copper and ammonia solution may be placed. It holds the liquid without risk of leakage; *e* is the object-stage of the microscope, *f*; *g*, the camera with its ground glass or sensitive plate.

In this manner I made many microscopic daguerreo-types. Some of them, such as were needed, were subsequently cut in wood to serve as illustrations for my Treatise on "Human Physiology." From that work I extract the specimens on the opposite page.

When the collodion process was invented, I made many of these photographs in that way. Some of my old original daguerreotypes were exhibited in the Centennial Exposition at Philadelphia, as illustrations of the early history of the art. Collodion photographs have, of course, the great advantage over daguerreotypes in this, that they may be magnified by the lantern and used for demonstration before a large class.

Fig. 61.—Spiracle of insect, magnified 75 diameters.

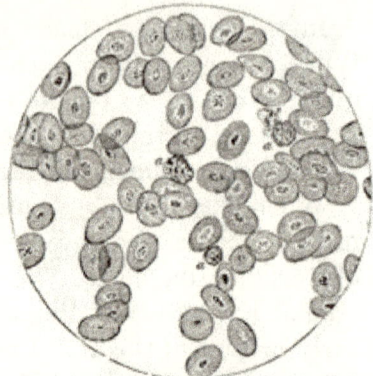

Fig. 62.—Elliptic blood-cells of frog, magnified 250 diameters.

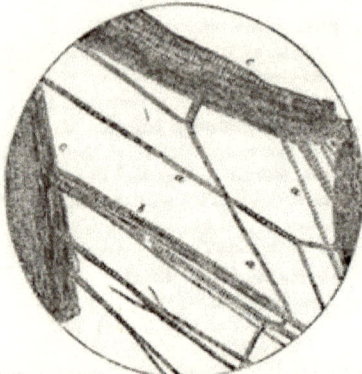

Fig. 63.—Ultimate muscular fibre, magnified 200 diameters.

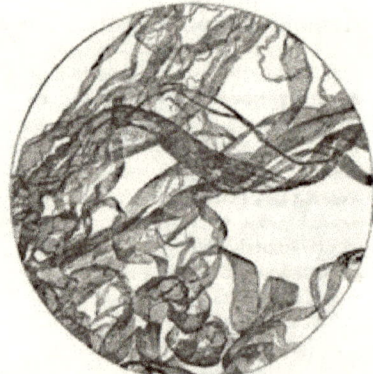

Fig. 64.—Spiral vessels of banana, magnified 50 diameters.

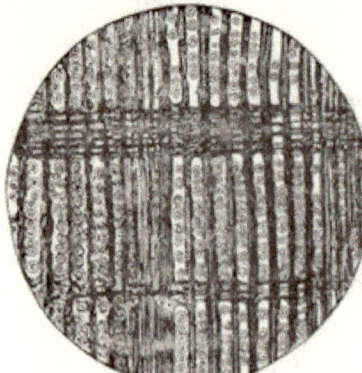

Fig. 65.—Woody fibre of pine, magnified 50 diameters.

Fig. 66.—Mucous membrane of the stomach of a carnivorous beetle, magnified 50 diameters.

•

MEMOIR XXVI.

ON CAPILLARY ATTRACTION AND INTERSTITIAL MOTIONS.
THE CAUSE OF THE FLOW OF SAP IN PLANTS AND
THE CIRCULATION OF THE BLOOD IN ANIMALS.

Extracted and condensed from various Memoirs as follows: On Capillary Attraction, Franklin Institute Journal, Philadelphia, September, 1834.   On the Tidal Motions of Movable Electric Conductors, Franklin Institute Journal, January, 1836.   Experiments on Endosmosis, Franklin Institute Journal, March, 1836. On Endosmosis through Water and Soap Bubbles, Franklin Institute Journal, July, 1836.   On Interstitial Movements, American Journal of Medical Sciences, May, 1836.   On the Physical Theory of Capillary Attraction, American Journal of Medical Sciences, February, 1838.   On the great Mechanical Force Generated by the Condensing Action of Tissues, American Journal of Medical Sciences, May, 1838.   On the Physical Theory of Endosmosis, American Journal of Medical Sciences, August, 1838.   Analysis of Some Coins, Silliman's American Journal, Vol. XXIX., 1836.   On the Circulation of the Blood, Silliman's American Journal, Second Series, Vol. II., 1846.   On the Constitution of the Atmosphere, London and Edinburgh Philosophical Magazine, October, 1838.   On Capillary Attraction, London and Edinburgh Philosophical Magazine, March, 1845.   On the Circulation of the Blood, London and Edinburgh Philosophical Magazine, March, 1846. A Treatise on the Forces that Produce the Organization of Plants, New York, 1844.   A Treatise on Human Physiology, New York, 1856.

Note.—These Memoirs are collectively so voluminous that I could not publish them in this volume in full.   I made the following abridgment of them, which was printed in Harper's Magazine, January, 1878.

Contents:—*Interstitial motions of solids.—Motions of metal in coins.— Movement of liquids in crevices.—Capillary attraction.—Conditions for a continuous flow.—Capillary attraction an electrical phenomenon. —Motions of liquid conductors.—Dutrochet's experiment of endosmosis. —Passage of gases through liquids.—Soap bubbles.—Passage against heavy pressure.—Distribution of sap in plants.—Circulation in plants due to sunlight producing gum.—Circulation of blood in animals explained.—The systemic, the pulmonary, and the portal circulation.*

It is necessary for the life of every organized being that a liquid should circulate through all its parts. In plants that liquid is called the sap; in animals, the blood.

When the distance through which such a liquid has to pass is small, there seems to be little difficulty in assigning the cause of its movement; but how shall we explain the rise of the sap in the great sequoia-trees of California, some of which attain a height of more than four hundred and thirty feet? To force a column of water to such an elevation would require a pressure of more than two hundred pounds on the square inch. Yet we cannot doubt that the power which overcomes these enormous resistances is the same as that engaged in the most insignificant transudations.

Even in the case of solid mineral substances the particles are not at rest. Boyle, in his tract on "The Languid Motions of Bodies," has collected several interesting instances.

He says: " But what is more extraordinary, a gentleman of my acquaintance had a turquois stone wherein were several spots of different colors, which seemed to him for many months to move slowly from one part of the stone to the other. And having the ring wherein it was set put into my custody, I drew pictures of the spots at different times; and by comparing several of the draughts together, it evidently appeared that they shifted their places, as if the matter whereof they consisted made its way through the substance of the stone. And as far as we observed, the motion of these spots was exceedingly slow and irregular. An experienced jeweller, likewise, assured me that in a few turquois stones he had observed two different blues in different parts of the same stone, and that one of these colors would, by slow and imperceptible degrees, invade and at length overspread that part of the stone which the other before possessed. And the same gentleman who lent me the spotted tur-

quois also showed me an agate haft of a knife, wherein
was a certain cloud which an ingenious person had for
some years observed to change its place in the stone."

Some years ago, having occasion to make an analysis
of certain Roman silver coins which had been long bur-
ied in the earth, I found that much of the alloying copper had made its way to the surface, constituting the
green patina of antiquarians, and that the silver had become comparatively pure.　An interstitial movement in
these denarii must therefore have taken place—a movement so slow that it had required many centuries to
yield the observed result.

If one end of a porous substance, such as a sponge, be
dipped into water, the liquid very soon percolates in all
directions through the mass, which becomes charged with
as much as it can hold.　If a piece of glass having a
crack in it be put into water so that the end of the
crack is immersed, the liquid instantly runs spontaneously to the other end.　And if from the crack other smaller
ones branch forth, along these also the water rapidly
finds its way.

These effects have long been studied by using slender
glass tubes, which operate in the same manner as cracks,
but permit the phenomena to be observed in a more convenient and exact manner.

Water will pass through a crevice the width of which
is less than one half of the millionth of an inch.　The
proof of this is readily
obtained experimentally.
If we take a convex lens,
$a a$ (Fig. 67), of long fo-

Fig. 67.

cus, and place it upon a glass plane, $b b$, there will be seen
at the point of apparent contact, c, on looking down, a
black spot surrounded by a series of variously colored
concentric circles, the appearance being well known

among optical writers under the name of Newton's colored rings. At the point *c* the lens and the plane are, as Newton has shown, a distance apart of about one half of the millionth of an inch; and from this centre, proceeding outwardly, the distance between the glasses of course increases. If anywhere at the outer portion a drop of water be introduced, it extends itself instantly across all the colored rings, reaching even across the central black spot.

If a tube of such diameter that it could lift water ten inches be broken off so as to be only six inches long, we might inquire whether the water would overflow from its top or simply remain suspended there.

Mathematical considerations, as well as direct experiments, prove that in such a case there would be no overflow. A capillary tube under these circumstances lifts the water, but does not produce a continuous current.

But if a removal of the liquid at the top of the tube take place in any manner, as by evaporation or by being dissolved in another liquid, a continuous current is produced.

As illustrating the production of such a continuous flow we may cite the case of a spirit-lamp, the wick of which may be regarded as a fagot of capillary tubes. Between the fibres of the wick there are interspaces that answer as tubes. If the cover of the lamp be taken off, all the spirit will eventually pass up the wick and escape from the reservoir by evaporation. Or, in an oil-lamp, the wick of which becomes readily saturated with the oil, but never exhibits an overflow, on the lamp being lighted, the oil is burned off, a current is established, and after a time the reservoir is emptied.

The phenomena of capillary tubes are connected with the adhesion of surfaces. Clairaut showed that if the mutual attraction of a solid and liquid amount to less than half the cohesion of the latter, the liquid will be depressed in a capillary tube made of the solid; if it be equal to half, the liquid will stand level in the tube; if it exceed half, the liquid will rise.

I published a paper in the Journal of the Franklin Institute of Philadelphia for September, 1834, its object being to show that the adhesion of surfaces, whether solid or liquid, to each other, and the rise or depression of liquids in capillary tubes, are strictly electrical phenomena.

When a glass plate is laid on the surface of quicksilver, a considerable force is required to separate them. On the separation being made, if the substances be examined by the electroscope, the glass will be found to be electrified positively, the mercury negatively. Their attraction or adhesion is, therefore, a necessary electrical result. So intense is this electrical development that if during the act of separation the mercury be in connec-

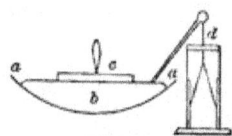

tion with a gold-leaf electroscope, the gold leaves are commonly torn asunder. In Fig. 68, $a\,a$ is a glass dish containing mercury, c the glass plate, d the gold-leaf electroscope.

Fig. 68.

In like manner, if some melted sulphur be poured into a conical glass and permitted to solidify, on making the separation the interior of the glass and the solid sulphur cone will be found to be in opposite electrical states. And the same occurs when surfaces of various kinds are parted from each other. There ought, therefore, to be adhesion. In Fig. 69, $a\,a$ is the conical glass, b the solidified sulphur, with handle for its removal attached.

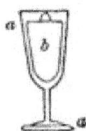

Fig. 69.

But if a glass plate be laid on a surface of water, there is no apparent development of electricity on separating them.   And the reason is obvious, for the glass has brought away with it a layer of water, and there has been no true separation of the solid from the liquid, but only of water from water.   The force of adhesion of the glass to the water has exceeded the cohesion of the water for itself.

If a plate of polished zinc be laid on mercury, there will, again, be no electrical development apparent on separating them.   For, owing to the conductibility of the zinc, there is nothing to prevent the opposite electricities from uniting, and all electrical manifestations must cease.

Whatever can disturb the electrical relations of a solid and a liquid will disturb their capillarity.   On wetting the interior of a glass tube, so as to form a temporary tube of water, and placing some mercury in it, the mercury will be depressed below the hydrostatic level.   But on connecting the mercury with the negative pole of a voltaic battery, and the water with the positive, the mercury at once rises, their mutual attraction being increased.

I derived these conclusions from the following experiments:

1. In a watch-glass (Fig. 70) place a quantity of pure mercury, $a\,b$, and upon it a drop of water, c.   Bring the water in contact with the positive platinum electrode of a voltaic battery, and touch the mercury with the negative.   The moment the contact is made the drop of water

Fig. 70.

loses its spherical form and spreads out into a thin circular disk, wetting the surface of the mercury.   The diameter of the disk seems to be greater in proportion as the battery is more powerful.

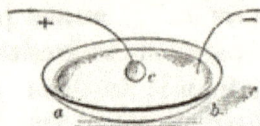

Under ordinary circumstances water does not wet

quicksilver. A drop of water remains on the surface of
quicksilver in the same manner that a drop of oil remains on water. As soon, however, as their electro-
chemical relations are disturbed by the aid of a voltaic
battery, the phenomenon of wetting at once occurs.

2. Take a tube, Fig. 71, in the form of an inverted
siphon, one branch of which, *a*, is about half an inch

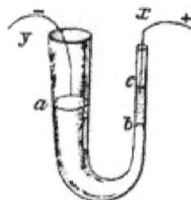

Fig. 71.

wide, and the other, *b*, not more than
the tenth of an inch. Fill the siphon to
a given height, *a b*, with mercury; the
metal, of course, does not rise in the narrow branch, *b*, to its hydrostatic level,
for mercury is depressed in a capillary
tube, inasmuch as it cannot wet glass.
Introduce a small column of water, *b c*.
The mercury may now be regarded as being in contact
with a tube of water, because that liquid wets the sides
of the glass intervening between it and the mercury.

Pass a slender platinum wire, *x*, down the tube, so as
to touch the water; let it be in communication with the
positive electrode of the voltaic battery; with the negative electrode, *y*, touch the mercury in the wide branch
of the siphon, *a*, and in an instant the metal will rise in
the narrow tube, and fall again to its former position as
soon as the current is stopped.

If into a watch-glass fifty or sixty grains of mercury
be poured, and over that as much water acidulated with
sulphuric acid as is sufficient to cover the surface of the
mercury, on the mercury being placed in contact with
the negative pole of a battery, and the water with the
positive, currents are produced both in the water and
the mercury, as was first observed by Erman and Ser-
rulas.

If the wires be on opposite sides of the mercury, as
shown in Fig. 72, the metal instantaneously elongates, as

indicated by the dotted line, and currents also are seen playing in the water. If the negative wire be introduced into the centre of the metallic globule, and the positive be brought on one

Fig. 72.

side, as in Fig. 73, the mercury will bulge out elliptically

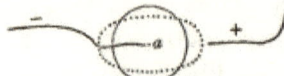

Fig. 73.

at both sides, nearest and farthest from the positive pole. If, now, the negative wire be cautiously raised from its position, so as to
be just out of contact with the surface of the metal, the mercury is immediately convulsed, its whole surface being covered with circular waves. On lowering the negative wire to its former position and advancing the positive, the moment it comes to the edge of the mercurial ellipsoid intense convulsions are produced, which increase until contact of the mercury and wire takes place.

At the same time that these movements are going on in the mercury, the surface of the water is ploughed by gentle currents, exactly resembling those that might be produced by directing a stream of air from a blow-pipe slantingly across the surface (Fig. 74).

The following experiment illustrates the nature of these effects:

Fig. 74.

A platinum needle, *a c*, Fig. 75, is suspended by a thread of unspun silk from a stand, *b b*, in a cup filled with acidulated water as high as *d d*. The needle hangs horizontally, its ends being about one fourth of an inch distant from the platinum polar wires, *p*, *n*, of a battery. Now the wire *p* being positive and *n* negative, the extremity *a* of the suspended needle would be negative and *c* positive

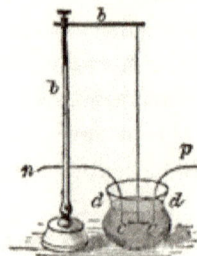

Fig. 75.

by induction. The conjoined effect of the forces thus brought to bear on the needle causes it to move on its centre, and take up a position of rest between the polar wires. To this, if it were turned aside, it would return after a few slow oscillations.

From this it appears that though the polar wires are plunged into a conducting medium and the current is passing, they still act as centres of attraction.

When a globule of mercury under the surface of some water is brought in presence of a point of attraction, p, situated at a short distance from its surface, two tides will be formed on the globule; one, a, Fig. 76, directly in front of the point of attraction, the other, b, 180° from it. In the quadrantal regions, c, d, there will be an ebb. If the point p be moved round the globule, both tides will follow it, keeping the same relative position that they had at first. These motions imitate on a small scale the effect which takes place by the action of the sun and moon in producing the tides of the ocean, and the explanation is the same as in the case of those tides.

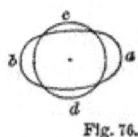
Fig. 76.

If a spring-tide were formed on a spherical ocean, and the sun and moon then annihilated, the elevation must sink, pressing the under waters aside, and causing them to rise where they were depressed. But the motion would not cease when the level was reached, for the water would arrive at that position with an accelerated velocity. It would, therefore, pass that position and form a high water where it had been low, and low water where it had been high. And this would be repeated again and again.

Now this theoretical case may be imitated with the globule of mercury, for on approaching the positive wire to it, a position will be reached at which contact will

take place between the protuberant tide on the mercury
and the wire.   At that moment the cause of attraction
is annihilated, the whole current of electricity now passing along perfect conductors, and fulfilling the supposed
case of an annihilation of the sun and moon at the time
of high tide.   And the same reasoning that held in one
case applies equally in the other—the mercurial tide falls
with an accelerated motion, and the line which before
was the transverse axis of the ellipse becomes the conju-
gate, tides being produced at right angles to the former
ones.   But here the strict comparison ends, for as the
mercury ebbs from its protuberant position the metallic
connection breaks, and the wire is again put in action as
a point of attraction.   The motion of the ebbing tide is
checked; it flows once more.   Once more the metallic
contact is complete; and when the tide falls, it is only to
flow again as long as the battery current passes.   Tides
take place at right angles to each other, in a series too
rapid to be counted, and the whole surface of the mercury is worked into those various and beautiful undulations which have been before referred to.

With respect to the currents that are observed, as if a
gentle wind were playing over the surface, the explanation is obvious.   We have seen that when a voltaic current is passed through mercury and water, the pressure
on the surface of contact is changed.   Newton has shown
("Principia," Vol. II., Bk. ii., Pr. 41) that if the particles
of a fluid do not lie in a right line, a pressure propagated
through that fluid will not be in a rectilinear direction,
but the particles that are obliquely
posited have a tendency to be urged
out of their position.   So the particles
a, a, Fig. 77, pressing on the particles
b, d, which stand obliquely to them by

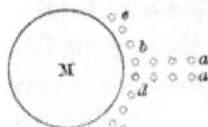

Fig. 77.

reason of the shape of the mass of mercury, M, have a

tendency to be urged from their places towards *e* and *c*
respectively, and the motion thus produced in a fluid
diverges from a rectilinear progress into the unmoved
spaces; and such a pressure taking place in a liquid free
to move continually returns the moving particles to
their position, after making them describe an elliptical
orbit.

Such is the nature of the evidence that may be brought
forward in proof of the hypothesis that the adhesion of
surfaces and capillary attraction are electrical results.
We may now pass to an exposition and explanation of
the more interesting instances of motion in the cases of
liquids and gases. If, in the lapse of centuries, metallic
copper can part itself from silver with which it has been
alloyed, coming forth from the interior of a coin to form
a new compound on the surface, such movements, it
might be expected, would more readily occur in liquids,
of which the cohesion is but small, and in gases, in which
cohesion perhaps does not exist at all.

And, first, as respects liquids:

The phenomena of endosmosis, first brought to general
notice in the case of liquid substances by M. Dutrochet,
may be explained as follows: If some alcohol be placed
in a bladder, the neck of which is tightly tied, and the
bladder be sunk into a vessel of water, percolation ensues, so that the bladder is distended to its utmost
capacity, and may even burst. Or if, instead of tying
the mouth of the bladder, a glass tube, open at both
ends and a foot or two long, be fastened into it without
leakage, as the water introduces itself through the pores
of the bladder to mingle with the alcohol the liquid rises
in the glass tube, and, when it has reached the top of it,
overflows. To express this inward passage of the water
the term endosmosis was introduced; and since a little of

the alcohol simultaneously passes outward to mix with the water, it is said to exhibit exosmosis.

In Fig. 78 is represented the endosmometer of Dutro-chet. It consists of a small blad-der, *a*, tightly tied to a tube, *d*, open at both ends, and bent as seen in the figure at *c;* the blad-der being completely filled with alcohol, and the tube to some such point as *d*, the arrangement is placed in a vessel of water, *e e;* almost immediately the level of the liquid will be seen to be ris-ing, the bend of the tube is reached, and one drop after an-

Fig. 78.

other falls from the open end into the receiver, *b*. And this continues until the liquids inside and outside of the bladder are uniformly commingled.

In these results there is nothing more than should take place on the ordinary principles of capillary action. The pores of a bladder are only short capillary tubes, into which water readily finds its way, because it can *wet* the substance surrounding the pores. If the blad-der be distended with air and sunk under water, al-though the water will fill the pores, it will not exude from them and accumulate in the interior of the blad-der; for, as we have seen, a capillary tube cannot estab-lish a continued current or flow. But the case becomes totally different when the bladder is filled with alcohol; for then as fast as the water presents itself on the inner end of each pore it is dissolved away by the alcohol, and the necessary condition for a continuous flow is complied with. Meantime through the pore itself a little alcohol passes in the opposite way by infiltrating through the incoming water, provided that the current be not too

Z

strong, and so the endosmosis of the water and exosmosis of the alcohol take place; the current of the former
greatly preponderates over that of the latter, and an accumulation of liquid in the interior of the bladder ensues.

That in all this there is nothing specially dependent
on the organic texture employed is obvious from the fact
that the same results arise when any inorganic porous
body is used. Vessels of unglazed earthenware, pieces
of baked slate or stucco, answer the purpose very well,
as will also a glass vessel with a minute fissure or crack
in it.

An incorrect representation of the conditions under
which endosmosis takes place is often made. It is said
to depend on the relative specific gravities of the liquid,
and that the lighter liquid always moves towards the
denser more abundantly than the denser towards the
lighter. But water endosmoses equally well to alcohol,
which is lighter than it, and to gum-water or salt-water,
which are heavier.

The force with which a liquid will thus pass through
a pore to mingle with another liquid beyond is, as we
shall presently see, very great.

It has sometimes been affirmed that living organic
textures possess a so-called selecting power, permitting
some substances to pass through them and refusing a
passage to others. But this so-called selecting power is
purely physical, as are the separations and apparent decompositions to which it gives rise.

If we take a glass tube, *a a*, Fig. 79, over the lower
end of which a piece of peritoneum or other delicate
membrane, *b b*, is tightly tied, and half fill it with litmus-
water, and then place it in a glass of alcohol, *c c*, the lev-
els of the liquids inside and outside being adjusted according to their specific gravities, so that there may be

no hydrostatic pressure either one way or the other through the pores of the peritoneum, as soon as the arrangement is completed, if the observer be so placed as to view it by transmitted light, he will see the water descending through the pores of the peritoneum in striæ and streams in a perfectly colorless state. The membrane has, therefore, absorbed and transmitted the water, but has refused to the coloring matter a passage. Such illustrations may, therefore, satisfy us that the selecting power of organic textures, like that of inorganic ones, is dependent on simple physical circumstances, and for these reasons we may exclude from the mechanism of animal ab

Fig. 79.

sorption the influence of any vital or other mysterious principles, and adopt the opinion of Abbé Hauy, that "those specious causes and imaginary powers to which in the Middle Ages all natural phenomena, even those of an astronomical kind, were referred, but which, through the genius of Newton and Laplace, have been banished from the celestial spaces, have taken their last refuge in the recesses of organic beings, and from these retreats science is preparing to expel them."

We may next turn to an examination of analogous phenomena in the case of gases. I found that on blowing a little bubble of melted shellac on the end of a glass tube, and putting some reddened litmus-water in its interior, ammoniacal gas, to which it was exposed, could almost instantly permeate the thin texture or barrier, and, gaining access to the inside of the bubble, turn the litmus blue.

When the pores of such barriers are of sensible size,

it is plain that the passing material is influenced by the
substance of which the pore consists only on those portions that come in contact therewith. Through the central parts of the pore the material will pass by mere
leakage. If, then, we desire to determine the physical
conditions under which these movements take place, we
must make use of barriers the pores of which are not of
sensible size.

I therefore closed the top of a glass tube, A, Fig. 80,
otherwise open at both ends, with a disk of
paper placed at such a distance down the tube
as to permit a stratum of water, a a, one eighth
of an inch thick, to be laid upon it. Conveying
the tube to the pneumatic trough, I filled it in
succession with various gases, and watched their
passage through the water roof, for so it might
be called, into the air. In such an experiment
a column of hydrogen gas half an inch in length

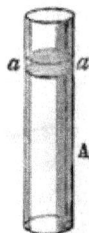

Fig. 80.

had escaped in twenty-four hours.

But experiments of this kind may be much shortened
by using very thin films instead of a thick stratum. A
glass bell-jar, a a, Fig. 81, was filled with
hydrogen gas, and by the side of it was
placed a small bottle containing atmospheric air. A finger dipped in soap-
water was passed over the mouth of the
bottle, so as to close it with a thin film,
and the glass bell-jar of hydrogen was

Fig. 81.

then placed over it, as at b. In the course of two minutes the film, which was at first horizontal, had become
convex, and eventually swelled into a large spherical
bubble, c. In sixteen minutes it had become so thin
that it was of a dark metallic lustre.

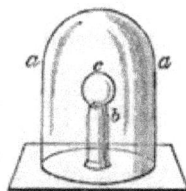

But the action is much more speedy if, instead of these
horizontal films, soap-bubbles are used. Such films are

at first too thick, they expose too small a surface to the atmosphere to which they are subjected, and it is not until the close of the experiment that the action becomes rapid. But a bubble at once exposes a large surface, and by using proper precautions there is no difficulty in preserving it for an hour, or even much longer.

The quickness of these motions is very well illustrated by the following experiment. If a glass flask, *a a*, be rinsed out with ammonia, and then by means of a bent glass tube, *b b*, narrowed at its projecting extremity, a soap-bubble, *c*, be blown therein, the air from the bubble being immediately drawn into the mouth without a moment's delay, the strong taste of the ammonia will be perceived. Or if a rod dipped in hydrochloric

Fig. 82.

acid be presented to the projecting end of the glass tube, as the bubble slowly collapsing presses out its contents, copious white fumes arise. Care must be taken that no drop of water obstructs the egress of the gas from the bubble, and that the cork fits the flask with a slight

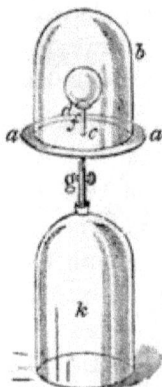

Fig. 83.

leakage. This, therefore, shows that vapors will pass through barriers having no proper pores, the transit taking place instantaneously.

I constructed an apparatus (Fig. 83) for exposing gases to each other with the intervention of a soap-bubble, and subsequently measuring and analyzing them. *a a* is a tin saucer about three inches in diameter and half an inch deep; into it water can be poured; it also serves as a platform to support a small bell-glass, *b*.

Through its centre at c passes a glass tube, f, one eighth of an inch in diameter, the upper extremity of which is cemented into a hole of the same size in a round, thin piece of copper, d, about half an inch in diameter; the other end of the pipe opens into another small bell-glass, k, through a perforation in its top, the communication being capable of being cut off by means of a stop-cock, g. The apparatus is used as follows: The upper bell, being taken off the platform, is filled with any gas to be tried —oxygen, for instance—and is placed aside on the shelf of a pneumatic trough. The lower bell-glass is then filled with water by depressing it in the trough; and the stop-cock being closed, five hundred measures of hydrogen, for instance, are thrown into it. After seeing that the copper plate d is free from moisture, a drop of water rendered viscid by soap is placed upon it, exactly where the orifice of the tube f opens. The upper glass containing the oxygen is now placed upon the tin saucer platform as in the figure. The lower glass is next depressed in the pneumatic trough, and as soon as the cock is opened a bubble of hydrogen containing five hundred measures expands, the spare oxygen escaping from the edge of the upper glass through the water in the tin saucer. The cock is next closed, and the apparatus placed on the trough shelf as long as the operator desires the experiment to continue. Keeping that position, when the cock is once more opened the gas passes into the lower bell until the bubble has entirely collapsed, when the cock is again closed, the contents of the bubble being now ready for measurement and analysis. As the gas was passing from the bubble into the lower bell, the water rose from the tin saucer into the upper bell, confining the gas that was outside of the bubble. This, by the common mode of manipulation, is to be transferred from the tin platform to the shelf of the trough for inspection.

By this apparatus it was found that one thousand measures of atmospheric air exposed to atmospheric air underwent no change either in volume or composition. The exposure in some cases lasted an hour.

One thousand measures of hydrogen, in the bubble, were exposed to atmospheric air in the bell. In five minutes there remained only four hundred and seventy-two. (It will be understood that the numbers quoted in this and other succeeding experiments are for the purpose of illustrating the general principle. They change with the relative proportion of gases inside and outside of the bubble.)

A reverse action ensues when nitrogen is substituted for hydrogen. The bubble swells instead of diminishing. Thus one hundred measures of nitrogen in half an hour became one hundred and seven and a half.

Oxygen decreases in bulk. Two hundred and fifty measures in ten minutes became one hundred and fifty-three. This gas passes more rapidly through the bubble than nitrogen.

Carbonic acid passes through the bubble very rapidly. When five hundred measures were used, the bubble collapsed almost as fast as it had expanded. Under a water roof half an inch thick and two inches in diameter, five thousand measures escaped into the air in forty-eight hours. In its place there were found two hundred measures of atmospheric air, which had passed in the opposite direction through the water roof.

By this apparatus it was proved that these motions through soap-bubbles continue until the gases on both sides of the bubble have the same chemical composition.

It has long been known that liquids and gases pass through porous structures though resisted by considerable force. Thus, if over the mouth of a cylindrical jar *b*, Fig. 84, a thin sheet of India-rubber be tied, and the

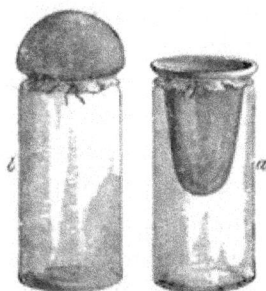

Fig. 84.

jar exposed to an atmosphere of carbonic-acid gas, that gas will force its way through the rubber and mingle with the atmospheric air in the jar, the rubber will be pressed outward, as at *b*, and eventually may be burst. I found that carbonic acid would thus force its way against a pressure of ten pounds on the square inch. If the conditions be reversed, the jar being filled with carbonic acid and then exposed to the atmosphere, the India-rubber will be depressed as at *a*, and stretch so as almost to sink to the bottom. Dr. Mitchell, of Philadelphia, has shown that this percolation will take place though resisted by many inches of mercury.

The same holds good as respects liquids. Thus water will readily pass through animal membrane to mix with alcohol against a pressure of fifteen pounds on the square inch.

In making experiments for determining the effect of such pressures it is to be borne in mind that there are certain disturbing circumstances which may vitiate the results. Among these is that general leakage which happens through the open pores of all tissues. Thus in the experiment first referred to (Fig. 78) it might be supposed that the force with which water passes through animal membrane into alcohol is not greater than one atmosphere, whereas in truth it is much more; but as soon as the pressure within the vessel had amounted to about one atmosphere, the alcohol escaped from the vessel by general leakage from the whole surface of the membrane as rapidly as the water entered. It is obvious that in a pore of sensible size those parts alone of a passing liquid in contact with its substance are subjected

to its influence, and those situated in its central regions
are ready to be influenced by any extraneous pressure.

In an experiment made on the passage of ammonia
into atmospheric air through India-rubber, it was found
that, though the passage of the gas was resisted by a
pressure of seventy-five inches of mercury, or upwards of
two atmospheres and a half, it took place apparently as
readily as if no such resistance had been opposed. The
question at once arises, Whence is this powerful impul-
sive force derived? Clearly not from the action of one
gas on the other. To the porous tissue or barrier alone
we must refer the seat of this power.

It is well known that porous solids of all kinds and
fluids absorb gaseous matter very readily, in volumes
varying according to circumstances. Water, for example, absorbs its own volume of carbonic acid, and four
hundred and eighty times its volume of hydrochloric-
acid gas. In the latter case, therefore, an extremely
great condensation takes place. So, too, a fragment of
porous charcoal absorbs nearly ten times its volume of
oxygen, and ninety times its volume of ammonia. These
gases, therefore, exist in the absorbing substance in a
state of very high compression. And the reasoning
which here applies, applies also in the case of two gases
separated by a tissue. If, for example, we separate by a
medium of this kind a certain volume of ammonia from
a like volume of nitrogen gas, though at the outset of
the experiment both the gases might be existing under
the same pressure, this equality would very rapidly be
lost. Ammonia, being absorbed more rapidly than nitrogen, would be presented to this latter gas not under
an equal pressure, but in a state of great condensation.
Under such circumstances, the transit of a gas is not
analogous to the case in which it flows under common
pressure into a vacuum, or into another gas; but the tis-

sue, continually acting as a perpetual condensing engine, brings the two media in contact with each other under extremely different conditions—the one in a compressed state, but ready to exert the whole of its elastic force; the other in a state perhaps little varying from its normal condition.

Moist membranes and films of water, by reason of their affinity for gaseous substances and their consequent condensing action, become the origin of great mechanical power. I have seen carbonic acid pass into atmospheric air through India-rubber against a pressure of ten atmospheres, and sulphuretted hydrogen against a pressure of twenty-five atmospheres.

The apparatus with which these results were obtained may be thus described. A strong glass tube, $a\ b$, Fig. 85, seven inches or more in length and half an inch in diameter, is hermetically closed at one end, through which a pair of platinum wires, b, c, pass to the interior, parallel but not touching. The other end, $a\ a$, has a lip or rim turned on it. Between the platinum wires a gauge tube, d, is dropped.

Fig. 85.

On the top of the gauge tube a small test glass, f, is placed, to contain a reagent suited to the gas under trial, as lime-water for carbonic acid, acetate of lead for sulphuretted hydrogen, litmus water for sulphurous acid. Sometimes, instead of this test tube, a piece of paper soaked in the proper reagent was employed. The large tube was then filled with water to the height $e\ e$, and over its lip a thin sheet of India-rubber was tightly tied, and over this again, to give strength, a very stout piece of silk. Everything being thus arranged, the projecting wires b, c, were connected with a voltaic battery; decomposition of the water ensued, oxygen and hydrogen being disengaged, and a condensed mixture of atmospheric air and those gases accumulated in the space $a\, a, e\, e$, the gauge tube showing the extent to which the condensation had gone. Now if the little cup f had been filled previously with lime-water, and the whole arrangement introduced into a jar of carbonic-acid gas, the upper part of the lime-water presently became milky, and after a time a copious precipitate of carbonate of lime subsided. This would readily take place when the gauge was indicating a pressure of ten atmospheres. In like manner, when a piece of paper covered with carbonate of lead had been introduced, and a pressure of twenty-four and a half atmospheres accumulated, on introducing the instrument into a vessel of sulphuretted hydrogen the paper quickly became brown. So sulphuretted hydrogen can pass through a sheet of India-rubber, and diffuse into an atmosphere of oxygen, hydrogen, and atmospheric air beyond, though it be resisted by a pressure equal to that of eight hundred feet of water.

The method of condensation here employed, because of its freedom from mechanical concussions, enabled me to continue these researches up to pressures of fifty atmospheres, without leakage, in comparatively slender tubes, and even under these circumstances gaseous diffusion seemed to take place without any restraint.

In passing, it may be remarked that when water is enclosed hermetically in a vessel, and a voltaic current passed through it, decomposition ensues, a portion of the gases making their appearance in the gaseous form, filling the small space occupied by the decomposed water, and the remainder being absorbed by that liquid as fast as it is given off. When the pressure is high, the dimensions of the vessel become sensibly greater, and the little bubble of air accumulated exceeds in bulk the volume of the decomposed water.

If, as it thus appears, no pressure we can command be
sufficient to restrain one gas from passing into another,
we next inquire what obstacle the condensed gas presents. There is abundant evidence to show that this
medium bears the same relation to the percolating gas
that a vacuum would do, inasmuch as the rate of discharge into it is the same as it is into a vacuum. If the
particles of different gases possess no repulsive tendency
as respects each other, if the presence of one makes no
difference in, nor produces any retardation in, the particles of the other, then it is immaterial how many of
such particles are condensed together in a given space.
The vacuum is not less a vacuum because it is contained
under smaller dimensions, any more than a torricellian
vacuum is less perfect when the mercury is made to rise
nearly to the top of a barometer tube than it was when
there was a vacant space several inches in length. This
would therefore indicate that these diffusions will take
place under all pressures, provided the gaseous condition
subsists; and this conclusion is abundantly borne out
by the experiments herein detailed.

The explanation we thus give of the action of condensing barriers rests upon a fundamental principle of
dynamics, that when the moving force and the matter
to be moved vary in the same proportion, the resulting
velocity will always be the same. Thus, if a cylinder
filled with air and fitted with a piston communicate with
a vacuum through an aperture, it is immaterial whether
the air be allowed to flow into the void without any
pressure, or whether it be urged by a direct action on
the piston—its velocity, as it goes into the void, will be
the same in both cases; for if it be compressed, the action of the piston is to reduce the air to such a density
that its elasticity is equal to the compressing force; and
because the elasticity varies as the density, the density

of the air increases with the impelling force. The matter to be moved is increased, therefore, in the same proportion as the pressure, and therefore the final velocity
is the same, and the same takes place in the case of a
tissue which is compressing a gas.

Such is the case while the gases are engaged with
each other in the tissue; but as soon as they pass from
it, the condensed gas, being no longer under its compression, expands freely, and, when measured, gives a result
differing from that which it would have been had not
the tissue compressed it.

Accordingly, when carbonic acid and air are separated
by a screen of stucco, which absorbs each to a very small
extent, they diffuse, according to the law of the square
roots of their density, one volume of air replacing 0.8091
of carbonic acid, the volume on that side of the screen
where the carbonic acid was increasing in quantity.
But if a thin sheet of India-rubber be used as a screen,
since it can condense one atmosphere of carbonic acid
while it does not act upon the air, though the same rate
of exchange ensues, there is a diminution of the gaseous
matter on the side containing the acid, and one volume
of air replaces 1.6182 of the acid.

In every plant two prominent operations are carried
forward: (1) the production of organic matter; (2) its
distribution through the various parts of the vegetable
system.

In 1844 I published a work under the title of "A
Treatise on the Forces that Produce the Organization of
Plants." It was a monograph chiefly devoted to the illustration or investigation of those operations, and com-
bated the existence of the Vital Force of physiologists
as a homogeneous and distinct power.

The progress of science shows plainly that living

structures, far from being the products of one such homogeneous power, are rather the resultants of the action
of a multitude of material forces. Gravity, cohesion,
elasticity, the agency of the imponderables, and all other
powers which operate both on masses and atoms, are
called into action, and hence it is that the very evolution
of a living form depends on the condition that all these
various agents conspire. Organized beings and organized bodies spring forth in those positions only to which
the rays of the sun have access. They are, therefore,
limited to the atmosphere, the sea, and the surface of the
earth.

If we expose some spring-water to the sunshine,
though it may have been clear and transparent at first,
it presently begins to assume a greenish tint, and after a
while flocks of green matter collect on the sides of the
vessel in which it is contained. On these flocks, whenever the sun is shining, bubbles of gas may be seen,
which, if collected, prove to be a mixture of oxygen and
nitrogen, their proportion being variable. Meantime
the green matter rapidly grows, its new parts, as they
are developed, being all day long covered with air-bells,
which disappear as soon as the sun is set. If these observations be made on a stream of water the current of
which runs slowly, it will be found that the green matter serves as food for thousands of aquatic insects which
make their habitations in it. These insects are indued
with powers of rapid locomotion, and possess a highly
organized structure. In their turn they fall a prey to
the fishes which frequent such streams.

Thus by the influence of the sunlight organic matter
is added to vegetable systems, the action being accompanied by a variety of chemical decompositions and interstitial diffusions. The substances arising are such as
are necessary for the uses of the plant; and in order to

distribute them, mechanical motion has to take place. This, in the more highly organized plants, goes under the designation of the flow of the sap.

The flow of the sap in plants and the circulation of the blood in animals are probably due to the same physical cause. And bringing into view the experiments already related respecting capillary attraction, I considered, in the work above referred to, the conditions necessary for producing a continual flow, such as evaporation, decomposition, solution, developing the general law of those movements, and illustrating the great force with which they are accomplished. I showed that these motions depend on this physical principle: "That if two liquids communicate with one another in a capillary tube, or in a porous or parenchymatous structure, and have for that tube or structure different chemical affinities, movement will ensue; that liquid which has the most energetic affinity will move with the greatest velocity, and may even drive the other liquid entirely before it;" and that this is due to common capillary attraction, which, in its turn, is due to electric excitement.

Applying this principle to the case of plants, the liquid of which the ascending sap is constituted is derived from the ground by the action of the spongioles, and consists of water holding in solution the different saline bodies necessary to the plant, along with carbonic acid, etc. This passes upwards by the woody fibre and ducts of the alburnum, making its way to the leaf, on the upper surface of which, in common cases, a change in its chemical constitution occurs through the influence of the sunlight. It obtains a quantity of carbon. This elaborated sap, or latex, now returns to the bark, and descends through its cellular tissue and intercellular spaces, finding its way by the route of the medullary rays to all parts of the plant. During its descent the

different vegetable principles necessary for the economy
of the plant are removed from it, and a certain quantity
goes down to the roots, partly to aid in their growth,
and partly to throw new quantities of ascending sap
into the tree.   In this descent the elaborated sap moves
through a system of vessels which anastomose with one
another in the same manner as the capillary vessels of
animals.

There are, therefore, two points in this circulation
which require attentive consideration — the spongiole
and the leaf.   The spongioles are nothing but the young
succulent extremities of the roots, which have been recently formed from portions of the descending sap, and
that sap is itself a species of mucilaginous solution.   Precisely, therefore, as water will pass through the tissue of
a bladder the interior of which is filled with gum-water,
so will moisture from the ground flow through the spongiole.   There is no difficulty in thus accounting for the
rise of the ascending sap on the principles of capillary
attraction, and indeed this is the explanation generally
received by vegetable physiologists.

Guided by the principle above laid down, I offered
the following as an explanation of the action of the leaf.
The ascending sap, which we may assume to be a weak,
watery solution, rises to the upper face of the leaf.   It
there obtains carbonic acid from the air; of this the sunlight effects the decomposition, with the production of
gum, the result being a change from water to a mucilaginous solution.   In the tissue of the leaf we have,
therefore, two liquids engaged—water and a mucilaginous solution.   On the principle above indicated, the
water will drive the mucilaginous solution before it, and
force it back along its proper vessels into the stem.

What, then, is the reason that the light of the sun
controls the rapidity with which the ascending current

comes? Because it controls the amount of carbonic acid reduced, and, therefore, the amount of elaborated sap formed. Why is it that the upward flow diminishes when changes are befalling the leaves, and why does it stop in the winter? Because the mucilaginous solution made by light diminishes in quantity, or ceases to be formed altogether.

There are, therefore, two sources of force in a flowering plant—the spongiole and the leaf; and they derive their power from ordinary physical principles. Whatever has been said respecting the movements of sap in exogenous plants applies also to the case of endogenous, and indeed to flowerless plants too.

It has been clearly established by the researches of comparative anatomists that the presence of a circulatory mechanism is determined by the centralization of the nutritive and respiratory apparatus. In exogenous and endogenous plants, from the circumstance that liquid and solid materials are introduced at distant points, channels of communication from one to the other, and indeed to every part, are required, and hence the introduction of a circulatory apparatus. In the lower tribes of vegetable life, where the separation of function does not exist, the circulatory mechanism is correspondingly absent. Sea-weeds absorb on their whole surface, and nutrition is directly carried forward at the points of reception. In lichens there is the first appearance of a transfusory mechanism, arising from the circumstance that on those parts which are shaded from the light, absorption most rapidly takes place: here, probably, however, the channels of movement are the interspaces between the cells, and the cause simple capillary attraction. In mushrooms there is a closer approximation to the mechanism more fully developed in the higher plants, for in them the rootlets absorb nutrient matter from the soil, from

A A

which it passes by capillary action to every part of the system.

The cause of the movement of the sap in flowering plants, both of the rise of the crude sap and the descent of the elaborated sap, is the light of the sun, which effects the decomposition of carbonic acid.

From this explanation of the causes of the movement of sap in plants, I turn to the circulation of the blood in animals.

In man there are three chief circulations — the systemic, the pulmonary, the portal. Bearing the above-mentioned general principle in mind, I presented the following explanation of these circulations:

1. *The Systemic.*—The arterial blood, which moves along the various aortic branches, and is distributed to every part of the system, contains oxygen which it has received during its passage through the lungs. Its color is crimson. As soon as it has reached its destination in the minute capillary vessels, it begins to carry on its proper process of oxidation, attacking in a measured way the tissues through which it is flowing. The direct result of this operation is an evolution of heat. But while this chemical change in the tissues is going forward, the arterial blood itself is also suffering a change in giving up its oxygen and gaining in exchange the results of combustion. From being crimson, it turns dark; from being arterial, it changes into venous blood.

Now, under these circumstances, what must take place in every capillary or each small portion of a porous structure? On the arterial side, we have the crimson arterial blood; on the venous side, dark venous blood—two different liquids. What, then, is the relation that obtains between each of these liquids and the walls of the tube, or the substance of the parenchyma in which they are

placed? Must it not be that the arterial blood bearing
its oxygen has an intense affinity for those structures,
but, those affinities being satisfied, that which was arterial passes into the condition of venous blood? The affinities it had for the structures with which it was in
contact are satisfied and have come to an end. The
arterial blood presented a highly energetic force, which
in the venous has diminished to zero. Under these circumstances, in accordance with the general principle, the
arterial blood must press the venous blood before it, and
the flow must be from the artery to the vein.

2. *The Pulmonary.*—In this the venous blood presents itself in the air-cells to receive oxygen. The systemic circulation deoxidized arterial blood, the pulmonary oxidizes venous. The latter, therefore, is the converse of the former. The venous blood has an affinity
for the oxygen dissolved in the tissues with which it is
in contact, and the arterial blood has none. Movement,
therefore, must ensue; but as the conditions of the affinity are reversed, so also is the direction of the motion,
for now the venous blood drives the arterial before it,
and drives it to the heart.

3. *The Portal Circulation.*—In this the same physical
principles apply. The blood which flows towards the
liver along the portal vein has been obtained by that
vein from the chylopoietic viscera; it has, therefore, the
same relation to the blood furnished from the different
and corresponding aortic branches as has the general
systemic venous blood. The arterial blood, therefore,
drives it before it in the same way that the general systemic circulation takes place; and, passing along the portal vein, it is now distributed to the liver. In this organ it also receives the blood which has been brought
by the hepatic artery.

The process of biliary secretion now takes place, and

compounds of carbon and hydrogen with soda are separated as bile, and pass along the biliary tubes. In its
final effect, therefore, the chemical action of the liver closely resembles the chemical action of the lungs. Compared
with the blood which passes along the branches of the
hepatic veins and finds its way into the ascending vena
cava, the portal blood differs by containing the elements
of bile.

Two systems of forces now conspire to drive the portal blood out of the liver into the ascending cava.

First, the blood which is coming along the capillary
portal veins and that which is receding by the hepatic
veins, compared together as to their affinities for the
substance of the liver, obviously have this relation : the
portal blood is acted upon by the liver, and there are
separated from it the constituents of the bile; the affinities that have been at work in producing the result
have all been satisfied; and the residual blood, over
which the liver can exert no action, constitutes that
which passes into the hepatic veins. Between the portal blood and the substance of the liver there is an energetic affinity, indicated by the circumstance that a chemical decomposition takes place and bile is separated; but,
that change once completed, the residue, which is no
longer acted upon, forms the venous blood of the hepatic veins, and hence the portal blood drives before it
the inert blood which is in those veins.

But, in addition to this, the blood of the hepatic artery, after serving for the economic purposes of the liver,
is thrown into the portal plexus. Hence arises a second
force, which, conspiring in its resultant with the former,
produces movement in the same direction. The presence of the arterial blood in the hepatic capillaries is not
only sufficient to give a force towards that in the capillaries of the portal veins, but also to give it a pressure

towards that in the hepatic veins. No regurgitation can take place backwards through the portal vein upon the blood arising from the chylopoietic viscera, because along that channel there is a pressure propagated in the opposite direction arising from the arterial blood of the aortic branches. This pressure conspires with that of the portal blood, and both together join in giving rise to motion towards the ascending cava.

On the same principle we may explain the circulation of the blood in other types of life; for example, in the case of the model adopted in fishes, the aorta of which has long been recognized as bearing a strong resemblance to the portal vein of the mammalia. To any one, however, who reflects on the principles here laid down, there will arise no difficulty in explaining the circulation in any particular case, if this plain precept be constantly kept in mind: that, in consequence of the physical principle which has been assigned, a pressure will always be exerted by the liquid which is ready to undergo a change upon that which has already undergone it—a pressure which, as there is no force to resist it, will always give rise to motion in a direction from the changing to the changed liquid.

I then, in the work referred to and in my "Physiology," continued the investigation to a determination of the uses and action of the heart (heretofore considered as the sole cause of the circulation), the action in asphyxia, the case of obstructed trachea, local inflammation, etc.

By regarding the affinity between the blood and the tissues with which it is in contact as the primary cause of the circulation, we assign a reason for those various phenomena which cannot be accounted for by Harvey's doctrine: the motions in the embryo; the periodic and local variations; the portal circulation; the changes in

the current as seen under the microscope; the movement
in the capillaries after the heart is cut out; the empty
condition of the arteries after death; the phenomena of
acardiac monsters; local inflammations and congestions;
the gangrene of parts while their capillaries are pervi-
ous; the retardation of the current on the application
of cold or of carbonic-acid gas; the results of asphyxia
and death by drowning or hanging; the changes of pressure in the arteries and veins respectively during a check
on the respiration; the *vis a tergo* of the veins; the effects of a ligature on those vessels; the action of irres-
pirable gases when breathed, and the opposite conditions when oxygen gas or protoxide of nitrogen is used.

A doctrine which accounts with simplicity for such a
long list of miscellaneous facts commends itself to our
attention at once.   There are, however, considerations of
a still weightier character which must compel us to
adopt it.   The affinity between the blood and the parts
with which it is in contact is a chemical fact beyond
contradiction.   The pressures and motions I have been
speaking of follow as the inevitable consequences of that
affinity.   We therefore cannot gainsay their existence in
the living mechanism, and the only doubt we can enter-
tain is as to whether they are of competent power to
produce all the effects before us.   But after what has
already been said respecting the energy of endosmotic
movements against pressures of many atmospheres, we
may abandon those doubts; and since we have here a
force of universality enough and intensity enough, and
in every instance acting in the right direction, it would
be unphilosophical to look farther, since such a force
must, under these conditions, exist in the physical necessity of the case.

MEMOIR XXVII.

ON THE EXISTENCE AND EFFECTS OF ALLOTROPISM ON THE CONSTITUENTS OF LIVING BEINGS.

From the Philosophical Magazine, April, 1849.

CONTENTS :—*Allotropism of elementary substances.—Frankenheim's nomenclature.—Brought about by light, heat, electricity, and by the nervous principle.—Explanation of inflammation and congestion.*

IT has been completely established for the majority of elementary substances that there are several forms under which each may occur—forms differing entirely both in their physical and chemical relations.

Thus, in the case of carbon, many such forms are known. To three of them Berzelius has directed attention: 1st, ordinary charcoal; 2d, plumbago; 3d, diamond. They are three distinct modifications of the same element. They differ in specific gravity, in specific heat, and in conducting power, both for electricity and caloric. In their relations to light, the first perfectly absorbs it, and is black; the second reflects it like a metal; the third is transparent like glass. When crystallized, plumbago and diamond do not belong to the same system: their chemical relations are also strikingly different. Charcoal takes fire with facility, and some varieties of it are even spontaneously combustible in the air; but crucibles and furnaces are made of plumbago because of its incombustibility; and the diamond with difficulty is set on fire in pure oxygen gas.

It seems immaterial to what class elementary bodies belong, whether electro-negative or positive: they present analogous results. Silicon, sulphur, selenium, phosphorus, titanium, chromium, uranium, tin, iridium, osmium, copper, nickel, cobalt, iron, oxygen, chlorine, are cases in point; and the instances which appear as exceptions are rapidly diminishing in number.

As is well known, to these singular modifications Berzelius gave the designation of allotropic forms, and the whole phenomenon passes conveniently under the designation of allotropism. He shows that the peculiarity assumed is often of such a persistent nature that it is not lost, even though the substance affected should go into combination with others. Thus, there are two forms of silicon—one combustible, and the other remarkably incombustible. Each, by uniting with oxygen, gives rise to silicic acid; the acid in one case being soluble in water and in hydrochloric acid, and in the other the reverse. And, in like manner, metallic arsenic, which exhibits the same duality of condition, gives rise to two different arsenious acids. Of phosphorus there are at least two modifications; and, accordingly, we have two compounds of that body with hydrogen, one of which is spontaneously inflammable, and the other not; and at least two oxygen acids, the monobasic and tribasic, in which the essential difference rests in the state of the phosphorus they contain.

It is to be remarked that, so far as observation extends, the most common cause of producing these singular differences is the action of that class of agents which we term imponderable substances. In very many cases, change of temperature brings about allotropic change; in others it is the agency of light, as in chlorine and phosphorus; and again, in others, association with foreign bodies, which apparently establish new voltaic relations. Heat, light, and electricity seem to be the general modifying agents.

Berzelius, following the suggestion of Frankenheim, proposes a nomenclature for pointing out the peculiar form referred to in any special case. It depends on the use of Greek letters. Thus, we have the three forms of carbon just alluded to designated on these principles by $C\alpha$, $C\beta$, $C\gamma$. But in a Memoir republished in this work, p. 285, it is remarked that we may often with greater convenience use the simple expressions "active" and "passive." Thus, active chlorine is that which will decompose water in the dark, passive chlorine failing to do so. In this Memoir the same expressions will be employed.

Hitherto, allotropism has only been considered as affecting inorganic states of matter, but its influence can be plainly traced in the far more interesting case of organic beings; and this, when placed in a proper point of view, yields a remarkable explanation of some of the most obscure but important facts in physiology and pathology. These explanations I propose now to point out.

In the *Philosophical Magazine* (March, 1846, p. 178) there is a paper by me explanatory of the causes of the circulation of the blood in the capillary vessels. It is merely an abridgment of a lecture which for eight years past has been delivered in this university. The doctrine there set forth has been generally received in America, and introduced into some of the standard works on physiology published in England. The principle on which it essentially depends, and which has been abundantly confirmed by direct experiment, is briefly this— that if there be two fluids occupying a capillary tube, or a porous structure of any kind, under the condition that one of them has a stronger chemical affinity for the substance of that tube or structure than the other, a movement of the liquids will at once ensue, that which has

the stronger affinity driving the other before it. On this
principle a clear account of the systemic circulation of
animals may be given; for the arterial blood, an oxidizing liquid, having a stronger affinity for the soft tissues
with which it is in contact than the venous blood, the
affinities of which have been satisfied, and therefore no
longer exist, necessarily exerts such a pressure that motion must ensue, the arterial blood forcing the venous
before it.

An application of the same principles shows that in
the pulmonary circulation the motions must necessarily
be in the opposite direction, or from the venous to the
arterial side, as is actually the case. It also explains
clearly the conditions of the portal circulation, in which
the direct action of the heart could hardly be expected
to be felt. With the generality which ought to belong
to a true theory, it meets all the cases which occur in the
lower orders of animal life, such as the greater circulation in fishes, in which there is no systemic heart; the
movements which take place in the vascular system of
insects; and even the extreme case of the rise and descent of sap in plants.

In this doctrine everything depends on the relationship between the nutritive fluid, or blood, and the solid
parts with which it is brought in contact; and whatever
changes that relationship must impress a corresponding
change on the circulation itself.

From experiments which I made some time ago, I have
been led to suppose that the arterialization of the blood,
as it takes place on the cell-walls of the lungs, bears a
strong analogy to the oxidation of white indigo. The
loose hold which the coloring matter of the blood retains
on the oxygen, coupled but not combined with it, is not
unlike what is witnessed in other nitrogenized coloring
matters, such as indigo, which oxidizes and deoxidizes

with the utmost facility. Charged with the oxygen it has thus obtained, the arterial blood passes to all parts of the system; and now arises that striking but all-important physiological fact, that it does not attack indiscriminately all those parts of the soft solids which it first encounters, but, proceeding in a measured way, exerts its action on such particles alone as have become effete, and, accomplishing the great process of interstitial death, resolves those particles into other forms, so that they can be eliminated from the system by the lungs, the kidneys, and the skin.

Now why is it that things proceed in this way? What is it that regulates this interstitial death? Why is one atom preserved and another surrendered?

It is upon these obscure points that the recent discoveries in allotropism shed a flood of light.

The three leading neutral nitrogenized bodies, fibrine, albumen, and caseine, are characterized by exhibiting allotropism in a most remarkable degree, and that in a double sense. 1st. Though so different from one another in their physical and chemical relations, it is admitted on all hands that they are mutually convertible: the albumen of the egg, during incubation, gives rise to fibrine and other allied bodies; from caseine in the milk with which the young mammalia are nourished, the albuminous and fibrinous constituents of their systems arise; the nurse fed on fibrine and albumen secretes caseine from the mammary gland. Indeed, there is no more reason to regard these three bodies as essentially distinct substances than there is to apply the same conclusion to charcoal, plumbago, and diamond. Between the two cases there is the most complete analogy; and if charcoal, plumbago, and diamond are merely allotropic forms of one substance, the same holds good for fibrine, albumen, and caseine. But, 2d, each of these three com-

pounds betrays a disposition under trivial causes to assume new forms; as with silicic acid so with fibrine,
there are two varieties—one soluble in water, the other
not. A difference of a few degrees of heat turns transparent albumen into the porcellanous variety, and analogous observations might be made respecting caseine.

It may therefore be asserted that these proteine bodies exhibit a propensity to allotropism in a far more remarkable manner than any other substances known; not
only passing indiscriminately into one another, but also
exhibiting special variations under the influence of the
most trivial causes.

And now we may recall the fact that, of the agents
which in the inorganic kingdom bring about these
changes, the so-called imponderable principles are pre-
eminent. I transfer this observation to the case of organized beings, and infer that the nervous system has
the power of throwing organized atoms into the active
or passive state; that this is the fundamental fact on
which all the laws of interstitial death depend; and that
upon this principle—its existing allotropic condition—
an organized molecule either submits to the oxidizing
influence of arterial blood, or successfully resists that
action.

But it has been stated that there are certain pathological conditions which upon these views meet with
a clear explanation—conditions which, though long and
laboriously studied by physicians, remain involved in
contradictions and obscurity. The conditions to which
I refer are those known as inflammation and congestion.

It is agreed among chemists that during the preva-
lence of these conditions the urine assumes a peculiar
constitution. In inflammatory actions the relative quantity of urea and sulphuric acid is much above the normal standard, while in congestive cases the reverse holds

good, and the urea and sulphuric acid are below the standard. What is the interpretation of these remarkable facts? We shall find they are very significant.

The quantity of urea and sulphuric acid in the urine undoubtedly expresses the quantity of proteine matter that has undergone oxidation in the system. In all cases where that quantity is above the normal standard, the destruction of proteine matter has been correspondingly accelerated; and where it is deficient, the destruction has been reduced. The result of inflammations corresponds to the first of these cases, and of congestions to the second.

Recalling now what has been said respecting the cause of the capillary circulation, we see how all these apparently disconnected facts group themselves together in the attitude of dependent effects. In inflammation there has been that allotropic change in the soft solids involved that they have assumed a disposition for rapid oxidation—they are active. Their relations with arterial blood have become highly exalted; and the blood flows, on the principles I have set forth, to the affected part with energy. Redness of that part and a higher temperature are the result. Oxidation goes on with promptitude, and urea and sulphuric acid begin to accumulate in the urine.

But in congestive cases it is the reverse; the parts affected are thrown into a more passive state. Oxidation goes on in a reluctant way, the amount of tissue metamorphosed diminishes, the urea and sulphuric acid diminish in the urine; and, on the principles which I have endeavored to explain respecting the capillary circulation, we perceive that an immediate action must be exerted on the flow of the blood: the passive condition of the tissues and diminished capacity for oxidation restrain the flow from the arteries, and, there being now less pressure on the contents of the veins, engorgement of those vessels is the result; and this condition of things is what a physician designates as congestion.

In this manner, if we admit the existence of allotropism in organic atoms, we can give a very clear explanation of the condition of the circulation in the pathological states of inflammation and congestion, and also of the peculiarities which in those states belong to the constitution of the urine.

UNIVERSITY OF NEW YORK, *Feb.* 17, 1849.

MEMOIR XXVIII.

ON THE DISTRIBUTION OF HEAT IN THE SPECTRUM.

From the American Journal of Science and Arts for 1872, Third Series, Vol. CIV., No. 161; Philosophical Magazine, August, 1872.

CONTENTS:—*Early experiments seeming to prove that the maximum of heat is in the less refrangible spaces. — Comparison of the dispersion and diffraction spectra.—Effect of compression in the less refrangible regions and of dilatation in the more refrangible.—Measure of heat in the two halves of the visible dispersion spectrum.—Description of the apparatus employed.—The different colored spaces are equally warm.*

MANY experimenters, at various times, have occupied themselves with the problem of the Distribution of Heat in the Spectrum. At first it was supposed that there is a coincidence between the luminous and calorific radiations, and that the maximum of intensity in both occurs at the same point—that is, in the yellow space. This view was abandoned on the publication of . the well-known experiments of Sir W. Herschel, who showed that in certain cases the maximum is below the red. Subsequently Melloni, having discovered the singular heat-transparency of rock-salt, proved that when a prism of that substance is used, the maximum in question is as far below the red as the red is below the yellow; but that if the light has passed through flint-glass, the maximum approaches the red; if through crown-glass, it passes into the red; if through water or alcohol, it enters the yellow.

In the case of the sun's spectrum the distribution of heat was more closely examined by Professor Müller, whose results in a general manner confirmed the views then held, that the invisible radiation below the red greatly exceeds that in the visible spectrum; and still more recently Dr. Tyndall, examining the spectrum of the electric light through rock-salt, showed that the curve indicating the distribution "in the region of the dark rays beneath the red shoots suddenly upward in a steep and massive peak, a kind of Matterhorn of heat, which quite dwarfs by its magnitude the portion of the diagram representing the visible radiation." These investigations were made under unexceptionable circumstances; the beam of electric light had practically undergone no atmospheric absorption, and the optical refracting train was of rock-salt.

Sir J. Herschel had shown in 1840 that when the sun's rays are dispersed by a flint-glass prism, the distribution of the heat towards the less refrangible region is not continuous, but there are three maximum points. These points, as shown by Dr. Tyndall, do not exist in the spectrum of electric light, the decline of which is continuous; they are therefore to be attributed to the absorptive disturbance which the sun's rays have undergone. Quite recently (1871), M. Lamansky has succeeded in identifying these interruptions by the aid of the thermo-multiplier. In his Memoir he states that, with the exception of Foucault and Fizeau, in their well-known experiments on the interference of heat, no one has made reference to these lines, and that all experimenters describe the heat-curve as a continuous one (*Philosophical Magazine*, April, 1872).

I may therefore be excused for remarking at this point that the three lines in question were not only observed by me nearly thirty years ago, but that an engraving of them was published in that journal, in a memoir announcing the discovery of fixed lines in the invisible portions of the spectrum (May, 1843). It will be seen, from an inspection of that engraving, that these lines are marked a, β, γ,   They were impressed on daguerre-otype plates by resorting to the well-known processes for obtaining photographs of the less refrangible regions of the spectrum.   The paper is republished in the present volume, Memoir III.

In view of the preceding statement and others that might be given, it may, I think, be affirmed that the general opinion held at the present day as to the constitution of the spectrum is this, that there exists a heat spectrum in the less refrangible regions, a light spectrum in the intermediate, and a spectrum producing chemical action in the more refrangible regions. An experimental attempt to correct this view, and to introduce a more accurate interpretation, will not be without interest, especially as it is necessarily and directly connected with the important subject of photometry. In this Memoir I shall offer some experiments and suggestions respecting the heat of the spectrum, and in another, shortly to be published, shall consider the distribution of the so-called chemical rays. Among the numerous problems of actino-chemistry, there are none more important than these.

All the experiments hitherto made on the heat of the spectrum have been conducted on the principle of exposing a thermometer in the differently colored spaces. Such was Sir W. Herschel's method. Leslie used a differential with small bulbs. Melloni, Müller, Tyndall, a thermo-electric pile, the form preferred being the linear. This was advanced successively through all the radiations, and the deflections of the multiplier noted.

Is not this method essentially defective ? Does it not necessarily lead to incorrect results?

"There is an inherent defect in the prismatic spectrum —a defect originating in the very cause which gives rise to that spectrum itself—unequal refrangibility. Of two groups of rays compared together, one taken in the red, the other in the violet region, it is clear that in the same spectrum, from the very circumstance of their greater refrangibility, those in the violet will be relatively more separated from each other than those in the red. The result of this increased separation in the more refrangible regions is to give an apparent dilatation to them, while the less refrangible are concentrated. The relative position of the colors must also vary: the fixed lines must be placed at distances greater than their true distances as the violet end is approached." I am quoting from the fifth chapter of a work " On the Forces which Produce the Organization of Plants," published by me in 1844. In this chapter, one of the chief points insisted on is the necessity of using wave-lengths in the measurement and discussion of spectrum results—a suggestion which, I believe, I was the first to make, and which I renewed in a memoir in the *Philosophical Magazine* (June, 1845).

The importance of these remarks respecting the peculiarities of the prismatic or dispersion spectrum may perhaps be most satisfactorily recognized on examining such a spectrum by the side of a diffraction or interference one. By the aid of Fig. 86 this may be done.

Regarding the space between the fixed lines D and E as representing the central region, in each the fixed lines

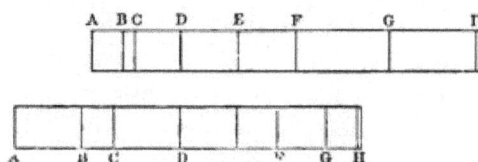

Fig. 86.

D and E are made coincident. The other lines are laid off in the prismatic as they appear through the flint-glass prism of the spectroscope; those of the diffraction are arranged according to their wave-lengths. It thus appears that in the prismatic, from the fixed line D to A, the yellow, orange, and red regions occupy but little more than half the space they do in the diffraction; while the green, blue, indigo, and violet, from the fixed line E to H, occupy nearly double the space in the prismatic that they do in the diffraction spectrum. The general result is, that in the prismatic the less refrangible regions are much compressed, and the more refrangible much dilated. And it is plain that the same will hold good in a still greater degree for any invisible rays that are below the red and above the violet respectively.

Now, if a thermometer of any kind were carried in succession from the greatly dilated, more refrangible regions to the greatly condensed, less refrangible, could the measures obtained be accepted as expressing the true distribution? The thermometric surface being invariable, would it not receive in the less refrangible spaces more than its proper amount of heat, and in the more refrangible less than its proper amount?

If we should admit that the distribution of heat in a correctly formed spectrum is uniform, it is plain that measures made by the use of a prism would not substantiate that admission. The concentration to which I have alluded as taking place in the less refrangible region would give an exaggerated, an increased heat for that region; and, on the contrary, the dilatation of the more refrangible would give an exaggerated diminution of heat for that space. But if it were possible to make satisfactory heat measures on the diffraction spectrum, in which the colored spaces and fixed lines are arranged according to their wave-lengths, the admission would be substantiated.

In view of these facts, I did attempt, many years ago, to make heat measures on the diffraction spectrum. But so small is the heat that, as may be seen in the *Philosophical Magazine* (March, 1857), the results were unsatisfactory. More recently I have tried another method of investigation, on principles which I will now explain.

For the sake of clearness, restricting our thoughts for the moment to the more familiar case of the visible spectrum, if we desired to ascertain the true distribution of heat, would not the proper method be to collect all the more refrangible rays into one focal group, and all the less refrangible into another focal group, and then measure the heat that each gave? If the view currently received be correct, would not nearly all the heat observed be found in the latter of these foci, and little, if indeed any, be found in the former? But if all the various regions of the spectrum possess equal heat-giving powers, would not the heat in each of these foci be the same?

Let us give greater precision to this idea. Using Angström's wave-lengths—the length at the line A is 7604, and that at H^2 3933, and these lines are not very far from the less and more refrangible ends of the *visible* spectrum respectively. The middle point of this spectrum is at 5768, which may therefore be called its optical centre. This is a little beyond the sodium line D, which is 5892. Now, if by suitable means we reunite all the rays between 7604 and 5768 into one focus, and all the rays between 5768 and 3933 into another focus, are we not in a position to determine the true distribution of the heat? Should the heat at these two foci be sensibly the same, must not the conclusion at present held be abandoned?

If in these investigations the rays of the sun be used, it is necessary to restrict the examination to the visible spectrum, excluding the invisible red and invisible violet radiations. On these the earth's atmosphere exerts not only a very powerful but a very variable action, and, what is still more, an action the result of which we cannot see, so that we are literally working in the dark.

There are days on which, owing to the excessive absorption taking place among the ultra-red rays, a rock-salt train has no advantage over one of glass. But if it is the visible spectrum alone that we are using, and the prisms are of a material *colorless* to the eye, we may be certain that they are exerting no selective absorption on any of the radiations of that spectrum, and that the indications they are giving are reliable.

This variable absorptive action of the atmosphere depends partly on changes in the amount of water vapor, and partly on the altitude of the sun. At midday and at midsummer it is at a minimum. The disturbance is not merely a thermochrose, for both ends of the spectrum are attacked. It is a matter of common observation that the horizontal sun has but little photographic power, owing to atmospheric absorption of the ultra-violet rays, and under the same circumstances his heating power is diminished, owing to the absorption of the ultra-red rays. But if the day be clear, and the sun's altitude sufficient, the visible spectrum may be considered as unaffected.

It should be borne in mind that the envelopes of the sun himself exert an absorptive action, which is powerfully felt in the ultra-violet region, as is indicated by the numerous fixed lines crowded together in that region. The force of this remark will be appreciated on examining the plate above referred to in the *Philosophical Magazine* for May, 1843.

It seems, then, that all the conditions necessary for the solution of this problem will be closely approached if we make use of prisms constituted of any substance which is *completely colorless to the eye*, and confine our measures to the *visible spectrum*, collecting all the radiations between the fixed line A and the centre of that spectrum just beyond D into one focus, and all the radiations between that centre and II^2 into another focus, and by the thermo-pile, or any other suitable means, measuring the heat of these foci.   •

Such is the method I have followed in obtaining the measures now to be presented: but before giving them, there are certain preparatory facts which I wish to submit to the consideration of the reader.

(1.) In the mode of experiment hitherto adopted, no special care has been taken to ascertain with accuracy the position of the " extreme red," yet that is held to be the point from which, on one side, we are to estimate the invisible, and, on the other, the visible spectrum. Different persons, perhaps because of a different sensitiveness of their eyes, will estimate that position differently. The red light shades off gradually—it is almost impossible to tell when it really comes to an end. A linear thermo-pile, such as is commonly used, is liable, under these circumstances, to give deceptive results, and any error in its indications counts in a double manner. It not only diminishes the value of one spectrum, but it adds that diminution to the value of the other. The force of this remark will be understood by considering the best experiments hitherto made on this subject, those of Dr. Tyndall, as related in his " Heat a Mode of Motion" (London edition, 1870, p. 420, etc.). In the case of the electric light, the result yielded by those experiments was that the heat in the invisible is eight times that of the visible region. But had there been an error in estimating the position of the extreme red by only two millimeters, so much would have been taken from the invisible and added to the visible that they would have been brought to equality, and then the slightest turn of the screw that carried the pile towards the dark space would have given a preponderance to the visible. It is obvious, therefore, that there cannot be certainty in such measures, unless the fixed lines be resorted to as standard points.

(2.) A ray which has passed through a solution of sulphate of copper and ammonia possesses no insignificant heating power. I took a stratum of a solution of that salt, of such strength that it only permitted waves to pass which are of less length than 4860. Seen in the spectroscope, the colors transmitted through it commenced with a thin green fringe, followed by blue, indigo, violet. It therefore gave rays in which, according to the accepted views, little or no heat should be detected. Yet I found that such rays produced one ninth of the heat of the solar beam. Does not this indisputably show that the more refrangible rays have a higher calorific power than is commonly imputed to them?

(3.) Again, by the use of the apparatus presently to be described, I found no difficulty in recognizing heat in the violet region. But in the mode of conducting the experiment heretofore resorted to, it could not be detected in rays more refrangible than the blue. It was this result which gave so much weight to the conclusion, that in the more refrangible regions the calorific power is replaced by chemical force, and strengthened the idea commonly entertained that the solar radiations consist of three distinct principles—heat, light, and actinism. In the Memoir above referred to, as soon to be published, I shall present some facts which apparently make this view indefensible.

(4.) If waves of light falling upon an absolutely black surface, and becoming extinct thereby, be transmuted into heat, if the warming of surfaces by incident light be nothing more than the conversion of motion into heat—an illustration of the modern doctrine of the correlation of forces, heat itself being only "a mode of motion"—it would seem extraordinary that the conversion should cease in the green or blue or in any middle ray. On the contrary, calorific effects ought to be traceable throughout the entire length of the spectrum. These views on the transference of motion from the ether to the particles of ponderable bodies, and conversely, I endeavored to explain in detail in a memoir on Phosphorescence inserted in the *Philosophical Magazine*, February, 1851, p. 98, etc. (Memoir VIII.). I had previously indicated them in the same journal, February, 1847 (Memoir V.). A given series of waves of red light impinging upon an extinguishing surface will produce a definite amount of heat, and similar series of violet waves should produce the same amount; for though an undulation of the latter may have only half the length of one of the former, and therefore only half its *vis viva*, yet, in consequence of the equal velocity of waves of every color, the impacts, or impulses, of the violet series will be twice as frequent as those of the red. The same principle applies to any intermediate color, and hence it follows that every color ought to have the same heating power.

Description of the Apparatus employed.

The optical arrangement I have employed for carrying the foregoing suggestions into practice is represented by Fig. 87, and in a horizontal section by Fig. 88.

A ray of sunlight reflected by a Silberman's heliostat comes into a dark room through a slit, *a*, one millimeter wide. It then passes through a prism, *b*. On the front face of this prism is a black paper screen, *c c*, having a rectangular opening, just sufficient to permit the light of the slit to pass. After refraction the dispersed rays fall as a spectrum on a concave metallic mirror, *d d*, nine inches in aperture and eleven in focus for parallel rays. I have sometimes used one of speculum metal, but more frequently one silvered on its front face. In front of this mirror there are therefore three foci. At a distance of eleven inches there is one, e, Fig. 88, giving a spectrum image of the sun. Still farther there is a second, f, which is a spectrum image of the slit, a, in which, if the prism be at its angle of minimum deviation, and the other adjustments be correctly made, will be seen the Fraunhofer lines. Again, still farther off, at g, is a focal image of the rectangular opening of the black paper, $c\,c$, on the front face of the prism. This image, arising from the recombination of all the dispersed rays, is consequently white. The second and third foci are at distances from the mirror depending on the distance of the slit, a, and the black paper, $c\,c$, respectively.

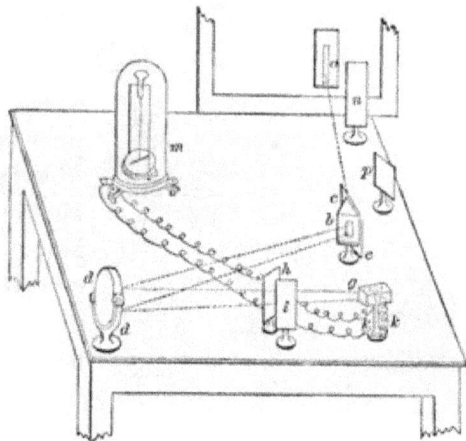

Fig. 87.

With the intention of being certain that the light coming through the slit, a, is falling properly on the rectangular opening in the prism screen, $c\,c$, a small looking-glass is placed at p. The experimenter, sitting near the multiplier, m, can then see distinctly the reflected image of that opening.

At the place where the second focal image, with its Fraunhofer lines, forms, two screens of white pasteboard, h, i, are arranged. By suitably placing the former of these, h, the more refrangible rays may be intercepted, and in like manner by the other, i, the less refrangible. In using these screens, and particularly h, care must be taken that no rays passing from the prism to the mirror be obstructed, a remark that applies especially to the invisible rays of less refrangibility than the red. For this reason the mirror, $d\,d$, must be placed at such an obliquity to its incident rays as to throw the focal images sufficiently on one side. Yet this obliquity must not be greater than is actually necessary for that purpose, or the purity of the second spectrum, with its Fraunhofer lines, will be interfered with. At the place of the third focus, arising

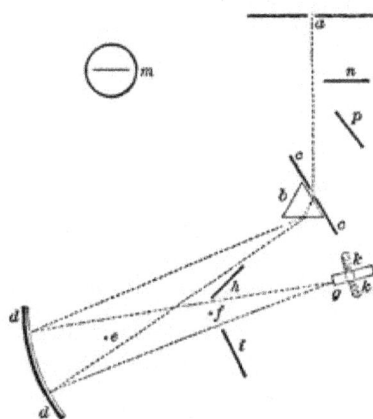

Fig. 89.

from the reunion of the dispersed rays, is the thermopile, g, connected by its wires, $k\,k$, with the multiplier, m.

Whenever any of the visible rays of the Fraunhofer spectrum are intercepted by advancing either of the screens h, i, the image on the face of the pile ceases to be white. It becomes of a superb tint, answering to the combination of the non-intercepted rays. A slip of white paper placed for a moment in front of the pile will satisfy the experimenter how magnificent these colors are. It is evident, therefore, that by this arrangement the pile will enable us to measure the heat of any particular ray, or of any selected combination of rays. The screens can be arranged so as to reach any designated Fraunhofer line.

The pile I have used is of the common square form; a linear pile would not answer. The focal image on the pile is of very much greater width than the slit a, on account of the obliquity of the front face of the prism.

By removing the screen h, and placing the screen i so that its edge coincides with the line A of the Fraunhofer spectrum, all the invisible heat radiations of less refrangibility than the red are cut off, except the contaminating ones arising from the general diffusion of light by the substance of the prism. Under these circumstances the image on the pile will be white, and the multiplier will give a deflection representing the heat of the visible and the extra-violet regions. If then the screen be advanced still farther, until it has intercepted all the less refrangible regions up to the sodium line D or a little beyond—that is, to the optical centre of the spectrum— the tint on the face of the pile will be greenish-blue, and the multiplier will give a measure of the heat of the more refrangible half of the visible spectrum, together with that of the ultra-violet rays: the latter portion may, however, be eliminated by properly using the other screen, h.

Besides the error arising from stray heat diffused through the spectrum in consequence of the optical imperfection of the prism, there is another which may be recognized on recollecting the relative positions of the prism, the concave mirror, and the face of the pile. It is evident that the prism, considered as a warm or a cool mass, is a source of disturbance, for the mirror reflects its image—that is, the image of the prism itself—to the pile. After the intromitted sunbeam has passed through the prism for a short time, the temperature of that mass has risen, and the heat from this source has become intermingled with the proper spectrum heat. But this error is very easily eliminated. It is only necessary to put a screen, n, in the path of the incoming ray, between the slit and the prism, and note the deflection of the multiplier. Used as we are here supposing, the multiplier has two zeros. The first, which may be termed the magnetic, is the position in which the needles will stand when no current is passing through the coil. The scale of the instrument should be set to this. The other, which may be termed the working zero, is formed by coupling the pile and the multiplier together, and introducing the screen n between the intromitting slit and the prism. On doing this it will probably be found that the index will deviate a few divisions. Its position should be accurately marked at the beginning and close of each set of measures, and the proper correction for them made. The disturbing influences of the mass of the prism, of the mirror, and of the pile itself, are thus eliminated. As respects the last, it should not be forgotten that it may be affected by changes in the position of the person of the experimenter himself.

With the intention of diminishing these errors, I have usually covered the upper and lower portions of the concave mirror, $d\,d$, with pieces of black paper, so arranged as to leave a band of sufficient width to receive and reflect the entire spectrum. $a\,a$, Fig. 89, is the upper paper, $b\,b$

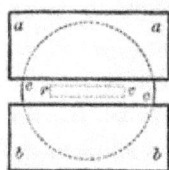

Fig. 89.

the lower, $c\,c$ the uncovered reflecting band, receiving the spectrum, $r\,v$. Had the spaces thus covered been permitted to reflect, they would have rendered more intense the image of the prism with its extraneous heat.

As regards the multiplier, care must be taken to avoid disturbance from aerial currents. I have one of these instruments of French construction, which could not be used in these delicate researches until proper arrangements were applied. It was covered with a glass shade. The slightest cause occasioned currents in its included air, which perpetually drifted and disturbed the needles. For this reason, and also for more accurate reading, it is best to view the position of the index through a small telescope.

The combination of needles being nearly astatic, attention must be paid to their magnetic perturbations, whether arising from local or other causes; and, since the vibrations are very slow, ample time must be given before the reading is ascertained.

The condition of the face of the pile is of importance. It must be such as to extinguish as completely as possible all the incident rays. To paint it with lamp-black, mixed with gum, will not answer—the surface so produced is too glossy and reflecting. The plan I have found best is to take a glass tube half an inch in diameter and six inches long, open at both ends, and use it as a chimney. A piece of camphor being set on fire at the lower end, and the face of the pile to be blackened being held for a moment at the upper, it is covered with a dense black film, without any risk of injury to the pile. Even at the best, when this has been done, there is an unavoidable source of error in the want of perfect blackness of the lamp-black. It is sufficient to inspect the face of the pile when receiving rays from the concave mirror to be satisfied how large a portion of light is reflected. The experiments of Dr. Tyndall show that this substance also transmits a considerable percentage of the heat falling on it. Its quality of transmitting light is well known to every one who has looked at the sun through a smoked glass.

The galvanometer I have used is calibrated according to the usual method. The numbers given in this Memoir do not represent the angles of deflection, but their corresponding forces.

The proper position of the intercepting screens h, i, can often be verified with precision by looking through blue cobalt glass. This glass insulates a definite red, an orange, and a yellow ray in the less refrangible regions, and then, commencing with the green, gives a continuous band to the end of the violet. Its red ray begins at the

less refrangible end of the spectrum, and ends near C;
it includes the fixed lines A, B, C. Its orange ray lies
wholly between C and D, including neither of those lines.
Its yellow ray begins near 5894, and ends about 5581;
the line D is therefore near its point of beginning. The
remaining continuous band begins about 5425; it therefore includes the lines E, b, F, G, H. I have found this
glass of much use in determining how far the screen i
has been pushed. It is convenient to select a light kind
of it, and by looking through one, two, or three pieces,
the depth of color can be regulated at pleasure.

The optical train which has acted on the sunbeam under examination is, therefore, (1) the sun's atmosphere;
(2) the earth's atmosphere; (3) the heliostat mirror of
speculum metal; (4) the prism; (5) the concave mirror
of silvered glass; (6) the blackened face of the thermo-
pile.

Results obtained by the Apparatus.

We are now ready to examine the results which this
optical apparatus yields, it having been of course previously ascertained that the reflecting band of the concave mirror, d d, is sufficient to receive all the radiations
coming from the prism, and that none are escaping past
its edges.

The operations required are as follows:

The heliostat is to be set, and its reflected ray brought
into the proper position. The optical train is adjusted,
the prism being at its minimum deviation, and the concave mirror giving a white image on the face of the
pile.

The screen h is then to be placed so that, without intercepting any rays coming from the prism to the mirror,
it cuts off all the Fraunhofer spectrum above H².

The screen i is so placed as to cut off all rays less refrangible than the sodium line D. More correctly, the

screen should be a little beyond D. The light on the face of the pile will now be greenish-blue.

The screen n is then placed so as to intercept the intromitted beam. When the needles of the multiplier come to rest, they give the working zero, which must be noted.

The intromitting screen n being now removed, the multiplier will indicate all the heat of the more refrangible rays; that is, from a little beyond D up to H^2. The force corrected for the working zero is to be noted.

The screen i is then removed to the line A, so as to give all the radiations between the lines A and H^2. The light on the face of the pile is white, and the multiplier gives the whole heat of the visible spectrum. By subtracting the foregoing measure from this, we have the heat of the less refrangible region; that is, from A to the centre of the spectrum.

As a matter of curiosity the experimenter may now, if he please, remove the screens h, i; the light on the face of the pile will still be white, and the multiplier will give the force of the entire radiations, except so far as they are disturbed by the thermochrose of the media. These measures, as they do not bear upon the problem under consideration, I do not give in the following tables.

Instead of advancing the screen i from the less towards the more refrangible regions, I have very frequently moved h from the more refrangible to the less. When it is brought down from H^2 to the centre of the spectrum, the light on the face of the pile is of an intense orange-red—it might perhaps be called a bromine-red. I need not give further details of this mode of experimentation, as I did not find that its results differed in any important degree from those obtained as just described.

The variation in different experiments may generally

be traced to errors in placing the screen i with exactness on the centre of the spectrum and on the line A.

For the sake of more convenient comparison, I have reduced all the different sets of experiments to the standard of 100 for the whole visible spectrum.

I have made use of four prisms: (1) rock-salt; (2) flint-glass; (3) bisulphide of carbon; (4) quartz, cut out of the crystal so as to give a single image.

All the observations here recorded were made on days when there was a cloudless sky.

TABLE I.—*Distribution of Heat by Rock-salt.*

		Series I.	Series II.
(1) Heat of the whole visible spectrum...............		100	100
(2) " more refrangible region..............		53	51
(3) " less " "		47	49

In this table the column marked Series I. gives the mean of four sets of measures, and that marked II. of three. At the beginning of each set the rock-salt was repolished.

TABLE II.—*Distribution of Heat by Flint-glass.*

		Series I.	Series II.
(1) Heat of the whole visible spectrum...............		100	100
(2) " more refrangible region..............		49	52
(3) " less " "		51	48

Series I. gives the mean of ten sets of measures, Series II. of eight.

TABLE III.—*Distribution of Heat by Bisulphide of Carbon.*

		Series I.	Series II.
(1) Heat of the whole visible spectrum...............		100	100
(2) " more refrangible region..............		52	49
(3) " less " "		48	51

The sulphide employed was devoid of any yellowish tinge; it was quite clear. Series I. is the mean of eight experiments, Series II. of ten.

TABLE IV.—*Distribution of Heat by Quartz.*

				Series I.	Series II.
(1) Heat of the whole visible spectrum				100	100
(2) "	more refrangible region			49	53
(3) "	less	"	"	51	47

Series I. represents twenty-seven experiments, Series II. twelve. In the former two quartz prisms were used to increase the dispersion; in the latter only one was employed.

Perhaps it may not be unnecessary for me to say that I have repeated these experiments many hundred times during a period of several months, including the winter and the summer, varying the conditions as to the hour of the day, arrangement of the apparatus, etc., as much as I could, and present the foregoing tables as fair examples of the results. Apprehending that the heliostat mirror, which was of speculum metal, might exert some disturbing influence on account of its faint reddish tinge, I replaced it with one of glass silvered on the front face, but could not detect any substantial difference in the results.

The important fact clearly brought into view by these experiments is, that if the visible spectrum be divided into two equal portions, the ray having a wave-length of 5768 being considered as the optical centre of such a spectrum, these portions will present heating powers so nearly equal that we may impute the differences to errors of experimentation. Assuming this as true, it necessarily follows that in the spectrum any two series of undulations will have the same heating power, no matter what their wave-lengths may be.

But this conclusion leads unavoidably to a most important modification of the views now universally held as regards the constitution of the spectrum. When a ray falls on an extinguishing surface, heat is produced; but that heat did not pre-exist in the ray. It arose from the stoppage of ether waves, and is a pure instance of the conversion of motion into heat—an illustration of the modern doctrines of the conservation and transmutation of force.

From this point of view, the conception that there exist in an incident ray various principles disappears altogether. We have to consider an incident ray as consisting solely of ethereal vibrations, which, when they are checked by an extinguishing substance, lose their *vis viva*. The effect that ensues depends on the quality of the substance. The vibrations imparted to it may be manifested by the production of heat, as in the case of lampblack, or by chemical changes, as in the case of many of the salts of silver. In the parallel instance of acoustics clear views have long ago been attained, and are firmly held. No one supposes that sound is one of the ingredients of the atmosphere, and it would not be more incorrect to assert that it is something emitted by the sounding body than it is to affirm that light or heat, or actinism, is emitted by the sun.

The progress of actino-chemistry would be greatly accelerated if there could be steadfastly maintained a clear conception of the distinction between the mechanism of a ray and the effects to which that ray may give rise. The evolution of heat, the sensation of light, the production of chemical changes, are merely effects—manifestations of the motions imparted to ponderable atoms; and these in their turn can give rise to converse results, as when we gradually raise the temperature of a substance the oscillating movements of its molecules are imparted to the ether, and waves of less and less length are successively engendered.

In the title of this Memoir I have employed the phrase "distribution of heat" in accordance with general usage; but if the conclusion arrived at be true, it is plain that this should be exchanged for "production of heat." The heat observed did not pre-exist in the incident rays: it is the result of their extinction.

The remark has been made that **these** results are essentially connected with photometry. In fact, any thermometer is converted into a photometer if its ball or other receiving surface be coated with a perfectly opaque non-reflecting substance.

•

MEMOIR XXIX.

ON THE DISTRIBUTION OF CHEMICAL FORCE IN THE
SPECTRUM.

From the American Journal of Science and Arts for December, 1872, and January,
1873; Philosophical Magazine, December, 1872.

CONTENTS:—*The curves of the calorific, luminous, and chemical spectra.* — *Their errors.* — *Inappropriateness of the term actinic rays.* — *There is no localization of chemical effects.*—*Every radiation can produce some specific effect.*—*Case of the silver compounds.*—*Bitumens and resins.*—*Carbonic acid.*—*Colors of flowers.*—*Law of Grotthus—Chlorine and hydrogen.*—*Bending of stems of plants.*—*Absorption essential to chemical action.*—*Decomposition of silver iodide.*—*Union of chlorine and hydrogen.* — *Angström's law.* — *General conclusions.* — *Chemical force exists in every portion of the spectrum.*—*Each radiation exercises chemical influences proper to itself.*

WITH scarcely an exception, the most recent works on the chemical action of radiations and spectrum analysis describe a tripartite arrangement of the spectrum, illustrated by an engraving of three curves, exhibiting the supposed relations of the calorific, the luminous, and the chemical spectra. This view, which by a mass of evidence may be shown to be erroneous, is exerting a very prejudicial effect on the progress of actino-chemistry.

I propose now to present certain facts which may aid in correcting this error. For this purpose it is necessary to show that chemical effects—decompositions and combinations—may take place in any part of the spectrum. The points to be established may be thus distinctly stated:

1st. That so far from chemical influences being restricted to the more refrangible rays, every part of the spectrum, visible and invisible, can give rise to chemical changes, of modify the molecular arrangement of bodies.

2d. That the ray effective in producing chemical or molecular changes in any special substance is determined by the absorptive property of that substance.

I may here remark that both these propositions were maintained by me many years ago: an example of the first will be found in the *Philosophical Magazine* (Dec., 1842), and of the second in a paper in the same journal, "On some Analogies between the Phenomena of the Chemical Rays and those of Radiant Heat" (Sept., 1841), Memoir XVII.

The opinion commonly held respecting the distribution of chemical force in the spectrum is mainly founded on the behavior of some of the compounds of silver. These darken when exposed to the more refrangible rays, and, unless correct methods of examination be resorted to, seem to be unaffected by the less refrangible. Hence it has been supposed that in the higher parts of the spectrum a special principle prevails, to which the designation of "actinic rays" is often applied—an inappropriate iteration. In these pages I use the derivatives of ἀκτίς not in this restricted sense, but as expressive of radiations of every kind. This is their proper signification.

Every part of the spectrum, no matter what its refrangibility may be, can produce chemical changes, and therefore there is no special localization of force in any limited region. Out of a large body of evidence that might be adduced, I select a few prominent instances.

1st.—*Case of the Compounds of Silver.*

Silver is the basis of the most important photographic sensitive substances. Its iodide, bromide, and chloride, darkening with rapidity under the influence of the more refrangible rays, have mainly been the cause of the misconception above alluded to respecting the tripartite constitution of the spectrum. It is necessary, therefore, to determine what are really the habitudes of these substances.

(1.) If a spectrum be received on iodide of silver, formed on the metallic tablet of the daguerreotype, and carefully screened from all access of extraneous light, both before and during the exposure, on developing with mercury vapor an impression is evolved in all the more refrangible regions. This stain corresponds in character and position to the blackening effect which under like circumstances would be found on any common sensitive silver paper. It is this which has given rise to the opinion that the so-called actinic rays exist only in the upper part of the spectrum. If, however, the action of the light be long continued, a white stain makes its appearance over all the less refrangible regions. It has a point of maximum to which I shall again presently refer.

(2.) But if the metallic tablet during its exposure to the spectrum be also receiving diffused light of little intensity, as the light of day or of a lamp, it will be found on developing that the impression obtained differs strikingly from the preceding. Every ray that the prism can transmit, from below the extreme red to beyond the extreme violet, has been active. The ultra-red lines a, β, γ are present. It must be borne in mind that the impression of these lines is a proof of proper spectrum action, and distinguishes it from that of diffused light, arising either from the atmosphere or from the imperfect transparency of the prism—a valuable indication. The resulting photograph shows two well-marked regions or phases of action. On its general surface, which, having condensed the mercury vapor, has the aspect of the high lights of the daguerreotype, and forms, as it were, the basis for the spectrum picture, there is in the region of the more refrangible rays a bluish or olive-colored impression, the counterpart of the result described in the foregoing paragraph. But in the region of the less refrangible rays no mercurial deposit has occurred, the place of those rays being depicted in metallic silver, dark, and answering to the shadows of the daguerreotype. This protected portion, which stands out in bold relief from the white background, reaches from a little below G to beyond the extreme red, and encloses the heat lines above named. They are in the form of white streaks. Though I speak of them as single lines, they are in reality groups, or perhaps bands.

The general appearance of the photograph at once suggests that the less refrangible rays can arrest the action of the daylight and protect the silver iodide from change. A close examination shows that there are three points—the extreme red, the centre of the yellow, and the extreme violet—which apparently can hold the daylight in check. There are also two intervening ones in which the actions conspire. The point of maximum protection corresponds to the point of maximum action referred to above in paragraph (1).

(3.) If the metallic tablet, previously to its exposure to the spectrum, be submitted for a few moments to a weak light, so that, were it developed, it would at this stage whiten all over, the action of the spectrum upon it will be the same as in the last case (2). But this change in the mode of the experiment leads to a very important conclusion. The less refrangible rays can reverse or undo the change, in whatever it may consist, that light has *already* impressed on the iodide of silver.

Now, bearing in mind the fact that the photographic

action of diffused light on this iodide is mainly due to the more refrangible rays it contains, we are brought by these experiments to the following conclusions :

1st. Every ray in the spectrum acts on silver iodide.

2d. The more refrangible rays apparently promote the action of daylight on that substance ; the less refrangible apparently arrest it.

3d. For the display of this arresting or antagonizing effect, it is not necessary that the less and more refrangible rays should be acting *simultaneously.* An interval may elapse, and they may act *successively.* Hence the effect is not due to the contemporaneous interference of waves of different periods of vibration with one another —the material particles of the changing substance of the silver iodide are involved.

I abstain for the moment from giving further details of these spectrum impressions. That has been very completely done by Herschel in the case of one I sent him many years ago. His examination of it, illustrated by a lithograph, may be found in the *Philosophical Magazine* (Feb., 1843). I shall have to return to the subject of the behavior of silver iodide in presence of radiations on a subsequent page of this Memoir.

The main point at present established is this, that the iodide under proper treatment is affected by every ray that a flint-glass prism can transmit, and therefore it is altogether erroneous to suppose that chemical force is restricted to the more refrangible portions of the spectrum.

2d.—*Case of Bitumens and Resins.*

These substances are of special interest in the history of photography, since in the hands of Niépce they probably were the first on which impressions in the camera were obtained and fixed. Their use has been abandoned in consequence, as it seems to me, of an incorrect opinion of their want of sensitiveness. Properly used, they are scarcely inferior to chloride of silver.

The theory of their use is very simple. Alcohol, ether, and various volatile oils, respectively dissolve certain portions of these substances. If such a solution be spread in a thin film upon glass, as in the collodion operation, and parts of the surface be then exposed to light, the portions so exposed become insoluble in the same menstruum. They may therefore be developed by its use. Practically, care has to be taken to moderate the solvent action and to check it at the proper time. The former is accomplished by dilution with some other appropriate liquid, the latter by the affusion of a stream of water.

The substance I have used is West India bitumen dissolved in benzine, and developed by a mixture of benzine and alcohol.

The bitumen solution being poured on a glass plate in a dark room, and drained off as in the operation of collodion, leaves a film sufficiently thin to be iridescent. This is exposed to the spectrum for five minutes, and then developed.

The beginning of the impression is below the line A, its termination beyond H. Every ray in the spectrum acts. The proof is continuous except where the Fraunhofer lines fall. A better illustration that the chemical action of the spectrum is not restricted to the higher rays, but is possessed by all, could hardly be adduced.

3d.—Case of Carbonic Acid.

The decomposition of carbonic acid by plants under the influence of sunshine is undoubtedly the most important of all actino-chemical facts. The existence of the vegetable world, and indeed it may be said the existence of all living beings, depends upon it.

I first effected this decomposition in the solar spectrum, as may be seen in Memoir X. The results obtained by me at that time from the direct spectrum experiment, that the decomposition of carbonic acid is effected by the less, not by the more refrangible rays, have been confirmed by all recent experimenters, who differ only as regards the exact position of the maximum. In the discussions that have arisen this decomposition has often been incorrectly referred to the *green* parts of plants. Plants which have been caused to germinate and grow to a certain stage in darkness are etiolated. Yet these, when brought into the sunlight, decompose carbonic acid, and *then* turn green. The chlorophyl thus produced is the effect of the decomposition, not its cause. Facts derived from the visible absorptive action of chlorophyl do not necessarily apply to the decomposition of carbonic acid. The curve of the production of chlorophyl, the curve of the destruction of chlorophyl, the curve of the visible absorption of chlorophyl, and the curve of the decomposition of carbonic acid, are not all necessarily coincident. To confound them together, as is too frequently done, is to be led to incorrect conclusions.

Two different methods may be resorted to for determining the rays which accomplish the decomposition of carbonic acid: 1st, the place of maximum evolution of oxygen gas in the spectrum may be determined; 2d, the place in which young etiolated plants turn green.

I resorted to both these methods, and obtained from them the same results. The rays which decompose carbonic acid are the same which turn etiolated plants green. They may be designated as the yellow with the orange on one side and a portion of green on the other. Though the form of experimentation does not admit of a close reference to the fixed lines, I think we are almost justified in supposing that the point of maximum action is in the yellow. It must be borne in mind that the rapidly increasing concentration of the rays occasioned by the peculiarity of prismatic dispersion towards the red end will give a deceptive preponderance in that direction. Without entering further into this discussion, it is sufficient for my present purpose to understand that the decomposition in question is accomplished by rays between the fixed lines B and F.

The two absorptive media, potassium bichromate and cupro-ammonium sulphate, so often and so usefully employed in actino-chemical researches, corroborate this conclusion. Plants cannot decompose carbonic acid, nor can they turn green in rays that have passed through a solution of the latter salt. They accomplish both those results in rays that have passed through the former.

The decomposition of carbonic acid, and the production of chlorophyl by the less refrangible rays of the spectrum, afford thus a striking illustration that chemical changes may be brought about by other than the so-called chemical rays.

4th.—*Case of the Colors of Flowers.*

The production and destruction of vegetable colors by the agency of light have, of course, long been a matter of common observation. Little has, however, been done in the special examination of the facts, and that little for the most part by Herschel.

We have only to examine his Memoir in the *Philosophical Transactions* (Part II., 1842) to be satisfied that nearly every radiation can produce effects. Thus the yellow stain imparted by the *Corchorus Japonica* to paper is whitened by the green, blue, indigo, and violet rays. The rose-red of the *Ten-weeks-stock* is in like manner changed by the yellow, orange, and red. The rich blue tint of the *Viola odorata*, turned green by sodium carbonate, is bleached by the same group of rays; that is,

by those less refrangible than the yellow. The green
(chlorophyl) of the *Elder* leaf is changed by the extreme
red.   •

It is needless to extend this list of examples. The
foregoing establish the principle that every part of the
spectrum displays activity, some vegetable colors being
affected by one, others by other rays. It is, however,
desirable that the general principle at which Herschel arrived—viz., that the *luminous* rays are chiefly effective—
should be more closely examined. Some important physiological explanations turn on that principle. These so-
called luminous rays are such as can impress the retina,
which, like organic colors, is a carbon compound. There
are strong reasons for inferring that carbon is affected
mainly by rays the wave-lengths of which are between
those of the extreme red and extreme violet, the maximum being in the yellow.

It is, however, to a former experimenter, Grotthus, that
we owe the discovery of the law under which these decompositions of the colors of flowers take place. This
law in repeated instances was verified by Herschel, and
more recently by myself. It may be thus expressed:
"The rays which are effective in the destruction of any
given vegetable color are those which by their union
produce a tint complementary to the color destroyed."
Even the partial establishment of this law, already accomplished, is sufficient to prove that chemical effects
are not limited to the more refrangible portions of the
spectrum, but can be occasioned by any ray.

5th.—*Case of the Union of Chlorine and Hydrogen.*

In Memoir XVIII. may be found the description of
an actinometer invented by me, depending for its indications on the combination of chlorine and hydrogen,
these gases having been evolved in equal volumes from

hydrochloric acid by a small voltaic battery. This instrument, modified to suit their purposes, was used by Professors Bunsen and Roscoe in their photometrical researches. Many of my experiments were repeated by them (*Transactions* of the Royal Society, 1856, 1857).

In Table III. of my Memoir above referred to, it is shown that this mixture is affected by every ray of the spectrum; but by different ones with very different energy. The maximum is in the indigo, the action there being more than 700 times as powerful as in the extreme red.

6th.—*Case of the Bending of the Stems of Plants in the Spectrum.*

It is a matter of common observation that plants tend to grow towards the light. Dr. Gardner was, however, the first to examine the details of this phenomenon in the spectrum. His Memoir is in the *Philosophical Magazine* (Jan., 1844). When seeds are made to germinate and grow for a few days in darkness, they develop vertical stems, very slender and some inches in length. These, on being placed so as to receive the spectrum, soon exhibit a bending motion. The stems in other parts of the spectrum turn towards the indigo; those in the indigo bend to the approaching ray. Removed into darkness, they recover their upright position. These movements are the most striking of all actinic phenomena. I have often witnessed them with admiration.

Dr. Gardner's experiments were repeated and confirmed by M. Dutrochet, who, in a report to the French Academy of Sciences (*Comptes-Rendus*, No. 26, June, 1844), added a number of facts respecting the bending of roots *from* the light, which he found to be occasioned by all the colored rays of the spectrum.

In Dr. Gardner's paper there are also some interesting

facts respecting the bleaching or decolorization of chlorophyl by light. He used an ethereal solution of that substance: •

"The first action of light is perceived in the mean red rays, and it attains a maximum incomparably greater at that point than elsewhere. The next part affected is in the indigo, and accompanying it there is an action from $+10.5$ to $+36.0$ of the same scale (Herschel's) beginning abruptly in Fraunhofer's blue. So striking is this whole result that some of my earlier spectra contained a perfectly neutral space, from -5.0 to $+20.5$, in which the chlorophyl was in no way changed, while the solar picture in the red was sharp and of a dazzling white. The maximum in the indigo was also bleached, producing a linear spectrum, as follows:

———————   ———————

in which the orange, yellow, and green rays are neutral. These, it will be remembered, are active in forming chlorophyl. Upon longer exposure, the subordinate action along the yellow, etc., occurs, but not until the other portions are perfectly bleached.

"In Sir J. Herschel's experiments there remained a salmon color after the discharge of the green. This is not seen when chlorophyl is used, and is due to a coloring matter in the leaf, soluble in water, but insoluble in ether."

I have quoted these results in detail because they illustrate in a striking manner the law that *vegetable colors are destroyed by rays complementary to those that have produced them*, and furnish proof that rays of every refrangibility may be chemically active.

At this point I abstain from adding other instances showing that chemical changes are brought about in every part of the spectrum. The list of cases here presented might be indefinitely extended, if these did not suffice. But how is it possible to restrict the chemical force of the spectrum to the region of the more refrangible rays, in face of the fact that compounds of silver such as the iodide, which have heretofore been mainly relied upon to support that view, and, in fact, originated it, are now proved to be affected by every ray from the invisible ultra-red to the invisible ultra-violet; how, when it is proved that the decomposition of carbonic acid, by far the most general and most important of the chemical actions of light, is brought about not by the more refrangible, but by the yellow rays? The delicate colors of flowers, which vary indefinitely in their tints, originate under the influence of rays of many different refrangibilities, and are bleached or destroyed by spectrum colors complementary to their own, and, therefore, varying indefinitely in their refrangibility. Towards the indigo ray the stems of plants incline; from the red their roots turn away. There is not a wave of light that does not leave its impress on bitumens and resins, some undulations promoting their oxidation, some their deoxidation. These actions are not limited to decomposition; they extend to combinations. Every ray in the spectrum brings on the union of chlorine and hydrogen.

The conclusion to which these facts point is, then, that it is erroneous to restrict the chemical force of the spectrum to the more refrangible, or, indeed, to any special region. There is not a ray, visible or invisible, that cannot produce a special chemical effect. The diagram so generally used to illustrate the calorific, luminous, and chemical parts of the spectrum serves only to mislead.

While thus we find that chemical action may take place throughout the entire length of the spectrum, the remarks that have been made in a previous Memoir (XXVI.) respecting the differences of calorific distribution in dispersion and diffraction, apply likewise to the chemical force. To be satisfied of this it is only necessary to compare photographic impressions given by a prism and a grating!

I published engravings of such photographs in 1844. They are referred to in Memoir VI. As they were obtained on silver plates made sensitive by iodine, bromine, and chlorine, they do not extend to the line F.

I had found that certain practical advantages arise from the use of a reflected instead of a transmitted spectrum. The ruled glass was therefore silvered upon its ruled face with tin amalgam, copying the surface perfectly. Of the series of spectra I used the first.

The fixed lines were beautifully represented in the photographs. They were, however, so numerous and so delicate that I did not attempt to do more than to mark the prominent ones. These were, I believe, the first diffraction photographs that had ever been obtained. The wave-lengths assigned were according to Fraunhofer's scale, which represents parts of a Paris inch.

The length of the photographic impression given by the prism I was then using, from the line H to the ultra-violet end of the spectrum, was about three times that from H to G; but in the spectrum by the grating, though the exposure was in one instance continued for a whole hour, the impression beyond H was not more than $1\frac{1}{2}$ times the length of that to G. In more moderate exposures the last fixed line in the photograph was about as far from H on one side as G was on the other. This, therefore, showed very clearly the difference of distribution in the diffraction and prismatic spectra.

Of the Chemical Action of Radiations on Substances.

Having offered the foregoing evidence in support of the first proposition considered in this Memoir, which was to the effect—

"That so far from chemical influences being restricted to the more refrangible rays, every part of the spectrum, visible and invisible, can give rise to chemical changes, or modify the molecular arrangement of bodies," I now pass to the second, which is—

"That the ray effective in producing chemical or molecular changes in any special substance is determined by the absorptive property of that substance."

This involves the conception of selective absorption, as I have formerly shown (*Philosophical Magazine*, Sept., 1841), Memoir XVII. A ray which produces a maximum effect on one substance may have no effect on another. Thus the rays which change chlorophyl are not those which change silver iodide.

In the examination of this subject I shall select two well-known instances, presenting the fewest elements and the simplest conditions. They are, 1st, the decomposition of silver iodide, the basis of so many photographic preparations; 2d, the production of hydrochloric acid by the union of its two constituents, chlorine and hydrogen—a mixture of these gases being exceedingly sensitive to light.

1st.—*Of the Decomposition of Silver Iodide.*

There are two forms in which the silver iodide has been used for photographic purposes: 1st, when prepared by the action of the vapor of iodine on metallic silver, as in the daguerreotype tablet; 2d, when nitrate of silver is decomposed by iodide of potassium, or other metallic iodides. These preparations differ strikingly in their ac-

tinic behavior, the former furnishing by far the most interesting series of facts.

When a polished surface of silver is exposed at common temperatures to the vapor of iodine, it speedily
tarnishes, a film of silver iodide forming. This passes
through several well-marked tints as the exposure continues and the thickness increases. They may be thus
enumerated in the order of their occurrence: 1, lemon-
yellow; 2, golden-yellow; 3, red; 4, blue; 5, lavender;
6, metallic; 7, deep yellow; 8, red; 9, green.

All these films are sensitive. Under the influence of
radiations they exhibit two phases of modification: 1st,
an invisible modification, which, however, can be made
apparent or developed, as Daguerre discovered, by exposure to the vapor of mercury, the iodide turning white
by the condensation of mercury upon it wherever it has
been exposed to light, but remaining unacted upon in
parts that have been in shadow; 2d, a visible modification, which arises under a longer exposure, the iodide
passing through various shades of olive and blue, and
eventually becoming dark gray.

But though all the variously tinted films of silver
iodide are impressionable, they differ greatly in relative
sensitiveness when compared with each other. This may
be very satisfactorily shown by producing on one silver
tablet bands of all the above-named colors—an effect
readily accomplished by suitably unscreening successive
portions of the tablet during the process of iodizing, and
then exposing all at the same time to a common radiation.
It will be found, on developing with mercury vapor, that
the bands of a yellow color have been the most sensitive,
those of a metallic aspect have been scarcely acted on,
and those of other tints intermediately. It is to be particularly remarked that the second yellow, numbered 7
in the above series, is equally sensitive with the first
yellow, numbered 2.

From this it appears that the sensitiveness of this form of iodide depends not merely on its chemical constitution, but also on its optical properties. The explanation of this different sensitiveness in different films of iodide becomes obvious when we cause a tablet prepared, as just described, with tinted bands to reflect the radiations falling on it to another tablet iodized to a yellow color, and placed in a camera. After due exposure and development of both with mercury, it will be found that the image of the first tablet formed on the second consists of bands of different shades of whiteness. The yellow parts of the first tablet have scarcely affected the second, but its metallic and blue parts have acted very powerfully. On comparing the first plate and its image on the second together, it will be perceived that the parts that have been affected on the one are less affected on the other.

It may therefore be inferred that the yellow films are sensitive because they absorb the incident radiation, and the metallic and blue are insensitive because they reflect it.

The effect, in whatever it may consist, which occurs during the invisible modification is not durable: it gradually passes away. If tablets that have received impressions be kept for a time before developing, the images upon them gradually disappear. On these tablets there is no lateral propagation of effect, nothing answering to conduction.

On examining the operation of a radiation continuously applied to one of these sensitive films, it will be discovered that a certain time elapses—that is, a certain amount of the radiation is consumed—before there is any perceptible effect. When that is accomplished, the radiation affects the film to a degree proportional to its quantity, until a second stage is reached; there is then another pause, followed by the second stage, in which visible modification or chemical decomposition sets in. The film begins to darken; it passes through successive tints —brown, red, olive, blue—and eventually becomes dark gray.

I have described in some of the foregoing paragraphs the action of the spectrum on silver iodide as presented on the tablet of the daguerreotype, showing the difference in the impression obtained: 1st, when extraneous light has been excluded; and, 2d, when it has been permitted simultaneously or previously to act.

In the latter case, in all that region of the spectrum from the more refrangible extremity to somewhat below the line G, the usual darkening effect manifested by silver compounds is observed; but beyond this, and to the extreme less refrangible rays, with certain variations of intensity, the action of the extraneous and simultaneously acting light is checked, and the effect of previously acting light is destroyed.

It happened that in 1842 I obtained two very fine specimens of the latter spectra; one of these I sent to Sir J. Herschel, the other is still in my possession.

In the *Philosophical Magazine* (Feb., 1843), Herschel gave a detailed description of these spectrum impressions. He was disposed to refer the appearance they present to the phenomena of thin films, but at the same time pointed out the difficulties in the way of that explanation. He also sent me three proofs he had obtained on ordinary sensitive paper, darkened by exposure to light, then washed with a solution of iodide of potassium, and placed in the spectrum. He described them as follows:

(1.) " Blackened paper from which excess of nitrate of silver has not been abstracted, under the influence of an iodic salt. Produced by a November sun. N. B.—View it also transparently against the light."

(2.) " Blackened paper under the influence of an iodic

salt when no excess of nitrate of silver exists in the paper."

(3.) "Action of spectrum under iodic influence when very little nitrate of silver remains in excess in the paper. To be viewed also transparently."

These paper photographs I still preserve. They are as perfect as when first made. The different colored spaces of the spectrum are marked upon them with pencil. The appearances they respectively present are as follows: (1) is bleached by the more refrangible rays, and blackened deeply from the yellow to the ultra-red; (2) is bleached from the ultra invisible red to the ultra-violet (a maximum occurs abruptly about the blue); (3) has the same upper spectrum as the others, a bleached dot in the centre of the yellow, and a darkened space on the extreme red. The action has reached from the ultra-red to the ultra-violet.

In Herschel's opinion, these effects in the less refrangible region are connected with the drying of the paper. It is well known that paper in a damp condition is more sensitive than such as is dry. But obviously this condition does not obtain in the case of the daguerreotype operation, which is essentially a dry process.

In 1846 MM. Foucault and Fizeau, having repeated the experiment originally made by me, presented a communication to the French Academy of Sciences to the effect that when a silver tablet which has been sensitized by exposure to iodine and bromine, and then impressed by light, is exposed to the spectrum, the effect is greatly increased in all the region above the line C, and is neutralized in all that below C. They remarked the distinctness with which the atmospheric line A comes out, and saw the ultra spectrum heat-rays α, β, γ described by me some years previously.

The interpretation given by them is, that the more refrangible rays promote the previous action of light; the less neutralize it. The curve representing the chemical intensity of the different rays would cross the axis of abscissas about the boundary of the red and orange; below that point to the ultra-red the ordinates would have negative values; above it to the ultra-violet those values would be positive (*Comptes-Rendus*, No. 14, tome xxiii.).

Hereupon M. Becquerel communicated to the same Academy a criticism on this interpretation, the opinion maintained by him being that while the more refrangible rays excite sensitive surfaces, the less refrangible, far from neutralizing, continue the action so begun. To the former he gave the designation "rayons excitateurs," to the latter "rayons continuateurs" (*Comptes-Rendus*, No. 17, tome xxiii.).

In 1847 M. Claudet communicated a paper to the Royal Society, subsequently published in the *Philosophical Magazine* (Feb., 1848), on this subject. His attention had been drawn to it by observing that the red image of the sun, during a dense fog, had destroyed the effect previously produced on a sensitive silver surface, and that this destruction could be occasioned at pleasure by the use of red and yellow screens. A surface which has been impressed by daylight, and the impression then obliterated by the less refrangible rays, had recovered its primitive condition. It was ready to be impressed again by daylight, and again the resulting effect might be destroyed. Claudet found that this excitation and neutralization might be repeated many times, the chemical constitution of the film remaining unchanged to the last.

These facts seem to be inconsistent with Herschel's opinion, that positive and negative pictures may succeed each other by the continued action of a radiation, on the principle of Newton's rings.

On a collodion surface such negative neutralizing or

reversing actions cannot be obtained by the less refrangible rays. The spectrum impression developed in the usual manner by an iron salt presents a sudden maximum about the line G, and continues thence to the highest limit of the spectrum. In the other direction it extends below F. From E to the ultra-red not a trace of action can be detected. The lines a, β, γ cannot be obtained on collodion. There is, therefore, a difference between its behavior under exposure to light and that of a daguerreotype tablet.

The reversals that are obtained on collodion by the use of certain haloid compounds are altogether different from the reversals on the thin films of a silver tablet. They are produced by the *more* refrangible rays.

On exposing a collodion surface (prepared in the usual manner) to daylight long enough to stain it completely, then washing off the free nitrate, and in succession dipping the plate into a weak solution of iodide of potassium, exposing it to the spectrum, washing, again dipping it into the nitrate bath, and finally developing, a reversed action is obtained. The daylight is perfectly neutralized, but not after the manner in a daguerreotype. In the region about G, the place of maximum action in collodion, the impression of the light is totally removed by an exposure of five seconds. In twelve seconds the protected space is much larger; in thirty seconds it has spread from F to H. It is, however, to be particularly remarked that the less refrangible rays show no action.

The results are substantially the same when, instead of iodide of potassium, the chloride of sodium, corrosive sublimate, bromide of potassium, or fluoride of potassium is used. In all these the reversing action is from F to H, and has its maximum somewhere about G. That is, the reversing action coincides with the direct action: there is no protection in the lower portion of the spectrum, as in the daguerreotype. The effect is altogether due to the change of composition of the sensitive film. Ordinarily it contains free nitrate; now it contains free iodide, chloride, etc.

The silver compounds of collodion absorb the radiations falling on them which are capable of producing a photographic effect. Yet, sensitive as it is, collodion is very far from having its maximum sensitiveness, as is shown by the following experiment, which is of no small interest to photographers: I took five dry collodion plates, prepared by what is known as the tannin process, and, having made a pile of them, caused the rays of a gas-flame to pass through them all at the same time. On developing, it was found that the first plate was strongly impressed, and the second, which had been behind it, apparently quite as much. Even the fifth was considerably stained. From this it follows that the collodion film, as ordinarily used, absorbs only a fractional part of the rays that can affect it. Could it be made to absorb the whole, its sensitiveness would be correspondingly increased.

Radiations that have suffered complete absorption can bring about no further change. Partial absorption, arising from inadequate thickness, may leave a ray possessed of a portion of its power. There must be a correspondence between the intensity of the incident ray and the thickness of the absorbing medium to produce a maximum effect.

Though the silver iodide is affected by radiations of every refrangibility, it is decomposed so that a subiodide results only by those of which the wave-length is less than 5000. If in presence of metallic silver, as on the daguerreotype tablet, the iodine disengaged unites with the free silver beneath. The rays of high refrangibility occasion in it chemical decomposition; those of less refrangibility, physical modification. In the language of the older theories of actino-chemistry, this substance may be said to exert a selective absorption. In this it illustrates the general principle, that it depends on the nature of the ponderable material presented to radiations which of them shall be absorbed.

2d.—Of the Union of Chlorine and Hydrogen.

An interesting experiment, illustrating the fact that chlorine gas absorbs the radiations which bring about its combination with hydrogen, may be made by covering a test-tube containing an explosive mixture of equal volumes of those gases with a large jar filled with chlorine. This arrangement may be exposed in the open daylight without risk of exploding the mixture; but if the experiment be made with a covering jar containing atmospheric air instead of chlorine, the gases immediately unite, and commonly with an explosion.

I placed a mixture of equal volumes of chlorine and hydrogen in a vessel made of plate-glass, the edges of the pieces being cemented together. This vessel was so arranged on a small porcelain trough, containing a saturated solution of common salt, that it could be used as a gas-jar. The radiations of a lamp were caused to pass through it, so as to be submitted to the selective absorption of the mixture. They were then received on a chlor-hydrogen actinometer.

Successive experiments were then made: 1st, with the radiations of a lamp after passing through the absorption vessel; 2d, with the same radiations after the vessel had been removed.

Two facts were now apparent: 1st, the mixture of chlorine and hydrogen in the absorption vessel began to unite under the influence of the rays of the lamp; 2d, the rays which had passed through that mixture had lost very much of their chemical force. It was not totally extinct, but the actinometer showed that it had undergone a very great diminution. .

From this it follows that, on its passage through a mixture of chlorine and hydrogen, the radiation had suffered absorption, and as respects the mixture under trial had become deactinized. Simultaneously the mixture itself had been affected, its constituent gases uniting. And thus it appears that the radiation had undergone a change in producing a change in the ponderable matter.

The following modification of this experiment shows the part played by the chlorine and hydrogen respectively when they are in the act of uniting:

(a) The glass absorption vessel above described was filled with atmospheric air, and the chemical force of the radiation passing from the lamp through it was determined. It was measured by the time required to cause the index of the actinometer to descend through one division. This was 12 seconds.

(b) The absorption vessel was now half filled with chlorine, obtained from hydrochloric acid and peroxide of manganese. The chemical force of the ray after passing through it was determined as before. It was now represented by 25½ seconds.

(c) To the chlorine an equal volume of hydrogen was added, the absorption vessel being consequently full of the mixture. The radiation was now passing through a stratum of chlorine diluted with hydrogen, and the point to be determined was whether it had undergone the same, or a greater, or less loss than in the preceding case, since the chlorine was now uniting with the hydrogen. On measuring the force, it was found to be represented by 19 seconds.

(d) Lastly, the first (a) of these measures was repeated with a view of ascertaining whether the intensity of the lamp had changed. It gave 12 seconds as before.

From these observations it may be concluded that the addition of hydrogen to chlorine does not increase its absorptive power. Moreover, it is obvious that the action of the radiation is expended primarily on the chlorine, giving it a disposition to unite with the hydrogen, and that the functions discharged by the chlorine and by the hydrogen respectively are altogether different. The ray itself also undergoes a change; it suffers absorption and loss of a part of its *vis viva*.

As to the ray which is thus absorbed. In 1835 I found that a radiation which had passed through a solution of potassium bichromate failed to accomplish the union of chlorine and hydrogen; but one which had passed through ammonia sulphate of copper could do it energetically. This indicates that the effective rays are among the more refrangible. On exposing these gases in the spectrum, the maximum action takes place in the indigo rays (*Philosophical Magazine*, Dec., 1843), Memoir XVIII.

Recently (1871) some suggestions have been made by M. Budde respecting the action of light upon chlorine. Admitting the correctness of the theorem that the molecules of most elementary gases consist of two atoms, he conceives that the effect of light on chlorine is to tend to divide, or actually to divide, its molecule into isolated atoms. These atoms, if the gas be kept in the dark, may reunite into molecules.

The chlorine molecule cannot unite with hydrogen; the chlorine atom can; hence insolation brings on combination. But if the chlorine be unmixed, there will, as a consequence of insolation, be a certain proportion of uncombined atoms; and from this, together with Avogadro's theorem, is drawn the conclusion that this gas through insolation increases in specific volume. Moreover, as the reunion of the chlorine atoms probably produces heat, rays of high refrangibility will cause chlorine to expand; but it will contract to its original volume when no longer under the influence of light.

In corroboration of this conclusion, Budde found that a differential thermometer filled with chlorine showed a certain expansion when placed in the red or yellow rays, but it gave an expansion six or seven times greater when in the violet rays. With carbonic acid and ether no such effect took place.

It should not be forgotten, however, in considering the bearing of these experiments, that chlorine merely because it is yellowish-green will absorb rays of a complementary—that is, of an indigo and violet—color and become heated thereby.

It has next to be determined whether the points of maximum action—that is, the points of maximum absorption—correspond to the rays of emission of either or both these gases, as they apparently ought to do under Angström's law: "A gas when luminous emits rays of light of the same refrangibility as those which it has the power to absorb."

Of the four rays characteristic of hydrogen there is one the wave-length of which is 4340. It is in the indigo space.

Plücker gives for chlorine a ray nearly answering to this. Its wave-length is 4338, and also another, 4346, the latter being one of the best marked of the chlorine lines.

There are, therefore, rays in the indigo which are absorbed both by hydrogen and by chlorine. The place of these rays in the spectrum corresponds to that in which the gases unite—the place of maximum action for their mixture.

But the absorptive action of chlorine is not limited

to a few isolated lines. The gas removes a very large portion of the spectrum. Subsequent experiments must determine whether each of these lines of absorption is also a line of maximum chemical action.

The chlor-hydrogen actinometer furnishes the means of ascertaining many facts respecting the combination of those substances advantageously, since it gives accurate quantitative measures.

By referring to my papers in the *Philosophical Magazine* (Dec., 1843; July, 1844; Nov., 1845; Nov., 1857) it will be found that chlorine and hydrogen do not unite in the dark at any ordinary temperature or in any length of time; but if exposed to a feeble radiation, such as that of a lamp, they are strongly affected. The phenomena present two phases: 1st, for a brief period there is no recognizable chemical effect, a preliminary actinization, or, as Professors Bunsen and Roscoe subsequently termed it, photo-chemical induction, taking place (it is manifested by an expansion and contraction of the mixture); 2d, the combination of the gases begins, it steadily increases, and soon acquires uniformity. In obtaining measures by the use of these gases we must, therefore, wait until this preliminary actinization is completed. That accomplished, the hydrochloric acid arising from the union of the gases is absorbed so quickly that the movements of the index-liquid over the graduated scale give trustworthy indications.

As regards the duration of the effect produced on the gases by this preliminary actinization, I found that it continued some time—several hours (*Philosophical Magazine*, July, 1844), Memoir XIX. Professors Bunsen and Roscoe, however, in their Memoir read before the Royal Society, state that it is quite transient (*Transactions of the Royal Society*, 1856).

This preliminary actinization completed, the quantity

of hydrochloric acid produced measures the quantity of the acting radiation. This I proved by using a gas-flame of standard height, and a measuring lens (page 265) consisting of a double convex, five inches in diameter, sectors of which could be uncovered by the rotation of pasteboard screens upon its centre, the quantity of hydrochloric acid produced in a given time being proportional to the area of the sector uncovered. The same was also proved by using a standard flame and exposing the gases during different periods of time. The quantity of hydrochloric acid produced is proportional to the time.

The following experiment illustrates the phenomena arising during the actinization of a mixture of chlorine and hydrogen, and substantiates several of the foregoing statements.

The diverging rays of a lamp were made parallel by a suitable combination of convex lenses. In the resulting beam a chlor-hydrogen actinometer was placed, there being in front of it a metallic screen, so arranged that it could be easily removed or replaced, and thus permit the rays of the lamp to fall on the actinometer or intercept them.

On removing the screen and allowing the rays to fall on the sensitive mixture in the actinometer, an expansion amounting to half a degree was observed. In 60 seconds this expansion ceased.

The volume of the mixture now remained stationary, no apparent change going on in it. At length, after the close of 270 seconds, it was beginning to contract and hydrochloric acid to form.

At the end of 45 seconds more a contraction of half a degree had occurred: the volume of the mixture was therefore now the same as when the experiment began, this half degree of contraction compensating for the half degree of expansion.

The rate of contraction of the gaseous mixture—that is, the rate at which its constituents were uniting—was then ascertained.

From these observations it appeared that when chlorine and hydrogen unite, under the influence of a radiation, there are four distinct periods of action:

1st. For a brief period the mixture expands.

2d. For a much longer period it then remains stationary in volume, though still absorbing rays.

3d. Contraction arising from the production of hydrochloric acid begins. At first it goes on slowly, then more and more rapidly.

4th. After that contraction is fully established, it proceeds with uniformity, equal quantities of hydrochloric acid being produced in equal times by the action of equal quantities of the rays.

The prominent phenomena exhibited by a mixture of chlorine and hydrogen are a preliminary absorption and a subsequent definite action.

It may be remarked, since a similar preliminary absorption occurs in the case of other sensitive substances, that there is in practical photography an advantage, both as respects time and correctness in light and shadow, gained by submitting a sensitive surface to a brief exposure in a dim light, so as to pass it through its preliminary stage.

The expansion referred to as taking place during the first of these periods may be advantageously observed when the disturbing radiation is very intense. It is well seen when a Leyden-jar is discharged in the vicinity of the actinometer. Though this light lasts but a very small fraction of a second, it produces an instantaneous expansion, followed by an instantaneous contraction. Not unfrequently the gases unite with an explosion. I have had several of these instruments destroyed in that manner.

It might be supposed that this instantaneous expansion is due to a heat disturbance arising from the absorption of rays that are not engaged in producing the chemical effect. But this interpretation seems to be incompatible with the instantaneously following contraction. Though it is admissible that heat should be instantaneously disengaged by the preliminary actinization, it is difficult to conceive how it can so instantaneously disappear.

When the radiation is withdrawn and the hydrochloric acid absorbed, there is no after-combining. The action is perfectly definite. For a given amount of chemical action, an equivalent quantity of the radiation is absorbed.

The instances I have cited in this discussion of the mode of action of radiations are, one of decomposition, in the case of the silver iodide; and one of combination, in the case of hydrochloric acid. I might have introduced another—the dissociation of ferric oxalate, which I have closely studied—but it would have made the Memoir of undue length. From the facts herein considered the following deductions may be drawn:

When a radiation impinges on a material substance, it imparts to that substance more or less of its *vis viva*, and therefore undergoes a change itself. The substance also is disturbed. Its physical and chemical properties determine the resulting phenomena.

(1st.) If the substance be black and undecomposable, the radiation establishes vibrations among the molecules it encounters. We interpret these vibrations as radiant heat. The molecules of the medium do not lose the *vis viva* they have acquired at once, since they are of greater density than the ether. Each becomes a centre of agitation, and heat-radiation and conduction in all directions are the result. The undulations thus set up are commonly of longer waves; and as the movements gradually decline, the shorter waves of these are the first to be extinguished, the longer ones the last. This, therefore, is in accordance with what I found to be the case in the gradual warming of a solid body, in which the long waves pertain to a low temperature, the short ones arising as the temperature ascends (Memoir I.).

In some cases, however, instead of the disturbing undulation giving rise to longer waves, it produces shorter ones, as is shown when a platinum wire is put into a hydrogen flame, or by Tyndall's experiments, in which invisible undulations below the red give rise to the ignition of platinum.

(2d.) If the substance be colored and undecomposable, it will extinguish rays complementary to its own tint. The temperature will rise correspondingly.

(3d.) If the substance be decomposable, those portions of the radiation presented to it which are of a complementary tint will be extinguished. The force thus disappearing will not be expended in establishing vibrations in the arresting particles, but in breaking down the union of those which have arrested them from associated particles. No vibrations, therefore, are originated; no heat is produced; there is no lateral conduction.

In actinic decompositions the effects may be conveniently divided into two phases: 1st, physical; 2d, chemical. The physical phase precedes the chemical. It consists in a preliminary disturbance of the group of molecules about to be decomposed. Up to a certain point the dislocation taking place may be retraced or reduced, and things brought back to their original condition. But that point once gained, decomposition ensues, and the result is permanent.

I may perhaps illustrate this by a familiar example. If a sheet of paper be held before a fire, its surface will

gradually warm; and if the exposure be not too long or
the fire too hot, on removing it the paper will gradually
cool, recovering its former condition without any permanent change. One could conceive that the laws of absorption and radiation might not only be studied, but
again and again illustrated by the exposure and removal
of such a sheet. But a certain point of temperature or
exposure gained, the paper scorches—that is, undergoes
chemical change—and then there is no restoration, no re-
covery of its original condition. Hence it may be said
of such a sheet of paper that it exhibits two phases, in
the first of which a return to the original condition is
possible; in the second such a return is impossible, because of the supervening of the chemical change.

An investigation of the facts produced by a ray presents, then, these two separate and distinct phases—the
physical and the chemical.

General Conclusions.

The facts presented in the preceding and the present
Memoir suggest the following conclusions:

1st. That the concentration of heat heretofore observed
in the less refrangible portion of the prismatic spectrum
arises from the special action of the prism, and would not
be perceived in a diffraction spectrum.

2d. From the long-observed and unquestionable fact
that there is in the prismatic spectrum a gradual diminution in the heat-measures from a maximum below the
red to a minimum in the violet, coupled with the fact
now presented by me (that the heat of the upper half of
the spectrum is equal to that of the lower half), it follows
that the true distribution of heat throughout the spaces
of the spectrum is equal. In consequence of the equal velocity of ether-waves, they will, on complete extinction by
a receiving surface, generate equal quantities of heat, no

matter what their length may be, provided that their extinction takes place without producing any chemical effect.

3d. That it is incorrect to restrict to the upper portion of the spectrum the property of producing chemical changes. Such changes may be produced by waves of any refrangibility.

4th. That every chemical effect observed in the spectrum is in consequence of the absorption of a specific radiation, the absorbed or acting radiation being determined by the properties of the substance undergoing change.

5th. That the figure so generally employed in works on actino-chemistry to indicate the distribution of heat, light, and actinism in the spectrum serves only to mislead. The heat curve is determined by the action of the prism, not by the properties of calorific radiations; the actinic curve does not represent any special peculiarities of the spectrum, but the habitudes of certain compounds of silver.

•

MEMOIR XXX.

ON BURNING GLASSES AND MIRRORS—THEIR HEATING AND CHEMICAL EFFECTS.

Collected and condensed from the Philosophical Magazine, May, 1851; Harper's Monthly Magazine, No. 329.

CONTENTS:—*Can concentrated rays produce new chemical decompositions? —Effects of amplitude, frequency, and direction of vibration in the ether-waves.—Clock lenses for long exposures.—Decomposition of water by chlorine.—Attempt to decompose it by bromine and iodine.—Use of absorbing troughs.—Dry silver iodide not decomposed by light.—Decompositions under water.—Decompositions in a spherical concave.— Effect of extraneous mixtures.—They do not make collodion more sensitive.—Antagonization of radiations.—Case of electric spark.—Effects of polarized light.—Attempts to polarize light by an electro-magnet.— Mechanical cause of decompositions by light.*

WHEN, many years ago, I commenced an experimental examination of the chemical action of light, I entertained great expectations of what might be accomplished by the use of burning-glasses. It seemed reasonable to suppose that if the direct sun-rays could occasion so many decompositions, their chemical force would be incomparably greater when their brilliancy was exalted by a mirror or a lens. Of the two, a concave metallic mirror should produce a more characteristic effect, since it returns the rays as it receives them, but a special and very important portion of them is absorbed by the selective action of the lens.

I had not, however, at that time the means of making these experiments in a satisfactory manner, and, though very much disappointed with the result, postponed the prosecution of them to a more favorable opportunity. Obtaining from time to time several isolated facts, I was led, in meditating upon them, to what seemed to be some general conclusions respecting the chemical action of radiations. Several of these were published, in a desultory manner, in various periodicals; but it was not until May, 1851, that they were collected in the *Philosophical Magazine*, under the title of a "Memoir on the Chemical Action of Light." Of this, the following is an abstract.

The general discussion of the problem of the chemical action of a ray involves the following considerations:

1. In what manner does the ray act, and what are the changes it undergoes?

2. What is the nature of the impression made on the material group, the decomposition of which ensues?

Many facts justify the supposition that the parts of all material substances are in a state of incessant vibration. To each particular thermometric degree there belongs a particular frequency of vibration. As soon as these motions approach four hundred billions in a second, red light is emitted, and the temperature is near 1000° Fahr. As the frequency increases, rays of a higher refrangibility are in succession evolved, and the temperature correspondingly rises. On the other hand, when these oscillatory movements decline, the temperature of the body falls.

These principles lead to a ready explanation of the nature of the exchanges of heat and the cause of the equilibrium of temperature. The vibratory molecular motions are necessarily propagated to the ether, through which medium they are again transferred to the particles of other bodies, on which the ethereal waves impinge, as a vibrating string excites undulations in the air, and these, in their turn, can give birth to analogous motions in other strings at a distance.

There is an analogy between the relations of a hot

and cold body and those of two strings, one of which is
emitting a musical sound and compelling the other to
execute synchronous movements. The ether in the one
case and the air in the other are the media through which
their motions pass.

Equilibrium of temperature takes place when the molecules of the substances concerned are in synchronous
and equal vibration. A hot body in presence of a cold
one compels the latter to hasten its rate of motion, its
own rate all the time declining, and this continues until
both have the same frequency; then equilibrium of temperature results. The theory of the exchanges of heat is,
therefore, only an expression for the exchanges of vibrations through the ether.

But temperature in thermotics is the equivalent term
for *brilliancy* in optics. Both refer to compound qualities, depending not only on *frequency* of vibration, but
also on its *amplitude*. As the degree of heat of a mass
rises, the mass expands, the increase in its volume indicating that not only do its parts vibrate more swiftly,
but also that their individual excursions are increased.
It follows, therefore, that every mass will have a determinate volume for every degree of heat, the volume increasing as the temperature rises. On this view the explanation of the expansion of bodies by heat is that their
parts are not only vibrating more quickly, but also that
the individual excursions are greater.

The atoms of the chemical elements differ in weight.
We therefore should not expect that the ethereal vibrations would throw them into movement with equal facility, but that some would yield more readily than others.
Is not this what we express in chemistry by the term
specific heat?—a body, the capacity of which is great, re-
quiring a prolonged application of ethereal pulses before
a consentaneous motion is reached, and in its turn im-

pressing on the ether during cooling a correspondingly prolonged series of motions.   And is not this the cause of that remarkable relation between the atomic weights of elementary bodies and their specific heats, discovered by Dulong and Petit?

These considerations may lead us to inquire whether the general cause of the decomposition of compound bodies by radiations is due to the circumstance that all the atoms of which their molecules are composed take on the vibratory motion with unequal facility.   Thus if a certain compound molecule be submitted to the influence of an intense radiation, some of its constituent particles may vibrate consentaneously at once, and others more tardily. Under these circumstances, the continued existence of the group may become impossible, and decomposition ensue in the necessity of the case.

In entering upon the experimental analysis of the action of a ray upon a decomposable body, there are three different points to be considered, so far as the ray itself is concerned: 1. To what extent and in what manner is the result affected by the *intensity* of the ray, or by the *amplitude* of the vibrating excursions?   2. How is it affected by the *frequency* of the pulsatory impressions? and, 3. How by the *direction* in which the vibrations are made, as involved in the idea of polarization?   I shall now examine these in succession.

1. *To what extent and in what manner is the decomposition of a compound body affected by the* intensity *of a ray or the amplitude of the vibrating excursions?*

If the different degrees of facility with which atoms receive the impression of ethereal vibrations be the true cause of decomposition by light, we should expect that many such changes would become possible under the influence of a burning-lens which are not so in the direct rays of the sun.

This idea is favored by what we find in the case of heat. The burning-glass has long had celebrity in that respect, and in former times was the most powerful means of reaching a high temperature.

The effect of the glass is due to the rapidity with which it can supply caloric, contrasted with the loss by conduction, radiation, etc. Thus an object of any kind exposed to the sun receives heat at a certain rate; but it is simultaneously experiencing a loss by conduction, radiation, and currents in the air. Exposed to the focus of a lens, the supply becomes, in a given time, greater than before; and the temperature rising, great effects are the necessary result.

But changes brought about by light are in a different predicament. Here conduction is entirely absent, as is also loss by currents in the air. The cumulative effects of a long exposure give the same action as a highly concentrated ray furnishes in a brief period of time. In this case, therefore, everything will depend on the absorptive power of the substance.

When a piece of polished silver is placed in the focus of a burning-lens, it remains quite cold, because of its high reflecting power; but if blackened, it melts in an instant. And so with chemical changes. A body which, like chlorine, can exert an absorptive action on the ray becomes modified, and induces changes; but if, like oxygen, it has not that property, it will remain indifferent and unaffected by the most intense radiation.

Considering, however, that the calorific effects of the converged solar rays are so striking, we may reasonably inquire whether, in like manner, the chemical action can be increased. There is a very general impression that the intense radiation of tropical climates accomplishes changes which cannot be imitated by the feebler light of higher latitudes, and perhaps decompositions may be

brought about by a large convex lens which the direct
rays of the sun are wholly inadequate to produce.

A very brilliant beam may possibly break up a given
combination, which a far greater quantity of light, acting
through a long period, might be inadequate to touch.
Sir R. Kane states that he, with M. Dumas, could remove
two atoms of hydrogen from acetone by the action of
chlorine in the sunshine at Paris, but in Dublin only
one.

In Fig. 90, a is a convex burning-lens supported in
ribbed frame, $b\,b\,;$ there is at c a second lens to hasten

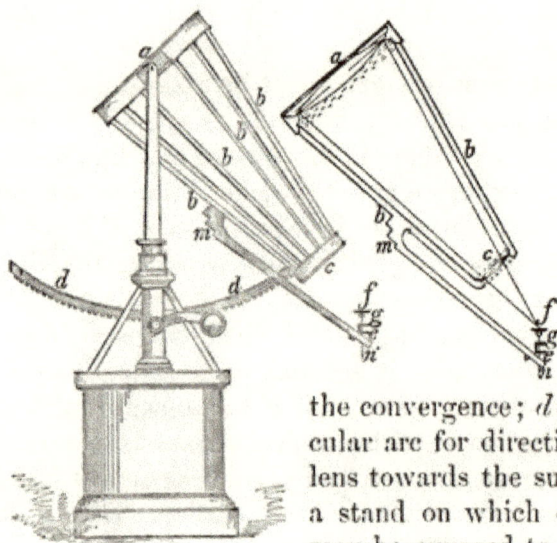

Fig. 90.

the convergence; $d\,d$, a circular arc for directing the
lens towards the sun; $e\,g$,
a stand on which objects
may be exposed to the fo-
cal point, $f.$ It is carried
by a stout bar, $m\,n$, attached to the frame.

By this instrument I endeavored to collect a series
of facts which might set this part of the question in
its true light. The lens a was of very fine and thin
French plate-glass, twelve inches in diameter in the
clear. Its goodness was such that on a fine day platinum might be melted in its focus. It was ground

and polished for me by the late Mr. Fitz, whose skill
was shown in the large and excellent telescopic objec-
tives that he made. He mounted it on a suitable support; it required, however, to be guided by the hand as
the sun moved. When the college building of the Medical Department of the University of New York was destroyed by fire in 1865, I had to regret the loss of this
instrument, with much other apparatus, and many documents that were of unappreciable value to me. Mounted
as the lens was, its use was attended with considerable
risk to the eyes, on account of the excessive brilliancy of
the focus. Screens and dark spectacles were found to be
very unsatisfactory, and an illness which I consequently
contracted admonished me either to abandon the subject
or pursue it in some other way.

The following experiments were made with a smaller
glass, consisting of a combination of two similar lenses,
their diameter being five inches and focal distance eight.
It was, in fact, the large lens of an old-fashioned lucernal
microscope, such as was made in London a century ago.
I had it fixed on a polar axis, as shown in Fig. 91, and
by the aid of a clock it could follow the motion of the
sun with such accuracy that, when once set in the morning, an object might be exposed in its focus, if desirable,
for a whole day. It had a contrivance on the frame carrying the lens for supporting small crucibles, glass matrasses (Fig. 92), charcoal supports, etc., at the proper
point, which might be either at the focus or at any other
distance from the lens, as the circumstances of the experiment required. Among these instruments were thermometers, blackened or otherwise so arranged as to exercise any desired selective absorption. At the outset
of any experiment, the whole face of the lens could be
covered with a blackened pasteboard screen, with a hole
half an inch in diameter. Through this a sufficient

amount of light could be transmitted to enable one to arrange the various details of the proposed experiment; and when everything was ready, the screen was removed, and in the concentrated and brilliant focus the action went on. I found that this simple contrivance was an invaluable relief to the eyes.

In Fig. 91, a a is the heliostat clock; b, its polar axis; d d, a frame carrying the lens, c, and having an arrangement at f for supporting flasks, crucibles, or other apparatus. This turns on a double joint at e, so that the lens may be directed to the sun.

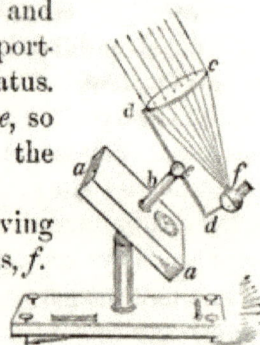

In Fig. 92, a is a small flask receiving the converging rays, b, at their focus, f.

The lens being five inches in diameter, and the space covered by the solar focal image, owing to

Fig. 91.

want of achromaticity and spherical aberration, one fifth of an inch, the multiplying effect would be 625 times, if the glass were perfectly transparent, and there were no loss by reflection from its surfaces. On a summer day of average brightness, with the thermometer at 68° in the shade, and the bulb, not being blackened, at 108° in the sun,

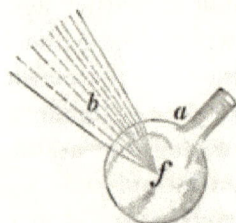

Fig. 92.

this lens could fuse copper instantly, the bead oxidizing only superficially, and cutting readily after fusion. Black oxide of copper in a little crucible of platinum foil melted into a slaty-looking substance at once. Wrought iron did not melt alone; but if exposed on a charcoal support in a globule of microcosmic salt, previously fused by the lens, it gave a clear, round bead, which readily extended when beaten upon an anvil. The globule of flux turned

black. The specimen employed was cut from a piece of
good iron wire; and though it might be thought that exposure on the charcoal would tend to turn it into cast
iron, its subsequent complete malleability seems to dis-
prove this. Spongy platinum did not melt alone, nor
even if enclosed in a globule of fused microcosmic salt.
We may therefore estimate the working power of this
lens on a substance placed in its focus as somewhat
above the point of fusion of wrought iron, and lower
than the point of fusion of platinum. This refers to temperature only. The power of the lens as to light must
be enormously greater.

We may now examine the chemical effects produced
by this lens.

Two small glass matrasses, the bulbs of which were
about half an inch in diameter, were filled with chlorine
water, the one being exposed to the direct rays of the
sun, the other to the converging rays of the lens. Decomposition of the water occurred in both, but with far
more activity in that placed in the focal point. The difference was at once so striking to the eye that I made no
attempt to measure it. It is plain that the greater the
quantity of incident light, the more rapid the decomposition; though, after the first moment of action, the solutions being no longer the same in constitution, the quantities of gas disengaged are no longer proportional to the
incident light.

There is thus no difficulty in effecting rapidly the decomposition of water by chlorine under the influence of
the sun, but under the same circumstances iodine and
bromine are inadequate to produce such an effect.

A solution of bromine in water was prepared, the water being first boiled to expel the air contained in it. It
was placed in a half-inch matrass (Fig. 93), and exposed
to the focus of the lens. As the temperature rose rap-

idly, the water was depressed in the bulb by the steam and bromine vapor which occupied the upper part, the bulb being placed uppermost, and the tube dipping into a phial which served as a reservoir. After the exposure had continued for two hours and a half, the matrass was removed from the lens and suffered to cool. There remained uncondensed a little bubble, measuring about $\frac{1}{100}$ of a cubic inch; but this was probably nothing more than the atmospheric air which had found access to the water, for on submitting the same specimen to another exposure for three hours, after the gas had been decanted from it, a little bubble, the diameter of which was estimated at one fiftieth of an inch, was all that could be procured.

In Fig. 93, *a* is the flask containing the bromine water; *b*, a phial serving as a reservoir. It is half filled with the same water.

In like manner I endeavored to decompose water by iodine, and with the same negative result, even when the exposure to the focal point lasted four hours. When proper care had been taken to remove from the solution all traces of air, no gas was evolved.

Fig. 93.

To reduce the heating effect of the lens, and allow the more refrangible rays alone to act, there was interposed between the lens and its focus a stratum of a solution of sulphate of copper and ammonia one third of an inch thick, included between two flat plates of glass, suitably arranged and carried along with the other parts by the movement of the clock. The cone of solar rays now passed through this absorbent medium (Fig. 94).

In Fig. 94, *a*, *b* is the flask and reservoir, but the converging rays pass through an absorbent trough, *c c*, shown in front view at *d*, *e* being the circular cell containing the blue solution.

In the focus of blue light thus formed there was exposed for two and a half hours (from 7¾ to 10¼ A.M., June 13, 1848) an inverted half-inch bulb containing iodine water, with a few particles of iodine. Temperature in the shade, 64°; in the sun, 86°. At the end of that time there was found an insignificant bubble of air, estimated at one thirtieth of an inch in diameter. It could, of course, be nothing but atmospheric air.

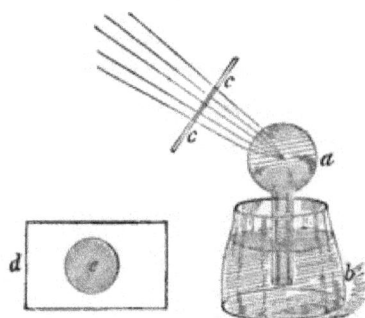

Fig. 94.

The absorbing medium was now removed, and the full rays of the sun permitted to converge on the matrass. The temperature of the water quickly ran up to the boiling-point, and the bulb was filled with steam and the purple vapor of iodine. Everything seemed favorable for the decomposition of the water to take place, if the iodine could accomplish it under so intense a radiation. At first I thought that the experiment had succeeded, for the color of the bulb became paler—a result that ought to have ensued if hydriodic acid was forming and oxygen being eliminated. The action, therefore, was kept up for four hours; but as soon as the sun was screened from the lens and the bulb began to cool, the water returned and filled it almost entirely. This, therefore, shows that under a most intense radiation iodine cannot decompose water.

A similar experiment was tried with bromine, and with the same result. It failed to decompose water.

Some silver chloride, carefully purified, was exposed in a little crucible of platinum foil (Fig. 95) so inclined that the cone of rays could come in at its mouth. The

absorbing trough was not used. Though the sun's rays
were not brilliant, the chloride at once melted, forming a
reddish-looking liquid. It was kept in that condition
all day. When cool it proved to be in the state of
horn-silver, easily cut by a knife. When the rays first
touched it a fume was disengaged, due probably to the
escape of vapor of water. It seems, therefore, that this
substance when perfectly dry is not decomposable by
sunlight, though so sensitive at common temperatures
when moist.

In Fig. 95, a is the platinum crucible; b, the place of the
material experimented upon, re-
ceiving at the focus the converging rays, c.

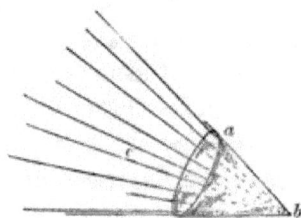

I must refer to the original
Memoir for the detail of numerous experiments on many metallic compounds, the general result of these being that, no mat-

Fig. 95.

ter how brilliant a ray may be, it cannot carry a decomposition farther than a feeble one acting for a corresponding longer period of time could do. Compounds
that can resist the force of an ordinary ray cannot be
broken down by the intense illumination of the focal
point of a burning-lens. That instrument cannot do
what the voltaic pile has done—effect decompositions
which had never been effected before.

To reduce the disturbing effect of heat as far as possible, and give every advantage to the condensed luminous focus, I received the cone of rays coming from the
twelve-inch burning-lens on a glass globe (Fig. 96) six
inches in diameter, filled with water. This increased
the converging of the rays, and brought them more
quickly to a focus. Then through the neck of the globe
was introduced to the focus, in a matrass, spoon, or other

suitable support, the substance to be experimented upon.
The mass of water kept the temperature down, and in
some cases the hot water was removed by an aspirator
and cold water introduced below. A spoon could be
used when powders were employed of so great a specific
gravity as not to drift too high from the focus in the
ascending current of hot water.

In Fig. 96, *a a* is a matrass filled with water, through
which come in the converging
rays, *c*. Through the neck at *d*
a spoon, *b*, may be passed down
to the focal point, *f*.

The result was, however, the
same as before. The focus of a
burning-lens cannot cause any
chemical change which the un-

Fig. 96.

converged sun-rays are incompetent to produce. It merely hastens the effect.

Upon the whole, we may therefore conclude that it is
not the *intensity* of a beam which determines its decomposing power, and that we cannot produce greater chemical effects by the action of converging mirrors and lenses
than we can by the application of the simple sunbeam,
continued for an equivalent period of time.

In estimating the influence of light on different solutions, we should constantly bear in mind that the maximum effect is never produced unless complete absorption
has taken place. When the color of a solution is pale,
it may require considerable thickness before complete
absorption is accomplished. Thus if two equal tubes,
containing equal quantities of the same solution of chlorine in water, be exposed to the rays, they will evolve
equal quantities of oxygen gas; but if behind one of
them a piece of looking-glass be placed, the effect is immediately increased. The rays that have passed through

the solution are compelled to cross it again, and, if not already exhausted, thus to act once more. The following illustrations are examples of the kind:

Two small bulbs of equal size containing chlorine water were exposed to the rays of the sun; behind one of them a concave hemispherical mirror was placed, so that the rays which had crossed the solution were compelled to cross it again. The amount of oxygen set free in this bulb was about one fourth greater than that in the other.

The same was repeated, the exposure being to the sky light instead of the sun-rays. The quantity of oxygen set free in the two bulbs was as 18 to 55.

It might be supposed that part of this increased effect is due to the rise of temperature, from the mirror obstructing radiation. To exert a cooling action the following modification was therefore tried. In a glass jar (Fig. 97) full of quicksilver a half-inch bulb containing chlorine water was placed in such a way that a small portion of its surface, about one eighth of an inch in diameter, projected above the surface of the liquid metal. On this part the solar focus from a burning-glass was thrown. The rays therefore gained access to the interior of the bulb, and were thrown about in all directions, crossing and recrossing the liquid in every way by the numerous reflections they underwent—the mercury, as it applied itself to the outer surface of the glass, acting like a spherical concave mirror, and, from its mass and high conducting power, effectually keeping the temperature down. The quantity of oxygen emitted in a given time was measured. The same experiment was then repeated with the bulb removed from the mercury. After the close of the same time, on measuring the oxygen set free, it was found that the reflecting action of the mercury had nearly tripled the effect.

F F

In Fig. 97, $a\,a$ is the vessel filled with mercury; $b\,d$, the glass flask immersed in it, but having at its upper part a small portion un-covered, through which the converging rays, c, may come in.

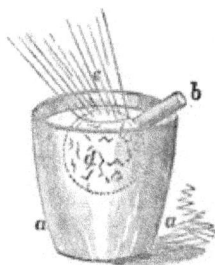
Fig. 97.

The power of a ray thus depending on the degree of absorption exerted upon it, I was led to inquire whether, by admixture with other suitable sub-stances in a solution undergoing de-composition, the effect could be increased. Chlorine wa-ter decomposes more rapidly as its yellow tint is deeper. Four equal bulbs were therefore taken—a, containing chlorine water; b, the same, deepened with chloride of gold; c, chlorine water, with commercial hydrochloric acid of a yellow tint; d, chlorine water, with tincture of iodine. These were all exposed together to the sun. It was at once obvious that a was giving off most oxygen, and eventually it was found that b yielded a much smaller quantity, and c and d none at all. The presence of these bodies, therefore, exerted a prejudicial effect.

A system of vibrating molecules will solicit an adja-cent one to execute similar motions through the medium of the intervening ether. A rise of temperature is due to an increased rapidity or intensity of the oscillations of the groups of vibrating molecules, but chemical de-composition is due to the dislocation of their parts. It, of course, by no means follows that when a compound molecule is undergoing entire disruption, those in the neighborhood should be compelled to pass into a similar state. For the very reason that chemical decomposition takes place is because the group that receives the pro-voking ray cannot vibrate consentaneously with it; and if that group cannot assume the motion in question, how can it possibly transmit it to any other?

Any artificial coloration by the addition of extraneous bodies does not increase the rate of decomposition, but retards it. This is precisely what ought to be expected. A compound atom has its grouping destroyed by the action of light upon its own parts, and is in no manner concerned in what is taking place in other atoms around. They therefore cannot increase the effect on it; but, on the contrary, they may greatly diminish the action on the mass by exerting a special absorption themselves. Thus the chloride of gold retards the decomposition of chlorine water, when mixed therewith, in the same manner as if it were placed in a trough in front of the water, and intercepted the impinging beam.

Experiments similar to the foregoing were made with a solution of ferric oxalate mixed with alcohol, ammonia citrate of iron, tincture of turmeric, sodic chloride, etc. In every instance it was clear that the action of the light is strictly molecular; that it is impressed on the group of atoms, and not on the mass; and that when various bodies are conjointly exposed to the sun, each one undergoes its own specific change, independently of and unaffected by all the rest.

These experiments, with others of a like kind, made many years ago, have an important bearing on some recently published by Professor Vogel, Captain Abney, Captain Waterhouse, and others on imparting increased sensitiveness to collodion by mixing it with variously colored substances. I repeated their experiments as carefully as I could, and should have thought that my want of success was due to unskilfulness had I not borne in mind the foregoing considerations.

2. We may next inquire, *To what extent and in what manner is the decomposition of a compound body affected by the* FREQUENCY *of vibration of a ray?*

From the beginning of optical chemistry, investigations have been made for the purpose of determining the action of rays of different refrangibilities. Almost a hundred years ago it had been shown, in special cases, that there is an antagonism between the opposite ends of the spectrum. Thus the phosphorescence excited in Canton's preparation by the violet end of the spectrum is extinguished by the red. As respects colored compounds, Grotthus showed that the active ray is very commonly of the tint complementary to that which it destroys.

More recently this branch of the subject has been examined to a great extent, and the behavior of all kinds of substances in the solar spectrum made known. The general result is this, that on wave-length—or, what is the same thing, frequency of vibration, the number of impulses it can communicate in a given period of time— depends the power of a ray to break down the union of any group of atoms. A compound that may resist a slowly recurring motion may be unable to maintain itself when the impulses increase in frequency.

So numerous and well known are the photographic and other changes brought on by light that I need not occupy space with a description of them here. I shall only refer to some curious instances of antagonism or interference, the details of which will be found in the original Memoir. Hitherto they have been very much overlooked.

Two rays may be so placed in relation to each other that their motions may conspire or may antagonize; and as one or other of these conditions ensues, the chemical result will correspond.

When iodine vapor is permitted to have access to a surface of polished silver, the silver tarnishes, the tarnished film increasing in thickness as the exposure to the iodine is prolonged. It assumes in succession colors

which undoubtedly arise from the interference of the incident light with the light reflected from the metal at
the back of the film. They are the colors of thin plates,
like those of a soap-bubble.

Now, there is a great difference in the action of light
upon these differently colored films, though chemically
they are the self-same silver iodide. Some have been unacted upon; in them the effect of the incident light has
been destroyed or reversed by the effect of the light reflected from the back of the film. Some have been powerfully acted upon; in them the chemical effect is at a
maximum — the incident and reflected rays have con-
spired.

If any proof were required that these maxima and
minima of chemical effect arise from the superposition of
similar or contrary motions, it is found in the relative
thickness of the films which have been acted or unacted
upon. Those in which there has been maximum action
have thicknesses as $2:1$; that showing the minimum
action is $1\frac{1}{2}$.

If the daylight and simple spectrum rays be permitted
to act together on a daguerreotype plate, the rays of
which the times of vibration are as 1, 2, etc., aid the daylight; but those of which the times of vibration are as
$\frac{1}{2}, 1\frac{1}{2}, 2\frac{1}{2}$, etc., interfere with it and destroy its effect.

In these numbers we may discern the suggestion of
some very important facts.

One of the most striking instances of this positive and
negative action I discovered in the case of the electric
spark. Let there be placed over a daguerreotype plate
(Fig. 98) two metal balls, connected respectively with
the inside and outside of a Leyden-jar in such a way that
the discharge may pass from one of the balls at about
half an inch distance to the sensitive plate, and from the
sensitive plate to the other ball at about the same dis-

tance. One spark is sufficient. The experiment should be made in a dark room.

In Fig. 98, *a a*, the metal photographic plate; *b, c*, the

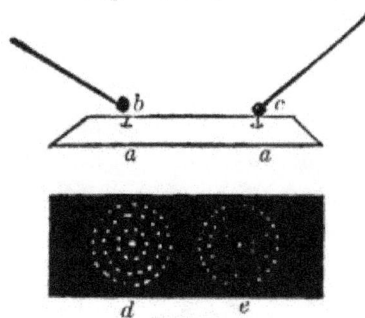

brass balls connected with the Leyden - phial. The spark passes between them and the metal plate. At *d, e*, the effect is shown, and again more plainly in Fig. 99.

On developing it will be found that on the point which received the spark there is a blue-white spot about one fortieth of an inch in diameter (Fig. 99). Immediately around this an annular space which is perfectly black, the rays of the spark having there had no action; then follows a white ring, and then another black one. Finally succeeds a whitish stain of an indistinct circular form, which can be traced by inclining the plate as having a diameter of about $1\frac{1}{2}$ inches.

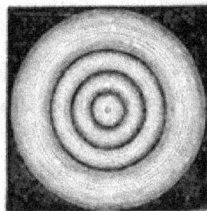

Fig. 99.

That part of the plate from which the spark escaped shows a repetition of the same phenomenon.

How shall we account for the production of these alternate white and black spaces—rings of action and inaction? Some persons might at first be led to suppose that this is only an interesting form of Priestley's experiment of "the fairy rings," formed by receiving the shock of a battery on a polished steel surface, when, by the oxidation that ensues, a film is formed of variable thickness, and giving the colors of thin plates. But a little consideration will show that this is impossible, and the facts are only to be explained on the principles of interference.

3. *In what manner is the decomposition of a compound body affected by the condition of* POLARIZATION *of the disturbing ray?*

A beam of light passing through a circular aperture one inch in diameter was received on the achromatic lens of a camera-obscura, and then fell on a doubly refracting prism, so placed as to give on the ground glass two circular images of the aperture, one third of an inch in diameter, and overlapping each other to a small extent. In these images the light was, of course, polarized at right angles respectively.

When paper rendered sensitive by being washed with ferric oxalate was placed so as to receive them—the light permitted to act nine minutes, and its effect developed by chloride of gold—the images (Fig. 100) were found of equal blackness, and the lenticular space formed by their overlapping of greater depth. This was repeated with several different photographic compounds, and always with the same result. It shows that plane polarized light acts precisely like common light, and with a rapidity proportional to its intensity.

In Fig. 100, *a*, *b*, the disks of plane polarized light, polarized at right angles to each other; at *c*, the place of overlapping.

While thus attempting to detect a difference in the decomposing action of common and polarized light, I made

Fig. 100.

some inquiries as to the possibility of polarizing light by a magnet.

A great many experiments have been made at different times for the purpose of producing disturbance on a ray of light by magnets. There are two methods which may be resorted to. The one hitherto followed has been to intercept the ray *in its course*, and submit it to magnetic action; but the principle on which my attempts have been founded is to attack it at its *origin*, and attempt to produce an impression on the shining body. These methods are essentially distinct. To borrow an illustration from acoustics, it is one thing to try to modify a sound on its passage through the air, and a very different thing to exert an influence on the sounding body.

When Bancalari's experiment on the influence of the poles of a powerful magnet on a flame was first published, I repeated it at once, expecting that the oscillations of the shining particles were constrained to take place in one plane by the magnetism, and that the light emitted would be polarized. The result, however, did not seem to prove this.

A similar experiment was then made with the electric spark from the prime conductor of a machine. It was compelled to cross between the poles of a powerful electro-magnet. But when the magnetism was on it did not seem that the light was polarized.

De la Rive has shown that the voltaic arc between charcoal points is greatly disturbed when it passes between the poles of a powerful electro-magnet. In the hope that this would produce the expected disturbance, I examined an arc formed between points of copper, platinum, and gas carbon; but though the sounds emitted were strong, resembling the sudden tearing of a piece of cloth, I could not perceive that the light was polarized.

In like manner the induction spark from a contact-breaker and the phosphorescent light from fluor-spar were tried without success. I still think, however, that with better means than those thus employed the experiment would succeed.

At the commencement of this Memoir it was stated that we should consider, 1st, the manner in which a ray of light acts in bringing about decomposition, and the changes it undergoes; 2d, the nature of the impression made on the material group, the decomposition of which ensues. The observations I proposed offering in relation to the former of these points being completed, I may pass to some remarks respecting the latter.

An examination of many cases of the decomposition of bodies by light has led me to the conclusion that its cause is to be attributed to the inability of the group of molecules affected to withstand the periodic impulses communicated to them. Of those molecules some, perhaps, take on a vibratory motion more readily than the others, and the continuance of a given group becoming impossible, a rearrangement ensues.

But in other cases the mechanism of decomposition is undoubtedly different: a change is impressed on one of the elements acted upon, which weakens its affinity for the others. Thus, under the influence of the sunshine, plants can decompose many bodies, such as carbonic, sulphuric, and phosphoric acids.

The nature of these changes may be best illustrated by tracing the complete course through which any one of these substances passes. The chief facts may be seen in the case of phosphorus. This substance, when freshly made, commonly exhibits a white waxy appearance, but when exposed to the sunshine, it turns to a deep mahogany-red. If the exposure has been long continued, or the effect hastened by the action of a burning-lens, the change of aspect is very striking. It is analogous to that which sulphur exhibits when heated to 400° or 500°. I have a specimen which has been kept for many years in an atmosphere of dry carbonic acid; the sides of the vessel are incrusted with crystals, which have slowly sublimed, and which in color resemble the ferrid cyanide of potassium.

The chemical properties of these two varieties of phosphorus are very different; indeed, there is scarcely a point in which they may not be said to be unlike. The common kind shines in the dark; the red does not. The common is soluble in a variety of menstrua, which do not act on the other: thus one of the methods of preparing red phosphorus is to expose a solution of the common in sulphuric ether to light—a red powder, the substance in question, precipitates. Compared together, the one displays a range of affinity which the other does not; nor do these properties seem to leave them when they are united with other bodies. Thus the active or white phosphorus, when united with hydrogen, yields a gas which is spontaneously combustible in the air; the red or passive variety yields a hydrogen compound of the same constitution, but devoid of the property of spontaneous combustibility.

It should be understood that though other agents—as a high temperature—can impress this remarkable change upon phosphorus, none can do it with more energy or more completely than the solar rays. I found by exposing a surface of white or active phosphorus to the prismatic spectrum that it is the more refrangible rays that are the most effective. Thus the rays most efficient in setting oxygen free from the bodies with which it is united have also the quality of impressing such a change on those bodies that they oxidize subsequently with difficulty. It follows that the true cause of such decompositions is the impression which the light makes on the elementary substance. Thus if phosphoric acid be decomposed by the solar rays, the decomposition is owing to the phosphorus being thrown into the red or passive state—a state in which its affinity for oxygen has almost entirely disappeared.

These considerations enable us to explain what takes

place in the economy of plants. The water of the soil is always charged with carbonic acid, which communicates to it the quality of dissolving bone-earth. The solution passing through the spongioles goes to the leaves as ascending sap. Here it is exposed to light, the effect of which is, aided by the cell-growth there taking place, to set the phosphoric acid free, and turn its phosphorus into the passive state. Its continued union with oxygen as an acid compound thus becomes impossible, and it is now associated with the proteine and oily bodies forming in the plant. Nor does it again unite with oxygen until it has passed into the systems of animals as a constituent of their nervous and muscular tissues. At the moment of activity of these, and especially of the former, it is oxidized, the change being apparently an immediate consequence of that activity, and, reverting to the acid state, it is finally dismissed from the system under the form of phosphate of soda and ammonia.

In the same manner might be explained the decomposition of carbonic acid by plants in the sunshine; for carbon, like phosphorus, and, indeed, like all other elementary bodies, has its active and passive states, as is exemplified in the contrast between diamond and lamp-black. The sunlight enables the leaves of plants to bring the carbon into the inactive state, and decomposition ensues as a secondary result. The carbon compounds arising form the food of various animals; nor does this element recover its active state until it has given rise to the processes of life, when it suddenly unites with oxygen brought by the arterial blood, and the compounds it then forms are dismissed from the system by the lungs and kidneys conjointly. It might seem that the mechanism of decomposition by vibratory movement is essentially different from that by these allotropic changes, but a more detailed examination will show that this is not necessarily so.

In this Memoir I have endeavored to examine how far the decomposing action of a radiation is dependent on the amplitude, the frequency, or the direction of its vibrations. The result arrived at is that decompositions are not determined by amplitude—that is, brilliancy— since a faint light continued long enough can produce precisely the same effect as the more concentrated ray of a burning-lens applied for a shorter time. Nor does the direction of motion, as involved in the idea of polarization, whether plane or circular, exert any effect, but it is the frequency of the periodic impulses that is the sole determining cause. And the phenomena of interference from the superposition of such small motions occur exactly as might have been predicted.

The immediate cause assigned for such decompositions is that a ray forcing the material particles on which it falls into a state of rapid vibration, it comes to pass in many compound molecules that their constituent atoms can no longer exist together as the same group, because of the impossibility of their being animated by consentaneous or conspiring motions, and dislocation, rearrangement, or decomposition is the result.

In this Memoir I have spoken of heat and light as though they were distinct agencies, and considered such facts as conductibility, etc., displayed by the one and not by the other. But if we recall what has been said in preceding Memoirs to the effect that these are only modes of motion, and that the difference of the effects they display turns on the character of the receiving surface or substance, there will be no difficulty in translating this commoner language into terms that are more exact, and in presenting the phenomena in question under a more rigidly scientific point of view. Familiar expressions very frequently convey to the mind clearer ideas than others which, perhaps, may be more strictly correct.

THE RUMFORD MEDAL.

APPENDIX.

"In 1796, Count Rumford presented to the Royal Society of London £1000, the interest of which was to be spent in striking two medals, both in the same die, one of gold and one of silver, of the value of the interest of the donation for two years, and to be given biennially for the most important discovery or improvement relating to light and heat that had been made during the preceding two years in any part of Europe."

At the same time Count Rumford made "a corresponding donation to the American Academy of Arts and Science of $5000, the interest of which was to be used in like manner as regards American discoveries. It was provided that if this term passed without any discovery or improvement being made that should be deemed worthy of the award, the accruing interest was to be added to the principal, and the augmented income thus arising was to be added to the medals when the next award was made."

The Royal Society accepted the trust, and made the first award to Rumford himself in 1802; in 1804, to Leslie; in 1806, to Murdock; in 1810, to Malus; in 1814, to Wells; in 1816, to Davy; in 1818, to Brewster; in 1824, to Fresnel; in 1834, to Melloni; in 1838, to Forbes; in 1840, to Biot; in 1842, to Fox-Talbot; in 1846, to Faraday; in 1848, to Regnault; in 1850, to Arago; in 1852, to Stokes; in 1854, to Arnott; in 1856, to Pasteur; in 1858, to Jamin; in 1860, to Clerk-Maxwell; in 1862, to Kirchhoff; in 1864, to Tyndall; in 1866, to Fizeau; in 1868, to Balfour Stewart, etc.

These awards covered the most important *discoveries* that had been made in relation to heat and light—the production of heat by friction, the correlation of forces, the effects of surfaces on radiation, the polarization of light, the theory of the formation of dew, the transverse vibrations of light, the application of the thermopile to the study of radiant heat, the refraction of dark rays, the invention of photography, the discovery of fluorescence, the fixed lines of the spectrum, the distribution of heat in the spectrum, the velocity of light, etc.

The American Academy permitted several years to pass without making any award. Meantime the fund had so greatly increased that the

Academy had to apply to the Legislature for authority to depart from
the strict letter of the endowment, and use the funds with more freedom
in the interest of advancing knowledge. In 1839, the Academy gave
from the interest of the fund the sum of $600 to Hare for the invention
of the oxyhydrogen blowpipe and improvements in galvanic apparatus.
In 1862, it awarded the medal to Ericsson for his caloric engine; in
1865, to Treadwell for improvements in the management of heat; in
1867, to Alvan Clark for improvements in the lenses of refracting tele-
scopes; in 1870, to Corliss for improvements in steam-engines. The fund
had now reached $42,000. It will be seen that the American awards
had been mainly for *inventions;* the English for *discoveries.*

At the six hundred and eighty-ninth meeting of the American Academy, held March 8, 1876, the chairman of the Rumford committee introduced the special business of the evening, and handed to the president, Hon. Charles Francis Adams, the Rumford medals (in gold and
silver), on each of which had been engraved the following inscription:
"Awarded by the American Academy of Arts and Science to John
William Draper for his Researches in Radiant Energy, May 25, 1875."

In presenting the medals, the president gave a brief history of the
fund, and announced that, "after a careful review of the services of Professor Draper in this great field of inquiry, the committee having the
subject in their charge have, for reasons given by them, recommended
through their chairman that the medals prescribed in the deed of trust
should be presented to him, as having fully deserved them."

The president then recapitulated some of these reasons:

"In 1840, Dr. Draper independently discovered the peculiar phenomena commonly known as Moser's images, which are formed when a medal
or coin is placed upon a polished surface of glass or metal. These images remain, as it were, latent, until a vapor is allowed to condense upon
the surface, when the image is developed and becomes visible.

"At a later period he devised the method of measuring the intensity
of the chemical action of light, afterwards perfected and employed by
Bunsen and Roscoe in their elaborate investigations. This method consists in exposing to the source of light a mixture of equal volumes of
chlorine and hydrogen gases. Combination takes place more or less
rapidly, and the intensity of the chemical action of the light is measured
by the diminution in volume. No other known method compares with
this in accuracy, and most valuable results have been obtained by its
use.

"In an elaborate investigation, published in 1847, Dr. Draper established experimentally the following important facts:

" 1. All solid substances, and probably liquids, become incandescent at the same temperature.

"2. The thermometric point at which substances become red-hot is about 977° Fahr.

" 3. The spectrum of an incandescent solid is continuous; it contains neither bright nor dark fixed lines.

"4. From common temperatures nearly up to 977° Fahr., the rays emitted by a solid are invisible. At that temperature they are red, and the heat of the incandescing body being made continuously to increase, other rays are added, increasing in refrangibility as the temperature rises.

" 5. While the addition of rays, so much the more refrangible as the temperature is higher, is taking place, there is an increase in the intensity of those already existing. Thirteen years afterwards Kirchhoff published his celebrated memoir on the relations between the coefficients of emission and absorption of bodies for light and heat, in which he established mathematically the same facts, and announced them as new.

" 6. Dr. Draper claims, and we believe with justice, to have been the first to apply the daguerreotype process to taking portraits.

" 7. Dr. Draper applied ruled glasses and specula to produce spectra for the study of the chemical action of light. The employment of ruled metallic specula for this purpose enabled him to avoid the absorbent action of glass and other transparent media, as well as to establish the points of maximum and minimum intensity with reference to portions of the spectrum defined by their wave-lengths. He obtained also the advantage of employing a normal spectrum in place of one which is abnormally condensed at one end and expanded at the other.

" 8. We owe to him valuable and original researches on the nature of the rays absorbed in the growth of plants in sunlight. These researches prove that the maximum action is produced by the yellow rays, and they have been fully confirmed by more recent investigations.

" 9. We owe to him, further, an elaborate discussion of the chemical action of light, supported in a great measure by his own experiments, and proving conclusively, and, as we believe, for the first time, that rays of all wave-lengths are capable of producing chemical changes, and that too little account has hitherto been taken of the nature of the substance in which the decomposition is produced.

" 10. Finally, Dr. Draper has recently published researches on the distribution of heat in the spectrum, which are of the highest interest, and which have largely contributed to the advancement of our knowledge of the subject of radiant energy.

"And now, in the absence of Dr. Draper, unable at this inclement season to execute a fatiguing journey, it gives me pleasure to recognize you, Mr. Quincy, as his worthy and competent representative.

"I pray you, in receiving these two medals on his behalf, in accordance with the terms of the original trust, to assure him, on the part of the Academy, of the high satisfaction taken by all its Fellows in doing honor to those who, like him, take a prominent rank in the advance of science throughout the world."

Mr. Quincy, on receiving the medals, said:

"MR. PRESIDENT,—In the name and on the behalf of Dr. Draper, I have the honor to receive the Rumford medals in gold and silver which the Academy has been pleased to award to him, and I will have them safely conveyed to him to-morrow, together with the assurances of the satisfaction of the Academy in this action which you wish me to communicate to him. In common with yourself, sir, and all the Fellows present, I regret that that eminent person is unable to attend this meeting and receive the medals himself. And, personally, I regret the absence of Dr. Wolcott Gibbs, who had promised to perform this grateful service for his friend, and who would have been able to make a more suitable reply to the able discourse with which you have accompanied the presentation of the medals, and to have done more justice to the claims of Dr. Draper to this distinction, than I can pretend to do. Dr. Gibbs having also been unavoidably prevented from being present this evening, I have now the honor to read a communication from Dr. Draper to the Academy, in acknowledgment of this testimony to his services to science."

Mr. Quincy then read the following letter:

"TO THE AMERICAN ACADEMY OF ARTS AND SCIENCES: Your favorable appreciation of my researches on radiations, expressed to-day by the award of the Rumford medals—the highest testimonial of approbation that American science has to bestow on those who have devoted themselves to the enlargement of knowledge—is to me a most acceptable return for the attention I have given to that subject through a period of more than forty years, and I deeply regret that through ill-health I am unable to receive it in person.

"Sir David Brewster, to whom science is under so many obligations for the discoveries he made, once said to me that the solar spectrum is a world in itself, and that the study of it will never be completed. His remark is perfectly just.

"But the spectrum is only a single manifestation of that infinite ether which makes known to us the presence of the universe, and in which whatever exists—if I may be permitted to say so—lives and moves and has its being.

"What object, then, can be offered to us more worthy of contemplation than the attributes of this intermedium between ourselves and the outer world?

"Its existence, the modes of motion through it, its transverse vibrations, their creation of the ideas of light and colors in the mind, the interferences of its waves, polarization, the conception of radiations and their physical and chemical effects—these have occupied the thoughts of men of the highest order. The observational powers of science have been greatly extended through the consequent invention of those grand instruments, the telescope, the microscope, the spectrometer. Through these we have obtained more majestic views of the nature of the universe. Through these we are able to contemplate the structure and genesis of other systems of worlds, and are gathering information as to the chemical constitution and history of the stars.

"In this noble advancement of science you, through some of your members, have taken no inconspicuous part. It adds impressively to the honor you have this day conferred on me that your action is the deliberate determination of competent, severe, impartial judges. I cannot adequately express my feelings of gratitude in such a presence, publicly pronouncing its approval on what I have done.

"I am, gentlemen, very truly yours,

"JOHN W. DRAPER."

The investigations and memoirs referred to by the committee of the Academy are contained in this volume.

INDEX.

THE END.

VALUABLE AND INTERESTING WORKS

FOR

PUBLIC AND PRIVATE LIBRARIES

PUBLISHED BY HARPER & BROTHERS, NEW YORK.

☞ *For a full List of Books suitable for Libraries, see* HARPER & BROTHERS' TRADE-LIST *and* CATALOGUE, *which may be had gratuitously on application to the Publishers personally,* **or** *by letter enclosing Nine Cents in Postage Stamps.*

☞ HARPER & BROTHERS *will send their publications by mail, postage prepaid (excepting certain books whose weight would exclude them from the mail), on receipt of the price.* HARPER & BROTHERS' *School and College Text-Books are marked in this list with an asterisk (*).*

FIRST CENTURY OF THE REPUBLIC. A Review of American Progress. 8vo, Cloth, $5 00; Sheep, $5 50; Half Morocco, $7 25.

Contents.

Introduction: I. Colonial Progress. By EUGENE LAWRENCE.—II. Mechanical Progress. By EDWARD H. KNIGHT.—III. Progress in Manufacture. By the Hon. DAVID A. WELLS.—IV. Agricultural Progress. By Professor WILLIAM H. BREWER.—V. The Development of our Mineral Resources. By Professor T. STERRY HUNT.—VI. Commercial Development. By EDWARD ATKINSON. —VII. Growth and Distribution of Population. By the Hon. FRANCIS A. WALKER.—VIII. Monetary Development. By Professor WILLIAM G. SUMNER.—IX. The Experiment of the Union, with its Preparations. By T. D. WOOLSEY, D.D., LL.D.—X. Educational Progress. By EUGENE LAWRENCE. —XI. Scientific Progress: 1. The Exact Sciences. By F. A. P. BARNARD, D.D., LL.D. 2. Natural Science. By Professor THEODORE GILL.—XII. A Century of American Literature. By EDWIN P. WHIPPLE.—XIII. Progress of the Fine Arts. By S. S. CONANT.—XIV. Medical and Sanitary Progress. By AUSTIN FLINT, M.D.—XV. American Jurisprudence. By BENJAMIN VAUGHAN ABBOTT.—XVI. Humanitarian Progress. By CHARLES L. BRACE. —XVII. Religious Development. By the Rev. JOHN F. HURST, D.D.

MOTLEY'S DUTCH REPUBLIC. The Rise of the Dutch Republic. A History. By JOHN LOTHROP MOTLEY, LL.D., D.C.L. With a Portrait of William of Orange. 3 vols., 8vo, Cloth, $10 50; Sheep, $12 00; Half Calf, $17 25.

MOTLEY'S UNITED NETHERLANDS. History of the United Netherlands: from the Death of William the Silent to the Twelve Years' Truce—1609. With a full View of the English-Dutch Struggle against Spain, and of the Origin and Destruction of the Spanish Armada. By JOHN LOTHROP MOTLEY, LL.D., D.C.L. Portraits. 4 vols., 8vo, Cloth, $14 00; Sheep, $16 00; Half Calf, $23 00.

MOTLEY'S LIFE AND DEATH OF JOHN OF BARNEVELD. The Life and Death of John of Barneveld, Advocate of Holland: with a View of the Primary Causes and Movements of "The Thirty-years' War." By JOHN LOTHROP MOTLEY, LL.D., D.C.L. Illustrated. In 2 vols., 8vo, Cloth, $7 00; Sheep, $8 00; Half Calf, $11 50.

*HAYDN'S DICTIONARY OF DATES, relating to all Ages and Nations. For Universal Reference. Edited by BENJAMIN VINCENT, Assistant Secretary and Keeper of the Library of the Royal Institution of Great Britain; and Revised for the Use of American Readers. 8vo, Cloth, $3 50; Sheep, $3 94.

HILDRETH'S UNITED STATES. History of the United States. FIRST SERIES: From the Discovery of the Continent to the Organization of the Government under the Federal Constitution. SECOND SERIES: From the Adoption of the Federal Constitution to the End of the Sixteenth Congress. By RICHARD HILDRETH. 6 vols., 8vo, Cloth, $18 00; Sheep, $21 00; Half Calf, $31 50.

HUME'S HISTORY OF ENGLAND. The History of England, from the Invasion of Julius Cæsar to the Abdication of James II., 1688. By DAVID HUME. 6 vols., 12mo, Cloth, $4 80; Sheep, $7 20; Half Calf, $15 30.

HUDSON'S HISTORY OF JOURNALISM. Journalism in the United States, from 1690 to 1872. By FREDERIC HUDSON. 8vo, Cloth, $5 00; Half Calf, $7 25.

JEFFERSON'S DOMESTIC LIFE. The Domestic Life of Thomas Jefferson: compiled from Family Letters and Reminiscences, by his Great-granddaughter, SARAH N. RANDOLPH. Illustrated. Crown 8vo, Cloth, $2 50.

JOHNSON'S COMPLETE WORKS. The Works of Samuel Johnson, LL.D. With an Essay on his Life and Genius, by ARTHUR MURPHY, Esq. 2 vols., 8vo, Cloth, $4 00; Sheep, $5 00; Half Calf, $8 50.

KINGLAKE'S CRIMEAN WAR. The Invasion of the Crimea: its Origin, and an Account of its Progress down to the Death of Lord Raglan. By ALEXANDER WILLIAM KINGLAKE. With Maps and Plans. Three Volumes now ready. 12mo, Cloth, $2 00 per vol.; Half Calf, $3 75 per vol.

LAMB'S COMPLETE WORKS. The Works of Charles Lamb. Comprising his Letters, Poems, Essays of Elia, Essays upon Shakspeare, Hogarth, &c., and a Sketch of his Life, with the Final Memorials, by T. NOON TALFOURD. With Portrait. 2 vols., 12mo, Cloth, $3 00; Half Calf, $6 50.

LAWRENCE'S HISTORICAL STUDIES. Historical Studies. By EUGENE LAWRENCE. Containing the following Essays: The Bishops of Rome.—Leo and Luther.—Loyola and the Jesuits.—Ecumenical Councils.—The Vaudois.—The Huguenots.—The Church of Jerusalem. —Dominic and the Inquisition.—The Conquest of Ireland.—The Greek Church. 8vo, Cloth, uncut edges and gilt tops, $3 00.

MYERS'S REMAINS OF LOST EMPIRES. Remains of Lost Empires: Sketches of the Ruins of Palmyra, Nineveh, Babylon, and Persepolis, with some Notes on India and the Cashmerian Himalayas. By P. V. N. MYERS. Illustrated. 8vo, Cloth, $3 50.

LOSSING'S FIELD-BOOK OF THE REVOLUTION. Pictorial Field-Book of the Revolution: or, Illustrations by Pen and Pencil of the History. Biography, Scenery, Relics, and Traditions of the War for Independence. By BENSON J. LOSSING. 2 vols., 8vo, Cloth, $14 00· Sheep or Roan, $15 00; Half Calf, $18 00.

LOSSING'S FIELD-BOOK OF THE WAR OF 1812. Pictorial Field-Book of the War of 1812: or, Illustrations by Pen and Pencil of the History, Biography, Scenery, Relics, and Traditions of the last War for American Independence. By BENSON J. LOSSING. With several hundred Engravings on Wood by Lossing and Barritt, chiefly from Original Sketches by the Author. 1088 pages, 8vo, Cloth, $7 00; Sheep or Roan, $8 50; Half Calf, ₹10 00.

MACAULAY'S HISTORY OF ENGLAND. The History of England from the Accession of James II. By THOMAS BABINGTON MACAULAY. 5 vols., 8vo, Cloth, $10 00; Sheep, $12 50; Half Calf, $21 25; 12mo, Cloth, $4 00; Sheep, $6 00; Half Calf, $12 75.

MACAULAY'S LIFE AND LETTERS. The Life and Letters of Lord Macaulay. By his Nephew, G. OTTO TREVELYAN, M.P. With Portrait on Steel. Complete in 2 vols., 8vo, Cloth, uncut edges and gilt tops, $5 00; Sheep, $6 00; Half Calf, $9 50. Popular Edition, 2 vols. in one, 12mo, Cloth, $1 75.

FORSTER'S LIFE OF DEAN SWIFT. The Early Life of Jonathan Swift (1667–1711). By JOHN FORSTER. With Portrait. 8vo, Cloth, $2 50.

*GREEN'S SHORT HISTORY OF THE ENGLISH PEOPLE. A Short History of the English People. By J. R. GREEN, M.A., Examiner in the School of Modern History, Oxford. With Tables and Colored Maps. 8vo, Cloth, $1 52.

HALLAM'S MIDDLE AGES. View of the State of Europe during the Middle Ages. By HENRY HALLAM. 8vo, Cloth, $5 00; Sheep, $2 50; Half Calf, $4 25.

HALLAM'S CONSTITUTIONAL HISTORY OF ENGLAND. The Constitutional History of England, from the Accession of Henry VII. to the Death of George II. By HENRY HALLAM. 8vo, Cloth, $2 00; Sheep, $2 50; Half Calf, $4 25.

HALLAM'S LITERATURE. Introduction to the Literature of Europe during the Fifteenth, Sixteenth, and Seventeenth Centuries. By HENRY HALLAM. 2 vols., 8vo, Cloth, $4 00; Sheep, $5 00; Half Calf, $8 50.

SCHWEINFURTH'S HEART OF AFRICA. The Heart of Africa. Three Years' Travels and Adventures in the Unexplored Regions of the Centre of Africa. From 1868 to 1871. By Dr. GEORG SCHWEINFURTH. Translated by ELLEN E. FREWER. With an Introduction by WINWOOD READE. Illustrated by about 130 Woodcuts from Drawings made by the Author, and with two Maps. 2 vols., 8vo, Cloth, $8 00.

M'CLINTOCK & STRONG'S CYCLOPÆDIA. Cyclopædia of Biblical, Theological, and Ecclesiastical Literature. Prepared by the Rev. JOHN M'CLINTOCK, D.D., and JAMES STRONG, S.T.D. 7 *vols. now ready.* Royal 8vo. Price per vol., Cloth, $5 00; Sheep, $6 00; Half Morocco, $8 00.

MOHAMMED AND MOHAMMEDANISM: Lectures Delivered at the Royal Institution of Great Britain in February and March, 1874. By R. BOSWORTH SMITH, M.A., Assistant Master in Harrow School; late Fellow of Trinity College, Oxford. With an Appendix containing Emanuel Deutsch's Article on "Islam." 12mo, Cloth, $1 50.

MOSHEIM'S ECCLESIASTICAL HISTORY, Ancient and Modern; in which the Rise, Progress, and Variation of Church Power are considered in their Connection with the State of Learning and Philosophy, and the Political History of Europe during that Period. Translated, with Notes, &c., by A. MACLAINE, D.D. Continued to 1826, by C. COOTE, LL.D. 2 vols., 8vo, Cloth, $4 00; Sheep, $5 00; Half Calf, $8 50.

HARPER'S NEW CLASSICAL LIBRARY. Literal Translations.

The following Volumes are now ready. 12mo, Cloth, $1 50 each.

CÆSAR. — VIRGIL. — SALLUST. — HORACE. — CICERO'S ORATIONS.— CICERO'S OFFICES, &c.—CICERO ON ORATORY AND ORATORS.— TACITUS (2 vols.).—TERENCE.—SOPHOCLES.—JUVENAL.—XENOPHON.—HOMER'S ILIAD,—HOMER'S ODYSSEY.—HERODOTUS.—DEMOSTHENES (2 vols.).—THUCYDIDES.—ÆSCHYLUS.—EURIPIDES (2 vols.).—LIVY (2 vols.).—PLATO [Select Dialogues].

LIVINGSTONE'S SOUTH AFRICA. Missionary Travels and Researches in South Africa: including a Sketch of Sixteen Years' Residence in the Interior of Africa, and a Journey from the Cape of Good Hope to Loanda on the West Coast; thence across the Continent, down the River Zambesi, to the Eastern Ocean. By DAVID LIVINGSTONE, LL.D., D.C.L. With Portrait, Maps, and Illustrations. 8vo, Cloth, $4 50; Sheep, $5 00; Half Calf, $6 75.

LIVINGSTONE'S ZAMBESI. Narrative of an Expedition to the Zambesi and its Tributaries, and of the Discovery of the Lakes Shirwa and Nyassa, 1858–1864. By DAVID and CHARLES LIVINGSTONE. With Map and Illustrations. 8vo, Cloth, $5 00; Sheep, $5 50; Half Calf, $7 25.

LIVINGSTONE'S LAST JOURNALS. The Last Journals of David Livingstone, in Central Africa, from 1865 to his Death. Continued by a Narrative of his Last Moments and Sufferings, obtained from his Faithful Servants Chuma and Susi. By HORACE WALLER, F.R.G.S., Rector of Twywell, Northampton. With Portrait, Maps, and Illustrations. 8vo, Cloth, $5 00; Sheep. $5 50; Half Calf, $7 25. Cheap Popular Edition, 8vo, Cloth, with Map and Illustrations, $2 50.

GROTE'S HISTORY OF GREECE. 12 vols., 12mo, Cloth, $18 00; Sheep, $22 80; Half Calf, $30 00.

RECLUS'S EARTH. The Earth: a Descriptive History of the Phenomena of the Life of the Globe. By ÉLISÉE RECLUS. With 234 Maps and Illustrations, and 23 Page Maps printed in Colors. 8vo, Cloth, $5 00; Half Calf, $7 25.

RECLUS'S OCEAN. The Ocean, Atmosphere, and Life. Being the Second Series of a Descriptive History of the Life of the Globe. By ÉLISÉE RECLUS. Profusely Illustrated with 250 Maps or Figures, and 27 Maps printed in Colors. 8vo, Cloth, $6 00; Half Calf, $8 25.

NORDHOFF'S COMMUNISTIC SOCIETIES OF THE UNITED STATES. The Communistic Societies of the United States, from Personal Visit and Observation; including Detailed Accounts of the Economists, Zoarites, Shakers, the Amana, Oneida, Bethel, Aurora, Icarian, and other existing Societies. With Particulars of their Religious Creeds and Practices, their Social Theories and Life, Numbers, Industries, and Present Condition. By CHARLES NORDHOFF. Illustrations. 8vo, Cloth, $4 00.

NORDHOFF'S CALIFORNIA. California: for Health, Pleasure, and Residence. A Book for Travellers and Settlers. Illustrated. 8vo, Cloth, $2 50.

NORDHOFF'S NORTHERN CALIFORNIA, OREGON, AND THE SANDWICH ISLANDS. Northern California, Oregon, and the Sandwich Islands. By CHARLES NORDHOFF. Illustrated. 8vo, Cloth, $2 50.

PARTON'S CARICATURE. Caricature and Other Comic Art, in All Times and Many Lands. By JAMES PARTON. With 203 Illustrations. 8vo, Cloth, Gilt Tops and uncut edges, $5 00.

*RAWLINSON'S MANUAL OF ANCIENT HISTORY. A Manual of Ancient History, from the Earliest Times to the Fall of the Western Empire. Comprising the History of Chaldæa, Assyria, Media, Babylonia, Lydia, Phœnicia, Syria, Judæa, Egypt, Carthage, Persia, Greece, Macedonia, Parthia, and Rome. By GEORGE RAWLINSON, M.A., Camden Professor of Ancient History in the University of Oxford. 12mo, Cloth, $1 46.

NICHOLS'S ART EDUCATION. Art Education applied to Industry. By GEORGE WARD NICHOLS, Author of "The Story of the Great March." Illustrated. 8vo, Cloth, $4 00.

BAKER'S ISMAILÏA. Ismailïa: a Narrative of the Expedition to Central Africa for the Suppression of the Slave-trade, organized by Ismail, Khedive of Egypt. By Sir SAMUEL WHITE BAKER, PASHA, F.R.S., F.R.G.S. With Maps, Portraits, and Illustrations. 8vo, Cloth, $5 00; Half Calf, $7 25.

BOSWELL'S JOHNSON. The Life of Samuel Johnson, LL.D., including a Journal of a Tour to the Hebrides. By JAMES BOSWELL, Esq. Edited by JOHN WILSON CROKER, LL.D., F.R.S. With a Portrait of Boswell. 2 vols., 8vo, Cloth, $4 00; Sheep, $5 00; Half Calf, $8 50.

VAN-LENNEP'S BIBLE LANDS. Bible Lands: their Modern Customs and Manners Illustrative of Scripture. By the Rev. HENRY J. VAN-LENNEP, D.D. Illustrated with upward of 350 Wood Engravings and two Colored Maps. 838 pp., 8vo, Cloth, $5 00; Sheep, $6 00; Half Morocco, $8 00.

VINCENT'S LAND OF THE WHITE ELEPHANT. The Land of the White Elephant: Sights and Scenes in Southeastern Asia. A Personal Narrative of Travel and Adventure in Farther India, embracing the Countries of Burma, Siam, Cambodia, and Cochin-China (1871–2). By FRANK VINCENT, Jr. Illustrated with Maps, Plans, and Woodcuts. Crown 8vo, Cloth, $3 50.

SHAKSPEARE. The Dramatic Works of William Shakspeare. With Corrections and Notes. Engravings. 6 vols., 12mo, Cloth, $9 00. 2 vols., 8vo, Cloth, $4 00; Sheep, $5 00. In one vol., 8vo, Sheep, $4 00.

SMILES'S HISTORY OF THE HUGUENOTS. The Huguenots: their Settlements, Churches, and Industries in England and Ireland. By SAMUEL SMILES. With an Appendix relating to the Huguenots in America. Crown 8vo, Cloth, $2 00.

SMILES'S HUGUENOTS AFTER THE REVOCATION. The Huguenots in France after the Revocation of the Edict of Nantes; with a Visit to the Country of the Vaudois. By SAMUEL SMILES. Crown 8vo, Cloth, $2 00.

SMILES'S LIFE OF THE STEPHENSONS. The Life of George Stephenson, and of his Son, Robert Stephenson; comprising, also, a History of the Invention and Introduction of the Railway Locomotive. By SAMUEL SMILES. With Steel Portraits and numerous Illustrations. 8vo, Cloth, $3 00.

SQUIER'S PERU. Peru: Incidents of Travel and Exploration in the Land of the Incas. By E. GEORGE SQUIER, M.A., F.S.A., late U. S. Commissioner to Peru, Author of "Nicaragua," "Ancient Monuments of Mississippi Valley," &c., &c. With Illustrations. 8vo, Cloth, $5 00.

STRICKLAND'S (Miss) QUEENS OF SCOTLAND. Lives of the Queens of Scotland and English Princesses connected with the Regal Succession of Great Britain. By AGNES STRICKLAND. 8 vols., 12mo, Cloth, $12 00; Half Calf, $26 00.

THE "CHALLENGER" EXPEDITION. The Atlantic: an Account of the General Results of the Exploring Expedition of H.M.S. "Challenger." By Sir WYVILLE THOMSON, K.C.B., F.R.S. With numerous Illustrations, Colored Maps, and Charts, from Drawings by J. J. Wyld, engraved by J. D. Cooper, and Portrait of the Author, engraved by C. H. Jeens. 2 vols., 8vo, $12 00.

BOURNE'S LIFE OF JOHN LOCKE. The Life of John Locke. By H. R. Fox BOURNE. 2 vols., 8vo, Cloth, uncut edges and gilt tops, $5 00.

www.ingramcontent.com/pod-product-compliance
Lightning Source LLC
Chambersburg PA
CBHW020902210326
41598CB00018B/1753